METHODS AND PROTOCOLS IN FOOD SCIENCE

Series Editor
Anderson S. Sant'Ana
University of Campinas
Campinas, Brazil

For further volumes:
http://www.springer.com/series/16556

Methods and Protocols in Food Science series is devoted to the publication of research protocols and methodologies in all fields of food science.

Volumes and chapters will be organized by field and presented in such way that the readers will be able to reproduce the experiments in a step-by-step style. Each protocol will be characterized by a brief introductory section, followed by a short aims section, in which the precise purpose of the protocol will be clarified.

Dairy Foods Processing

Edited by

Adriano Gomes da Cruz

Department of Food, Federal Institute of Education, Science and Technology of Rio de Janeiro (IFRJ), Rio de Janeiro, Brazil

Tatiana Colombo Pimentel

Department of Food, Federal Institute of Science, Paranavaí, Paraná, Brazil

Erick Almeida Esmerino

Department of Food Technology, Federal University of Rio de Janeiro, Rio de Janeiro, Brazil

Silvani Verruck

Department of Food Science and Technology, Federal University of Santa Catarina, Florianópolis, Santa Catarina, Brazil

Humana Press

Editors
Adriano Gomes da Cruz
Department of Food
Federal Institute of Education, Science and Technology of Rio de Janeiro (IFRJ)
Rio de Janeiro, Brazil

Tatiana Colombo Pimentel
Department of Food
Federal Institute of Science
Paranavaí, Paraná, Brazil

Erick Almeida Esmerino
Department of Food Technology
Federal University of Rio de Janeiro
Rio de Janeiro, Brazil

Silvani Verruck
Department of Food Science and Technology
Federal University of Santa Catarina
Florianópolis, Santa Catarina, Brazil

ISSN 2662-950X ISSN 2662-9518 (electronic)
Methods and Protocols in Food Science
ISBN 978-1-0716-4146-0 ISBN 978-1-0716-4144-6 (eBook)
https://doi.org/10.1007/978-1-0716-4144-6

This Humana imprint is published by the registered company Springer Science+Business Media, LLC, part of Springer Nature.
The registered company address is: 1 New York Plaza, New York, NY 10004, U.S.A.

If disposing of this product, please recycle the paper.

Preface to the Series

Methods and Protocols in Food Science series is devoted to the publication of research protocols and methodologies in all fields of food science. The series is unique as it includes protocols developed, validated, and used by food and related scientists as well as theoretical basis are provided for each protocol. Aspects related to improvements in the protocols, adaptations, and further developments in the protocols may also be approached.

Methods and Protocols in Food Science series aims to bring the most recent developments in research protocols in the field as well as very well established methods. As such the series targets undergraduates, graduates, and researchers in the field of food science and correlated areas. The protocols documented in the series will be highly useful for scientific inquiries in the field of food sciences, presented in such way that the readers will be able to reproduce the experiments in a step-by-step style.

Each protocol will be characterized by a brief introductory section, followed by a short aims section, in which the precise purpose of the protocol is clarified. Then, an in-depth list of materials and reagents required for employing the protocol is presented, followed by a comprehensive and step-by-step procedure on how to perform that experiment. The next section brings the dos and don'ts when carrying out the protocol, followed by the main pitfalls faced and how to troubleshoot them. Finally, template results will be presented and their meaning/conclusions addressed.

The *Methods and Protocols in Food Science* series will fill an important gap, addressing a common complain of food scientists, regarding the difficulties in repeating experiments detailed in scientific papers. With this, the series has a potential to become a reference material in food science laboratories of research centers and universities throughout the world.

Campinas, Brazil ***Anderson S. Sant'Ana***

Preface

The production, consumption, and trade of milk and dairy products have experienced substantial growth, outpacing global population expansion. According to the Food and Agriculture Organization of the United Nations (FAO), the world's population increased by 27% over the last two decades, reaching 7.75 billion people in 2020. In contrast, milk production surged by 56%, reaching 906 billion liters. This includes 81% cow, 15% buffalo, and 4% goat, sheep, and camel milk. Consequently, per capita consumption rose by 23%, reaching 117 kg of milk equivalent per person per year. Of the 906 billion liters of milk produced and traded globally in 2020, 51% was consumed as fluid milk and derivatives, 20% as cheese, 18% as butter, and 11% as powdered milk (both skimmed and whole), generating a gross production value close to US$ 433 billion. Europe (26%), North America (12%), and Oceania (4%), together with Asia (42%), accounted for 84% of the milk volume produced in 2020, with Latin America and Africa contributing 11% and 5%, respectively.

Given the high productivity and the essential role of milk as a nutrient source for health, this book aims to provide a comprehensive introduction to the methods and procedures involved in manufacturing dairy foods and beverages. This volume, a part of the successful Methods and Protocols in Food Science (MeFS) series, is meticulously curated by scientists from disciplines such as Food Engineering and Food Technology. Each chapter includes thorough introductions to the respective topics, lists of materials and reagents, step-by-step reproducible protocols, and notes on troubleshooting and avoiding pitfalls.

This volume contains 20 chapters, each focusing on specific aspects of dairy food processing.

Chapter 1, *Pasteurized and Ultra-High Temperature Treated Milk*, details the methodologies for producing safe and shelf-stable milk products.

Chapter 2, *Flavored Milk*, explores techniques to enhance milk with various flavors, appealing to diverse consumer preferences.

Chapter 3, *Mozzarella Cheese*, provides protocols for producing this popular cheese, emphasizing techniques to ensure consistency and quality.

Chapter 4, *Prato Cheese*, introduces the traditional methods for making this semi-hard Brazilian cheese.

Chapter 5, *Processed Cheese*, focuses on methods to create cheese products with improved shelf life and versatility.

Chapter 6, *Fresh Cheese*, covers the production of various types of fresh cheese, highlighting their unique properties and applications.

Chapter 7, *Parmesan Cheese*, offers detailed protocols for producing this hard, aged cheese renowned for its rich flavor.

Chapter 8, *Cheddar Cheese*, discusses the traditional and modern techniques for making one of the world's most popular cheeses.

Chapter 9, *Blue Cheese*, explores the production processes yielding distinctively flavored and textured blue-veined cheeses.

Chapter 10, *Ricotta Cheese*, details how this creamy, whey-based cheese is produced.

Chapter 11, *Mini-Cheese Models*, presents smaller-scale cheese production techniques suitable for experimental and educational purposes.

Chapter 12, *Concentrated, Isolated, and Hydrolyzed Whey Protein Obtained by Membrane Separation Processes*, delves into advanced techniques for extracting valuable proteins from whey.

Chapter 13, *Powdered Milk*, describes the production processes for creating stable, long-lasting milk powder products.

Chapter 14, *Set, Stirred, and Liquid Yogurts Obtained by Lactic Fermentation*, explores the fermentation processes that produce various types of yogurt.

Chapter 15, Traditional and Industrial Methods for Milk Kefir Production, covers the methodologies for creating this probiotic-rich fermented milk drink.

Chapter 16, *Labneh, Ayran, Amasi, and Tarag*, introduces the techniques for producing these traditional fermented dairy beverages.

Chapter 17, *Non-Fermented Dairy Beverages*, focuses on producing dairy drinks that do not undergo fermentation.

Chapter 18, *Milk Cream and Butter*, details the processes for producing these rich, dairy-fat products.

Chapter 19, *Ice Cream and Frozen Yogurt*, provides protocols for creating these popular frozen dairy desserts.

Chapter 20, *Dulce de Leche*, explores the techniques for making this sweet, caramel-like dairy product.

This volume is designed to serve as a foundational resource for graduate students, postdoctoral associates, and researchers who are beginning their academic careers or seeking new challenges in the field of dairy science. Combining well-established protocols and procedures used by numerous laboratories in academia and industry, this book offers a broad and detailed perspective on the methods used in dairy food production.

Rio de Janeiro, Brazil ***Adriano Gomes da Cruz***
Paranavaí, Brazil ***Tatiana Colombo Pimentel***
Rio de Janeiro, Brazil ***Erick Almeida Esmerino***
Florianópolis, Brazil ***Silvani Verruck***

Contents

Contributors

RANA MUHAMMAD AADIL • *National Institute of Food Science and Technology, University of Agriculture, Faisalabad, Pakistan*
MUHAMMAD AJMAL • *Department of Dairy Technology, University of Veterinary and Animal Sciences, Lahore, Pakistan*
MARZIA ALBENZIO • *Department of Agriculture, Food, Natural Resources, and Engineering (DAFNE), University of Foggia, Foggia, Italy*
LUAN AMARAL ALEXANDRE • *Graduate Program in Food Science and Technology, Federal University of Santa Catarina, Florianópolis, SC, Brazil*
ADRIANA TORRES SILVA E ALVES • *Institute of Food Technology ITAL Department of Dairy Technology and Lactic Acid Bacteria (Tecnolat), Instituto de Tecnologia de Alimentos, São Paulo, Brazil*
FERESHTEH ANSARI • *Razi Vaccine and Serum Research Institute, Agricultural Research, Education and Extension Organization (AREEO), Tehran, Iran; Research Center for Evidence-Based Medicine, Health Management and Safety Promotion Research Institute, Tabriz University of Medical Sciences, Tabriz, Iran; Iranian EBM Centre: A Joanna Briggs Institute Affiliated Group, Tabriz, Iran*
HANNAH CAROLINE SANTOS ARAUJO • *Post Graduate Program in Food Science and Technology—PROCTA—Federal University of Sergipe, São Cristóvão, Brazil; Laboratory of Flavor and Chromatographic Analysis, Federal University of Sergipe, São Cristóvão, Brazil*
GIORDANA DEMAMAN AREND • *Laboratory of Membrane Processes (LABSEM), Federal University of Santa Catarina, Florianopolis, SC, Brazil*
MUHAMMAD ASHRAF • *Department of Chemistry, Government College University, Lahore, Pakistan*
ANDRÉA CÁTIA LEAL BADARÓ • *Federal University of Technology – Parana (UTFPR), Campus Francisco Beltrão, Paraná, Brazil*
MARYAM BATOOL • *Department of Biosciences, COMSATS University Islamabad Sahiwal Campus, Sahiwal, Punjab, Pakistan*
CARINA V. BERGAMINI • *Instituto de Lactología Industrial (INLAIN-CONICET), Facultad de Ingeniería Química (FIQ-UNL), Santiago del Estero, Santa Fe, Argentina*
L. G. CAMPOS-MÚZQUIZ • *Facultad de Ciencias Químicas, Universidad Autónoma de Coahuila, Saltillo, Coahuila, Mexico*
ROSICLER COLET • *Federal Institute of Santa Catarina, São Miguel do Oeste, SC, Brazil*
RAFFAELE COPPOLA • *Department of Agricultural, Environmental and Food Science, University of Molise, Campobasso, Italy*
RENATA GOLIN BUENO COSTA • *Empresa de Pesquisa Agropecuária de Minas Gerais, Instituto de Laticínios Cândido Tostes, Juiz de Fora, Minas Gerais, Brazil*
D. E. CRUZ-CASAS • *Facultad de Ciencias Químicas, Universidad Autónoma de Coahuila, Saltillo, Coahuila, Mexico*
ADRIANO GOMES DA CRUZ • *Department of Food, Federal Institute of Education, Science and Technology of Rio de Janeiro (IFRJ), Rio de Janeiro, Rio de Janeiro, Brazil*

ALICE CRISTINA DA SILVA • *Graduate Program in Food Science and Technology, Federal University of Santa Catarina, Florianópolis, SC, Brazil*

RAMON DA SILVA ROCHA • *Department of Food Engineering (ZEA), College of Animal Science and Food Engineering (FZEA), University of São Paulo (USP), São Paulo, Brazil; Departamento de Alimentos e Nutrição Experimental, Faculdade de Ciências Farmacêuticas (FCF), Universidade de São Paulo (USP), São Paulo, Brazil*

FABIANE PICININ DE CASTRO CISLAGHI • *Federal University of Technology – Parana (UTFPR), Campus Francisco Beltrão, Paraná, Brazil*

MÔNICA SILVA DE JESUS • *Post Graduate Program in Food Science and Technology—PROCTA—Federal University of Sergipe, São Cristóvão, Brazil; Laboratory of Flavor and Chromatographic Analysis, Federal University of Sergipe, São Cristóvão, Brazil*

ANA PAULA ZAPELINI DE MELO • *Graduate Program in Food Science and Technology, Federal University of Santa Catarina, Florianópolis, SC, Brazil*

BIANCA CRISTINA ROCHA DE OLIVEIRA • *College of Veterinary, Department of Food Technology (MTA), Universidade Federal Fluminense (UFF), Rio de Janeiro, Brazil*

FABIANE ANDRESA DE OLIVEIRA • *Federal University of Technology – Parana (UTFPR), Campus Francisco Beltrão, Paraná, Brazil*

ADRIANA CRISTINA DE OLIVEIRA SILVA • *Department of Food Technology (MTA), College of Veterinary, Universidade Federal Fluminense (UFF), Rio de Janeiro, Brazil*

JUNIO CÉSAR JACINTO DE PAULA • *Instituto de Laticínios Cândido Tostes, Empresa de Pesquisa Agropecuária de Minas Gerais (EPAMIG), Juiz de Fora, MG, Brazil*

JAQUELINE DE PAULA REZENDE • *Food Science Department, Federal University of Lavras, Lavras, MG, Brazil*

LETÍCIA BRUNI DE SOUZA • *Department of Food Technology, Federal University of Viçosa, Viçosa, MG, Brazil*

ANA CLARISSA DOS SANTOS PIRES • *Department of Food Technology, Federal University of Viçosa, Viçosa, MG, Brazil*

NATÁLIA REGINA COLDEBELLA FERREIRA • *Graduate Program in Food Science and Technology, Federal University of Santa Catarina, Florianópolis, SC, Brazil*

ADRIANA GÁMBARO • *Departamento de Ciencia y Tecnología de Alimentos, Facultad de Química, Universidad de la República, Montevideo, Uruguay*

NABILA GULZAR • *Department of Dairy Technology, University of Veterinary and Animal Sciences, Lahore, Pakistan*

ELIARA ACIPRESTE HUDSON • *Department of Food Technology, Federal University of Viçosa, Viçosa, MG, Brazil*

USMAN MIR KHAN • *National Institute of Food Science and Technology, University of Agriculture, Faisalabad, Pakistan*

ANAM LATIF • *Institute of Food Science and Nutrition, University of Sargodha, Sargodha, Pakistan*

ESMAEEL MIRZAKHANI • *Dietary Supplements and Probiotic Research Center, Alborz University of Medical Sciences, Karaj, Iran*

MASOUD AMAN MOHAMMADI • *Department of Food Technology, Faculty of Nutrition Science and Food Technology, Nutritional, and Food Technology Research Institute, Shahid Beheshti University of Medical Sciences, Tehran, Iran*

MUHAMMAD NADEEM • *Department of Dairy Technology, University of Veterinary and Animal Sciences, Lahore, Pakistan*

NARENDRA NARAIN • *Post Graduate Program in Food Science and Technology—PROCTA—Federal University of Sergipe, São Cristóvão, Brazil; Laboratory of Flavor and Chromatographic Analysis, Federal University of Sergipe, São Cristóvão, Brazil*

FILOMENA NAZZARO • *Institute of Food Science, CNR, Avellino, Italy*

MARIA TEREZINHA SANTOS LEITE NETA • *Post Graduate Program in Food Science and Technology—PROCTA—Federal University of Sergipe, São Cristóvão, Brazil; Laboratory of Flavor and Chromatographic Analysis, Federal University of Sergipe, São Cristóvão, Brazil*

ANA FLÁVIA COELHO PACHECO • *Instituto de Laticínios Cândido Tostes, Empresa de Pesquisa Agropecuária de Minas Gerais (EPAMIG), Juiz de Fora, MG, Brazil*

PAULO HENRIQUE COSTA PAIVA • *Instituto de Laticínios Cândido Tostes, Empresa de Pesquisa Agropecuária de Minas Gerais (EPAMIG), Juiz de Fora, MG, Brazil*

LUIS A. PANIZZOLO • *Departamento de Ciencia y Tecnología de Alimentos, Facultad de Química, Universidad de la República, Montevideo, Uruguay*

GUILLERMO H. PERALTA • *Instituto de Lactología Industrial (INLAIN-CONICET), Facultad de Ingeniería Química (FIQ-UNL), Santiago del Estero, Santa Fe, Argentina; Facultad de Ciencias Agrarias (FCA-UNL), Esperanza, Argentina*

STEPHANIE SILVA PINTO • *Federal Institute of Santa Catarina, São Miguel do Oeste, SC, Brazil; State University of Santa Catarina – UDESC, Pinhalzinho, SC, Brazil*

HADI POURJAFAR • *Dietary Supplements and Probiotic Research Center, Alborz University of Medical Sciences, Karaj, Iran*

MARIANE CAROLINA PRONER • *Laboratory of Membrane Processes (LABSEM), Federal University of Santa Catarina, Florianopolis, SC, Brazil*

S. C. RANGEL-ORTEGA • *Facultad de Ciencias Químicas, Universidad Autónoma de Coahuila, Saltillo, Coahuila, Mexico*

KATIA REZZADORI • *Department of Food Science and Technology, Federal University of Santa Catarina, Florianopolis, SC, Brazil*

ANALÍA RODRÍGUEZ • *Departamento de Ciencia y Tecnología de Alimentos, Facultad de Química, Universidad de la República, Montevideo, Uruguay*

R. RODRÍGUEZ-HERRERA • *Facultad de Ciencias Químicas, Universidad Autónoma de Coahuila, Saltillo, Coahuila, Mexico*

SERGIO D. ROZYCKI • *Instituto de Tecnología de Alimentos, Facultad de Ingeniería Química, Universidad Nacional del Litoral, Santa Fe, Argentina*

RAFAEL DONIZETE DUTRA SANDES • *Post Graduate Program in Food Science and Technology—PROCTA—Federal University of Sergipe, São Cristóvão, Brazil; Laboratory of Flavor and Chromatographic Analysis, Federal University of Sergipe, São Cristóvão, Brazil*

ANTONELLA SANTILLO • *Department of Agriculture, Food, Natural Resources, and Engineering (DAFNE), University of Foggia, Foggia, Italy*

PATRÍCIA FERNANDA SCHONS • *Federal Institute of Santa Catarina, São Miguel do Oeste, SC, Brazil*

DENISE SOBRAL • *Empresa de Pesquisa Agropecuária de Minas Gerais, Instituto de Laticínios Cândido Tostes, Juiz de Fora, Minas Gerais, Brazil*

Maria Eduarda Marques Soutelino • *College of Veterinary, Department of Food Technology (MTA), Universidade Federal Fluminense (UFF), Rio de Janeiro, Brazil*
Leila Maria Spadoti • *Institute of Food Technology ITAL Department of Dairy Technology and Lactic Acid Bacteria (Tecnolat), Instituto de Tecnologia de Alimentos, São Paulo, Brazil*
Patrizio Tremonte • *Department of Agricultural, Environmental and Food Science, University of Molise, Campobasso, Italy*
Silvani Verruck • *Department of Food Science and Technology, Federal University of Santa Catarina, Florianópolis, SC, Brazil*
Patrícia Blumer Zacarchenco • *Institute of Food Technology ITAL Department of Dairy Technology and Lactic Acid Bacteria (Tecnolat), Instituto de Tecnologia de Alimentos, São Paulo, Brazil*

Chapter 1

Pasteurized and Ultrahigh-Temperature-Treated Milk

Muhammad Ajmal, Muhammad Nadeem, Nabila Gulzar, Muhammad Ashraf, and Maryam Batool

Abstract

Pasteurization, pioneered by Louis Pasteur in the nineteenth century, involves heating raw milk to a moderate temperature (72 °C for 15 s) followed by rapid cooling. Ultrahigh-temperature (UHT) processing, on the other hand, entails heating milk to ultrahigh temperatures (135 °C) for a short duration (2–5 s) and then rapidly cooling it. This intense thermal treatment destroys both pathogenic and spoilage microorganisms and spores, resulting in a product with an extended shelf life of several months when stored at ambient temperatures. Pasteurization and UHT processing are indispensable technologies in the dairy industry, playing vital roles in guaranteeing food safety, extending shelf life, preserving nutritional value, and meeting consumer demands for convenient and high-quality dairy products. They are essential steps in the production of various dairy products, including milk, cream, yogurt, and flavored milk drinks. Additionally, these thermal treatments help preserve the nutritional integrity of dairy products by minimizing the loss of vitamins and proteins during processing.

Key words Milk, Pasteurization, Ultrahigh temperature, Long shelf-life milk, Aseptic

1 Introduction

In the food industry, milk is one of the most precious, valuable foods and a major source of essential nutrients including protein, fat, vitamins, and minerals which are important for human growth [1]. Milk is a precious food which has a very short life about 4–6 hours after milking. The shelf life of milk and milk products can be extended through various methods including cooling, freezing, heating (pasteurization, UHT, sterilization), bactofugation, and microfiltration, and the main objective is to eliminate or inhibit the growth of microorganisms [2]. According to Walstra et al. [3] for milk, thermal processes are applied to eliminate, reduce, and inactivate microorganisms or enzymes and as a result enhance the shelf life of products.

Adriano Gomes da Cruz et al. (eds.), *Dairy Foods Processing*, Methods and Protocols in Food Science,
https://doi.org/10.1007/978-1-0716-4144-6_1,

In 1864 scientist Louis Pasteur first time introduced the term pasteurization, and it is a heat treatment technique to kill the pathogenic microorganisms in food, beverage, and milk—Pasteur's groundbreaking research on the role of microorganisms in food spoilage and fermentation process in the food industry. Pasteurization technique plays a vital role in ensuring the shelf life, quality, and safety and aims to inactivate or destroy vegetative (pathogen and spoilage) microorganisms [4]. Pasteurization heat technique is popular to extend the shelf of raw milk to 1 to 2 weeks weeks at refrigerator temperature without any changes in nutrients and sensory attributes [5]. In this technique heating and cooling parameters can be controlled for effective microbial safety and quality and ensuring consumer satisfaction. Milk free from pathogenic microorganisms is known as safe milk and heat-treated milk known as pasteurized milk.

In 2016 it was reported in the USA that 14.4 billion liters of fluid milk was sold and out of whole milk 99.9% was pasteurized milk [6]. Heat treatment is required to reduce spoilage microflora and improve the quality of products [7]. Pasteurization is a thermal technique that is mandatory for manufacturing of fluid milk and milk products [8]. The term pasteurization in the dairy industry means to heat the milk particles in properly designed equipment with suitable time–temperature combination. Two methods are commonly used for pasteurization processes which are low temperature long time (LTLT) and high temperature short time (HTST), and milk is treated at 65 °C for 30 min and 72 °C for 15 ss, respectively [9]. At pasteurized temperature alkaline phosphatase enzyme is completely denatured and negative alkaline phosphatase tests indicate the pasteurization process has been done [4]. The shelf life of pasteurized milk in developed countries is 10 days, as compared to 3–5 days in developing countries [10]. The shelf life of commercial pasteurized milk is limited due to low processing temperature, and some spp. of microorganisms such as psychrophilic, *Pseudomonas*, and *Bacillus* can survive at that temperature and continue to grow during storage of milk [11].

Ultra-pasteurization is a food preservation technique in which food is exposed to a high temperature (135–150 °C) for a very short period of time (2–5 s) to manufacturing a commercially sterile product to kill the vegetative and spore-forming bacteria followed by aseptic packaging and rapid cooling at 20 °C [12, 13]. In UHT treatment absolutely spore-forming bacteria are inactivated (killed) and milk is microbiologically safe and in superior quality. Ultrahigh-temperature (UHT) processing techniques are highly valuable for beverage, dairy desserts, juices, soups, and food industry to extend the shelf life, safety, and hygiene of products which are driving the market growth. Almond milk and soy milk are processed through UHT operation for a long shelf life [8]. The shelf life of UHT-treated products is between 3 and 6 months without refrigeration and this makes them convenient for customer consumption with higher demand [14].

For evaluating the performance of UHT processing, *Bacillus stearothermophilus* is used as an index organism, and at UHT it should be negative (inactive). Heating of milk at higher temperature also results in undesirable changes in chemical quality, browning, and cooked flavoring due to generation of sulfhydryl compounds from fat globules at that temperature. In this method sensory and physical properties of milk remain acceptable.

2 Materials and Steps Involved for Pasteurization and UHT Milk Process

2.1 Technical Steps Involved in Processing of Milk

Pasteurization and UHT process consist of several technical and sequential steps for effective heat treatment, maintaining product safety and quality. The detailed steps are given below according to the method by Kilara [15]:

2.2 Reception of Raw Milk

Milk is milked at farms by hand or machine milking and may be stored at a chilling center or directly transported to a milk processing unit with the help of sanitized tanks or containers. When milk is received at the processing unit, its temperature should be below 10 °C but ideal 4 °C. Milk should be clean, sweet, of pleasant flavor, and free from extraneous material such as dust, hair, and gases. Quality of final products depends on milk incoming milk from shed area (farm). When the milking process is completed at the farm, milk should reach the chilling or processing unit within 3–4 hours to prevent the deterioration of the raw milk. When milk reached the processing unit after thorough and gentle mixing, a sample of milk is collected from the tank/container for laboratory analysis to check the microbial load and quality of raw milk before accepting it for further processing. If the quality of milk (*see* **Notes 1–5**) is good, it will be accepted or otherwise rejected. After a decision is made for accepting the milk, then milk is received for processing [16].

2.3 Quality Analysis of Raw Milk

Accepted milk sample is to be tested in the quality control lab for final decision on product production. Product production totally depended on the quality of raw milk. Qualitative, quantitative, microbial, and adulteration tests for different adulterants (*see* **Note 6**) are performed including the following:

2.3.1 Alcohol Test (APT)

This method determines the stability of the protein of milk to coagulate due to developed acidity or high concentration of salt. This method is also used to detect the sustainability of protein for heat processing in industry [17].

Procedure A milk sample of 2 mL was mixed with equal amount of ethanol solution (55%, 60%, 68%) of desired strength in petri plate and well mixed without shaking.

Conclusion Appearance of precipitate indicates positive (+ve) APT.

2.3.2 Clot on Boiling (COB)

This test is used for the determination of sustainability of milk protein whether milk is suitable for processing. Milk acidity has a great role in stability of milk protein during processing and higher acidity of milk decreases the stability of milk. This test is performed when milk reached the processing unit; if this test is positive, then milk will be rejected.

Procedure Put the 4–5 mL sample in the test tube, then heat the sample on spirit lamp to boiling point, and after heating observe any clot on the wall of the test tube [16].

Result Appearance of any clot on the wall of the test tube indicates positive COB.

2.3.3 Titratable Acidity (%TA)

Natural acidity of fresh milk is between 0.08 and 0.14% and is due to casein, whey protein, citrates, phosphates, and CO_2. Bacteria convert lactose to lactic acid in milk and this acidity developed by microbes is known as developed acidity. This is a quantitative method used to determine the total titratable acidity of fresh milk in degree thornic (Th°) at reception at plant, reception in area and after processing of milk in industry.

Procedure Weigh 9 mL sample in a beaker. Add two to four drops of 1% indicator (phenolphthalein) and mix. Titrate the sample with NaOH 0.1 N with the help of a burette until a light pink color appears. Read the volume of 0.1 N NaOH used and calculate the result by the following formula:

$$\text{Acidity } \% = \frac{\text{NaOH } 0.1\text{N used in mL} \times 0.0009 \times 100}{\text{Weight of sample}}$$

or

$$\text{Acidity } \% = \text{mL of } 0.1 \text{ N NaOH used x } 0.1 \text{ [18].}$$

2.3.4 Determination of Fat% by Gerber Method

Fat is the most important component of milk because almost all industries purchase the milk on the basis of fat% in milk. Different methods are used for determination of fat%, but Gerber method is a quick, quantitative method to determine the fat contents of fresh milk at the reception and processing unit.

Procedure For determination of fat%, heat the milk sample at 45 °C and cool to 20 ° C. Take H_2SO_4 (10 mL) with the help of auto-measure in butyrometer. Add 10.94 mL milk sample and 1 mL isoamyl alcohol. Insert the stopper into the butyrometer with the help of a stopper key and well mix. Place the butyrometer in Gerber machine for 5 min. For raw and 10 min for UHT milk at 65 °C and 1100 rpm. After that remove the butyrometer from the Gerber machine and immediately read the lower meniscus of the fat column in the butyrometer [16].

2.3.5 Determination of Solid Not Fat (SNF)

Milk processors have serious concerns about total solid (TS) and fat of raw milk. After calculation of butter fat with the help of a Gerber machine, also calculate the TS% and SNF% of raw milk. This is a very important parameter to check the quantity and profit margin of final products.

Procedure For determination of lactometer reading, heat the milk sample at 45 ° C and cool to 20 ° C. Fill the glass cylinder completely with milk and place the lactometer over the cylinder and leave it. After the stability of the lactometer in the cylinder, note the reading on the lactometer scale [17].

Calculation

SNF = (LR/4) + 0.72 + (0.22 X Fat % of sample).

Specific gravity = LR/1000 + 1.

2.3.6 Methylene Blue Reduction (MBR) Test

The objective of this test is to know about the microbiological quality and overall hygienic condition of milk. According to this test, bacteria in milk reduce the methylene blue dye into colorless form by the process of microbial respiration. The rate of disappearance of dye is directly proportional to activity of bacteria in the sample. The greater the number of bacteria, the sooner the color will disappear and this indicates the large number of microbes present in the sample [19].

2.4 Filtration Process

Filtration of milk in the dairy industry is an initial, critical, and fundamental step to remove foreign particles from raw milk. Milk is heated before performing the filtration because preheating increases the efficiency of the filtration process. Preheating is mostly performed at 40 °C and at that temperature the fat is in liquid form and is easily removed. Preheating practice is not performed in every dairy industry and it depends on requirement. Raw milk is received from a dairy farm/collection center and brought to the dairy processing plant. Sample of raw milk is collected before filtration for quality analysis. Preliminary visual inspection is performed before the filtration process to assess the nature of contaminants and the need for further investigation. Filtration plays a vital role in ensuring the purity, safety, and quality of final products by removing visible impurities such as debris, dirt, hairs, and other contaminants. According to the nature of impurities, filtration is performed at the processing unit at membrane, mechanical, or centrifugal filtration. In mechanical filtration milk is passed through meshes, screen, or strainers to trap solid particles from liquid milk and this is most commonly practiced in dairy processing units. Membrane filtration is used selectively for microorganism's filtration based on pore size. Keeping records of filtration parameters and outcomes is necessary to maintain the quality of milk throughout the processing chain. When the filtration process is completed by the filtration system, it is sanitized to prevent cross contamination. For optimal performance and reliability of the filtration system, regular inspection and maintenance are necessary [17, 20].

2.5 Cooling and Storage of Milk

After undergoing filtration processing, the temperature of raw milk is more than 4 °C, and it can be reduced with the help of plate heat exchanger (PHE). In the dairy processing unit, milk is stored below 4 °C until used for products. To ensure the milk's freshness, bacteriological quality and remains safe for further various dairy products such as pasteurized milk, yogurt, ice cream, cheese or UHT milk. Cooling tanks are a pillar of the dairy industry to maintain the quality of raw milk before processing or converting into final products. The cooling process in a storage tank usually involves a refrigerated cooling system, plate heat exchanger, internal tubular cooler, or jacketed vat. When milk has been adequately cooled, it remains in tanks for temporary holding before further processing. Tanks are made from stainless steel and insulated to control the temperature of milk during storage period. Sensors and monitoring system is installed to utilization of storage capacity, to maintain the temperature and agitator to prevent the cream separation. Samples of storage milk are periodically analyzed for freshness, taste and quality specifications. Milk is transferred from tank to processing unit through a hygienic system (pipes) [20].

2.6 Balance Tank and Regeneration Suction

Balance tank is used to maintain a constant supply of milk to the system (pasteurizer). Raw milk is supplied with the help of a booster (centrifugal) pump into plate heat exchanger (regeneration 1 (R1)) where raw milk is pre-warmed by hot pasteurized milk returning from the holding tube. In the regeneration system, raw and pasteurized are flowing into two different channels in counterflow where heat is exchanged between pasteurized milk and raw milk. Regeneration unit is playing a vital role to recover the energy in industry by heat transfer method.

2.7 Cream Separation and Standardization

Cream separator is a mechanical device which is used to separate the cream from whole milk through a centrifugation process. Cream separator is designed for continuous operation of fat separation. Raw milk is passed through a cream separator with the help of a regulator at very high speed to separate lighter part of milk (fat $\rho = 0.90$) and heavier part (skim milk $\rho = 1.037$). Cream is light in weight and tends to move toward the center, whereas skim milk is heavier in weight and moves away from the center under the action of centripetal and centrifugal forces, respectively. Whole milk is converted into standardized milk, low fat milk, skim milk, and cream with the help of a cream separator. Cream and skim milk are separated in a bowl and go out through a separator. After cream separation from whole milk, the by-product is known as skim milk and it contains less than 0.5% fat. Later on, this skim milk is converted into skim milk powder (SMP), some soft type of cheese, and other flavoring drinks. After the standardization of milk, surplus cream is used for the production of butter or ghee. The efficiency of the separator depends on various factors including speed of bowl, temperature of milk, quality of milk, adjustment screw, and size of fat globules. The best temperature of separation of cream is 40–45 °C, and if the size of fat globules is more than 2 microns, then more cream will be separated. Cream separators have the ability to separate "slime," and slime contains foreign materials including red blood cells, leucocytes, bacteria, udder cells, and dust and dirt particles collected inside the bowl shell of the cream separator. The main objective of de-slugging is to remove insoluble particles from milk to prevent the sedimentation in milk during storage period of processed milk (UHT).

Standardization is the process of adjusting the fat content of milk to a desired level by removing or adding the cream. In the dairy industry, it is very important for legal requirements and consistency of fat in dairy products. Fat analysis of raw milk is important to know the need of standardization. To achieve desired level and accuracy, the adjustment of fat in dairy products is mostly controlled by standardization valves or flow meters and pumps that are installed within the cream separator. Standardization valve on the separator permits the operator to get separated milk of predetermined fat content. By blending of skim milk fractions and cream, various fluids of desired level milk fat are produced. Quality analysis is performed during the process to maintain consistency and accurate fat content of final products [21]

2.8 Homogenization

Homogenization is the mechanical process of breakdown of fat globules into smaller ones and ensures the uniform distribution throughout the milk at very high temperature. In 1904 it was used for the first time in the dairy industry by Gaulin and milk is known as "Gaulin milk." In this method milk is forced through a narrow orifice at very high pressure (2500 PSI) and velocity (210 m/s); as a result a sharing action between fat globules and layer of liquid under flow causes breakup of the fat globules. Before homogenization milk is heated up to 65 °C to inactivate the lipase enzyme because after the homogenization, presence of lipase causes degradation of fat into free fatty acid and aldehydes. Two types of homogenizers are used in the dairy industry for processing of milk. Single stage homogenizer (upstream) is used when milk fat is less than 6% and it is installed before final temperature gain (holding tube). Milk fat is more than 6% or only for UHT treatment; double stage homogenizer is used after single stage homogenizer (downstream) to prevent the clumping of fat and protein after final temperature. Some types of proteins are denatured at UHT and cluster with fat globules and cause age gelation during storage period. The main objectives of homogenization are prevention of cream separation, stable emulsion, maintenance of sensory attributes, consistency improvement, and product acceptance in the marketplace. Milk contains less or greater amounts of gases including carbon dioxide (3.5–4.8%), nitrogen (1%), and oxygen (0.1%), and this amount can be increased during transportation and cooling of milk. These gases exist in milk in the form of dispersion and cause serious problems during milk treatment and homogenization that is why deaeration is necessary before homogenization (*see* **Note 10**) [21].

2.9 Final Heating (Holding Tubes)

Holding tube is a specific length of tube where milk is holding for a time to attain desired temperature. Time of milk in the holding tube depends on the speed of the milk and that is controlled by a timing pump. Basically, a holding tube is a section of piping of sufficient length that provides the desired residence time for the final temperature of milk. For pasteurization and UHT milk, the time–temperature combination is 72 °C for 15 s and 142 for 2–4 s, respectively.

2.10 Aseptic Packaging

After cooling (25 °C) UHT milk is transferred into an aseptic balance tank of a packaging machine, followed by packaging of milk aseptically. Aseptic packaging is defined as packaging of product in a pre-sterilized container under a sterile environment and ensuring that there is no post-processing contamination of milk. Milk packages are stored at room temperature and later are distributed to the market or superstore or directly to consumers [22, 23].

2.11 Cleaning and Sanitization

Sanitizing means to clean, dry, and disinfect equipment, immediately after processing of products. Disinfection means destruction of microorganisms and cleaning means removal of soil (deposit fat, protein, salt) from the surface. During processing, residual products stick on the surface of the process line and are caused by growth of microorganisms. The aseptic intermediate cleaning (AIC), cleaning in place (CIP), and cleaning out place (COP) systems are used for cleaning and sanitization of dairy processing units. To maintain the quality of products, cleaning and sanitization of plant equipment are necessary after processing of products or in a specific interval of processing time according to requirements. The CIP program is operated in different operation conditions according to requirement. CIP cycle is automatically controlled according to program, and general sequence for UHT plant is prerinsing, caustic cleaning, hot-water rinsing, acid cleaning, and final rinsing [24].

3 Methods of Pasteurization

There are two general methods of pasteurization; one is known as "continuous flow method of pasteurization" or "high temperature short time system" and other as the "batch pasteurization" or "holding system" (*see* **Note 7**).

3.1 Continuous Flow Method of Pasteurization

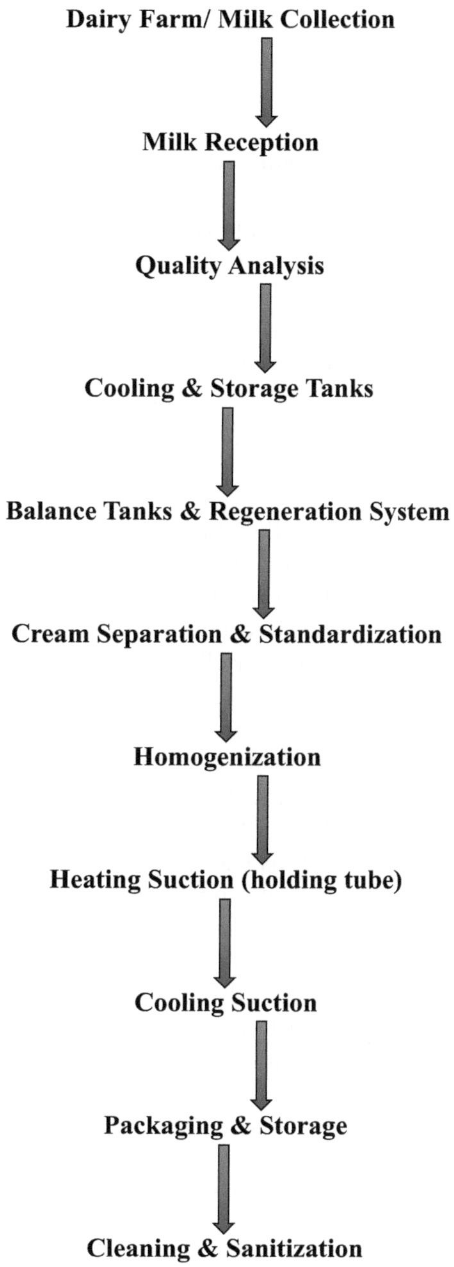

Continuous pasteurization is known as "high temperature short time (HTST) system." According to the HTST method in the dairy industry, milk is heat-treated at 72 °C for 15 s to eliminate pathogen microorganisms and preserve the quality, freshness, and sensory features of milk. The shelf life of milk is enhanced to 1 to 2 weeks at refrigerator temperature. Raw milk is transferred into a balance tank and it is located at the start of the PHE system. The objective of the balance tank is to maintain a constant supply of milk to the system (pasteurizer). Raw milk is pumped with the help of a booster (centrifugal) pump into regeneration 1 (R1) where raw milk is pre-warmed by hot pasteurized milk flowing from the holding tube to cold storage tank. In the regenerator raw and pasteurized milk are flowing in counterflow where heat is exchanged between hot milk and cold milk in a fast method. Fat and solid not fat (SNF) of raw may be adjusted to meet the specification of products through standardization before pasteurization. After preheating milk flows toward a heating suction where it is heated (65 °C) with steam or hot water in stainless steel in the opposite direction of the milk. Now finally the hot milk enters into a holding tube where it is held for at least 15 s to attain pasteurized temperature. Holding time of milk in a holding tube is dependent on the speed of milk and controlled by a timing pump. Basically, the holding tube is a section of piping in continuous flow pasteurizers of sufficient length that provides the desired residence time (15 s) for final heating (72 °C) of the milk. Recorder controller monitors the flow direction with the help of a diversion device. If legally standard temperature is not attained, a flow diversion device transfers the milk again into the balance tank; otherwise milk continues through the system known as pasteurized milk. Pasteurized milk is entered through regeneration 2 (R2) or pasteurized side and gives up heat to cold milk and in return pasteurized milk is partially cooled. After that partially cooled pasteurized milk passes through the cooling section where milk is cooled below 5 °C. Pasteurized milk is transferred into the balance tank of the packaging machine and followed by packaging of milk. Milk packages are stored in cold/blast rooms to maintain freshness and quality of product. Pasteurized products are distributed to the market or superstore or directly to consumers. To maintain the quality, safety, and shelf life of products, cold chain management is necessary during transportation of products.

3.2 The Holding System/Batch Pasteurization

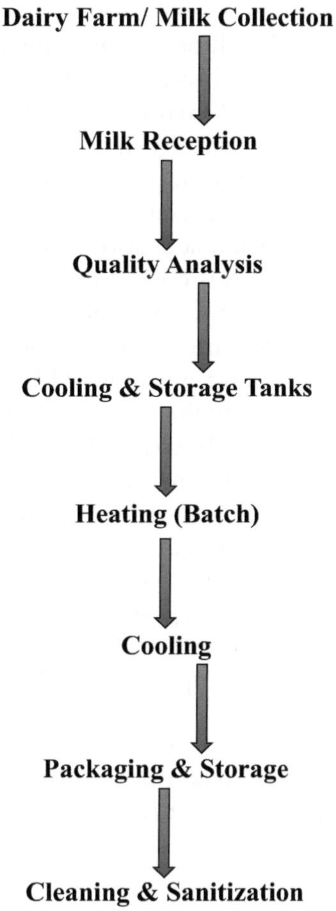

The holding system is also known as "low temperature long time (LTLT)" or "batch pasteurization" (*see* **Notes 8–9**). Batch pasteurization involved the following steps. First of all, milk is transferred to pasteurization vat where milk is heated by using steam or hot water at 65 °C (145–150 °F) for 30 min or according to product specification and then immediately cooled below 5 °C. All components (if needed) such as milk powder, stabilizer, flavorings, and vitamins are added prior to beginning the pasteurization. To ensure that each particle of milk is heated with desired time temperature, it is necessary that properly designated equipment is installed for pasteurization of milk. In batch pasteurization vats are used for heating and cooling in a stainless-steel jacketed vat. In batch processing time and temperature are monitored by a control panel to ensure effective microbial reduction and to prevent undesired changes in sensory attributes of milk or composition. For uniformity in temperature distribution and to prevent the fat separation in pasteurization vat, an agitator is installed. Milk is kept in pasteurization vat for a specific duration (30 min) to ensure microbial inactivation. After completing the holding period, pasteurized milk is cooled down immediately and transferred into a temporary storage tank before packing. The milk is packed into sterilized packing materials such as pouches or bottles using an automated filling machine. Milk packages are stored in cold/blast rooms to maintain freshness and quality of product. Pasteurized products are distributed to the market or superstore or directly to consumers. To maintain the quality, safety, and shelf life of products, cold chain management is necessary during transportation of products [10].

Batch pasteurization units have various advantages including minimal equipment requirements, ease of operation, flexibility capacity, and also consumer demand for minimal processes involved, despite its limitations such as small processing capacity as compared to flash pasteurization.

3.3 In the UHT Processing of Milk, the Following Steps (Units) Are Generally Involved

There are two general methods of ultrahigh-temperature treatment of milk; one is known as "direct heating method" and other as the "indirect heating method."

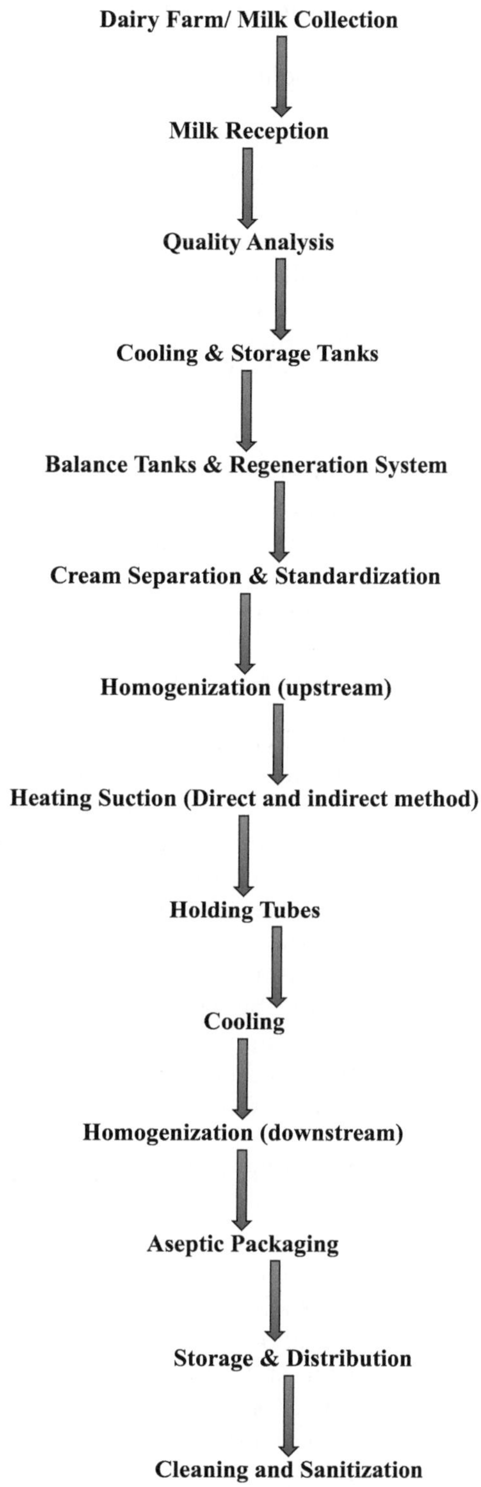
Dairy Farm/ Milk Collection
Milk Reception
Quality Analysis
Cooling & Storage Tanks
Balance Tanks & Regeneration System
Cream Separation & Standardization
Homogenization (upstream)
Heating Suction (Direct and indirect method)
Holding Tubes
Cooling
Homogenization (downstream)
Aseptic Packaging
Storage & Distribution
Cleaning and Sanitization

Plant is pre-sterilized before starting the production of UHT product in order to avoid reinfection of the treated product. Hot water is used for sterilization of dairy plant for 30 minutes and cools by cooling water according to desired processing temperature. In UHT treatment (*see* **Note 11**), raw milk is supplied toward balance tanks and to regeneration suction to raise the temperature up to 50 °C. After regeneration suction milk is transferred to standardization suction where milk is standardized according to legal requirement (fat 3.5% and SNF 8.5%). In case the concentration of milk components is less or greater than the desired level, then fat may be added or removed from raw milk by cream separator. After that milk goes to aseptic homogenizer for homogenization (if needed) and then moves to heating suction (direct or indirect heating). When heating is done, milk is transferred for final heating (143 °C/4 s) in holding tube, and at the end of the holding tube, a flow diversion valve (FDV) is installed for final temperature of milk; if temperature of milk is not attained, it diverts the milk flow toward balance tank. After going through the holding tube, the milk undergoes rapid cooling suction where it is cooled; the process is followed by step-down homogenization, aseptic packaging, storage, and distribution. The details of direct and indirect heating system are given below [22, 25]:

3.3.1 Direct Heating Method of UHT Milk

In a direct method, steam is directly injected into milk with the help of a steam injector. The steam is mixed with milk for a very short time. There is no medium between steam and milk and so there is almost instantaneous heating at 80–143 °C in 1–2 s; this is followed by flash cooling (Valero et al. 2001). Milk is heated by three direct methods: (a) injection heating (steam is directly injected into milk to heat the milk), (b) infusion heating (processing is through steam-into-milk arrangement), and (c) electric heating (milk is heated by electrical coil). In direct method heating is very rapid with minimal chemical changes. Direct UHT treatment is preferred than indirect UHT treatment because in this treatment there are less chances of browning (Maillard reaction) (*see* **Notes 12–15**).

Method

Raw milk (4 °C) supplied to balance the tank from storage tank and forwarded by feed pump to PHE for regeneration. After preheating milk is fed into a cream separator for standardization of fat followed by homogenizer (if needed) for homogenization of milk to improve the stability of fat and protein during storage period. After homogenization milk is forwarded to heating suction (heating regeneration) and into a pressure pump where milk pressure is increased by 4 bar (preventing the product from boiling) and the product continues in steam. Steam is injected by a nozzle injector and instantly raises the product temperature, and the product goes to the holding tube for final temperature (136–141 °C/4 s), and then it is flash cooled. To prevent the sedimentation and coagulation in milk during storage period and it is aseptic homogenization at high pressure then the product is cooled to 25 °C in PHE and milk is intermediate stored in an aseptic storage tank to direct the aseptic filling machine. [22, 26].

3.3.2 Indirect Heating Method of UHT Milk

In indirect method heating medium and milk to be heated never contact each other, separated by metal wall or plates. Milk and steam flow into two different channels and heat is exchanged by conduction method. In the dairy industry, tubular and plate heat exchangers are mostly used. Plate heat exchangers (PHE) have a large heating surface as compared to tubular heat exchangers. In PHE several rectangular stainless-steel plates with corrugations are arranged in sequences. Corrugation in plates produced turbulence for better heat transfer. One disadvantage of PHE is leakage of jackets which causes the mixing of raw and processed milk during the regeneration process in PHE. Tubular heat exchanger is majorly of two types: (a) concentric tube, where milk tube is surrounded by steam tubes, and (b) shell and tube, where milk tubes are surrounded by steam chamber. Tubular heat exchangers are best for high temperature and pressure of milk with no chance of mixing of milk. Indirect UHT treatment of milk is more susceptible to cooking and off-flavor due to fouling of milk with tubular walls. In this method changes in chemical quality are greater than direct method because heat load on product is greater and raises temperature very gradually.

Method

Raw milk (4 °C) supplied to balance the tank from storage tank and forwarded by feed pump to PHE for regeneration. After preheating, milk is fed into a cream separator for standardization of fat followed by homogenizer for homogenization at 180–250 bar pressure of milk to improve the stability of fat and protein during storage period. Non-aseptic homogenizer can be used in upstream homogenization in indirect heating systems. Preheated homogenized milk continues to heating suction of tube heat exchanger or plate heat exchanger where it is heated about 135 °C by indirect heating medium and to holding tube for final temperature (141 °C for 4 s); then rapid cooling is done. To prevent the sedimentation and coagulation in milk during storage period and it is necessary aseptic homogenization is performed then the product is cooled to 25 °C through PHE (*see* **Note 16**) [22].

4 Notes

1. Milk safety is started before, during, and after milking at farm. Milk is free from dirt, gases, heavy metal, residues of pesticides, antibodies, and mastitis. Milk from unhealthy teats is known as

mastitis milk. In mastitis milk somatic cell count and chlorine content are greater than normal milk.

2. Colostrum is a rich source of antibodies, proteins, fat, and minerals as compared to normal milk. Colostrum should not mix with processed milk. After calving the first 3–5 days' milk is known as colostrum.
3. Farmers and milking animals should be healthy and safe from internal disease for good quality of milk. It is necessary that milk from animals suffering with any disease or under treatment does not mix with healthy animal milk stream for processing.
4. Farmers try to transport milk as soon as possible to processing plants because at 4 °C some cold-tolerant bacteria produce spores (lipase, proteases) that spoil milk.
5. Milk containers should be made with stainless steel and not iron, copper, or non-food-grade material. Containers with wide mouth are used for easy shifting of milk and easy container cleaning.
6. In some countries, farmers or milk resellers adopted illegal practices to get maximum benefits by adding adulterants including whey powder, vegetable fat, preservatives, and neutralizer. Adulterated milk is harmful for human health, so at the time of receiving the milk at processing plant, these tests are performed for identification of any adulterants.
7. Pasteurized milk is safe milk without significantly altering its composition and nutritional profile. It's important to note that pasteurized milk effectively reduces the risk of foodborne illness in humans by pathogens in raw milk including *E. coli*, *Salmonella*, and *Listeria*. To check the proper pasteurization, phosphatase enzyme activity is analyzed because at pasteurized temperature phosphatase enzyme is destroyed.
8. Batch pasteurization can result in inconsistent processing and limitations arise because each batch must be processed individually and each product has different sensory attributes. Continuous pasteurization is energy saving processing, particularly when PHE is installed. In PHE heat is exchanged from hot milk to cool milk during the process.
9. Bactofugation is an important instrument used to remove bacteria from raw milk and also known as bactofuge or bacteria separator. It is typically required at modern dairy processing plants or cheese units to remove the microorganisms for better quality of products.
10. Homogenization is necessary for UHT milk because in unhomogenized milk cream is separated at the top and denatured proteins are deposited on the bottom. Milk for cheese and

yogurt production is preferred unhomogenized for better curd formation.

11. Ultrahigh-temperature (UHT) milk is known as long-life milk and undergoes very high temperature for a few seconds, and at that temperature all types of microorganisms (spore-forming, pathogen, vegetative) are destroyed. Despite its convenience for transport, storage, usage and lead to some drawback's loss of heat-sensitive nutrients including vitamins, antioxidants and enzymes.
12. In UHT milk color, cooked flavor, and taste are changed as compared to fresh or pasteurized milk.
13. Maillard reaction or browning reaction in UHT milk is due to reaction of milk sugar with protein (lysine) of milk at high temperature. The intensity of Maillard or browning reaction is dependent upon heat intensity and types of UHT processing.
14. In the direct method, milk is directly in contact with the heating medium, while indirect heat is exchanged with the help of any medium (plates, tubes) without direct contact with the heating medium. Both methods are effective for heat treatment, but they offer different advantages in terms of efficiency and quality control of milk preservation.
15. Cooked flavor, Maillard reaction, and denaturation of whey proteins in indirect UHT heating method are greater as compared to direct heating method.
16. In developing countries pasteurized milk is more acceptable as compared to UHT milk due to its affordable price. UHT processing is more expensive than traditional pasteurization because it requires high temperatures, specialized equipment, and the packaging materials (Tetra Pak cartons). In UHT milk higher production is expensive and that can enhance the retail price of final products.

References

1. Oftedal OT (2002) The mammary gland and its origin during synapsid evolution. J Mammary Gland Biol Neoplasia 7(3):225–252
2. Giribaldi M, Coscia A, Peila C, Antoniazzi S, Lamberti C, Ortoffi M (2016) Pasteurization of human milk by a benchtop high-temperature short-time device. Innov Food Sci Emerg Technol 36:228–233
3. Walstra P, Wouters JTM, Geurts TJ (2006) Dairy science and technology. Taylor & Francis, Boca Raton
4. Kelly AL, Datta N, Deeth HC (2006) Thermal processing of dairy products. In: Sun D (ed) Thermal food processing: new technologies and quality issues. CRC Press, Boca Raton, pp 266–292
5. International Dairy Federation (2022) Milk heat treatment Bulletin:516/2022
6. Dairy Management Inc (DMI) (2017) DMI quarterly retail snapshot: Milk Q1 2017. Accessed August 2017. https://mmistorebrands.com/wp-content/uploads/2017/11/DMI-Quarterly-Milk-Snapshot-Q1
7. Poushi MK, Sharifi D (2024) State-of-the-art in milk processing for improvement of the quality of pasteurized milk and UHT milk. Academic, pp 19–27

8. Sakkas L, Moutafi A, Moschopoulou E, Moatsou G (2014) Assessment of heat treatment of various types of milk. Food Chem 159:293–301
9. Holdsworth SD, Simpson R (2007) Sterilization, pasteurization and cooking criteria. In: Thermal Proce pack foods, 3rd edn. Springer, New York, pp 123–138
10. Ajmal M, Nadeem M, Imran M, Gulzar N, Batool M Tayyab M (2018) Impact of immediate and delayed chilling of raw Milk on chemical changes in lipid fraction of pasteurized Milk. Lip in Heal and Dise 17:190
11. Rauh PV, Yinghua Xiao Y (2021) The shelf life of heat-treated dairy products. Int Dairy J 125
12. Tomasula PM, Kozempel MF (2004) Flow characteristics of a pilot-scale high temperature, short time pasteurizer. J Dairy Sci 87: 2761–2768
13. O'Connor TP, O'Brien NM (2006) Lipid oxidation. In: Fox PF, PLH MS (eds) Advanced dairy chemistry Lipids, vol 2. Springer, New York, pp 557–600
14. Bimbo F, Bonanno A, Viscecchia B (2016) Hedonic analysis of the price of UHT treated milk in Italy. J Dairy Sci 99:1095–1102
15. Kilara A (2013) Principles of dairy processing. In: Chandan RC, Kilara A (eds) Manufacturing yogurt and fermented milks, 2nd edn. John Wiley & Sons, Chichester, UK, pp 95–113
16. Davis JG (1951) Milk testing: the laboratory control of Milk. Dairy Industries Ltd., London, UK. Page 128
17. Sommer HH (1938) Page 147 in market Milk and related products. Madison, Wisconsin
18. American public health Association (APHA) (1978) Standard methods for the examination of dairy products, 14th edn, Washington, DC, pp 355–357
19. Deeth H, Lewis MJ (2017) High temperature processing of milk and milk products. Wiley-Blackwell, Chichester
20. Walstra P, Walstra P, Wouters JTM, Geurts TJ (2006) Dairy science and technology, 2nd edn. CRC Press, p 782
21. Lee CL, Liao HL, Lee WC, Hsu CK, Hsueh FC, Pan JQ, Chu CH, Wei CT, Chen MJ (2018) Standards and labeling of milk fat and spread products in different countries. J Food and Drug Anal 20(2):469–480
22. Datta N, Elliott AJ, Perkins ML, Deeth HC (2002) Ultra-high-temperature (UHT) treatment of milk. Austr J Dairy Technol 57(3):211
23. Pereira JR, Tamanini R, Rios EA, Oliveira VHS, Yamamura AAM, Beloti V (2013) Microbiota mesófila aeróbia contaminante do leite UHT. Rev Inst Laticínios Cândido Tostes Juiz de Fora 68:25–31
24. Ribeiro-Furtini LL, Abreu LR (2006) Utilização de APPCC na indústria de alimentos. Rev Ciênc Agrotec 30:358–363
25. Vianna PCB, Walter EHM, Dias MEF, Faria JAF, Netto FM, Gigante ML (2012) Effect of addition of CO_2 to raw milk on quality of UHT-treated milk. J Dairy Sci 95(8): 4256–4262
26. Burton H (1988) Ultra-high-temperature processing of Milk and Milk products. Elsevier, London

Chapter 2

Flavored Milk

Maria Eduarda Marques Soutelino, Bianca Cristina Rocha de Oliveira, Ramon da Silva Rocha, and Adriana Cristina de Oliveira Silva

Abstract

Flavored milk's main characteristic is that it is a fluid product, with the addition of a few ingredients and a simple technological process. It is a product with great market acceptance, especially for children and has great potential in incorporating new ingredients and functional components, aiming to reach new audiences. This chapter presents the main technological approaches in flavored milk and discusses innovations and perspectives. Several technological modifications are applied to flavored milk to improve the texture and sensorial characteristics of the product. Furthermore, a major appeal of these products is the reduction of added sugar through sweeteners and fruits (pulps and extracts) to meet the demand for healthy products and reduce the incidence of chronic diseases. Finally, emerging technologies have been widely used in the formulation of flavored milk, which produces products similar to conventional pasteurization, in addition to reducing energy expenditure in processing and offering a greater quantity of bioactive compounds. In any case, it is a product with great potential and has become increasingly common in the consumer's diet.

Key words Dairy products, Flavored milk, Milk drink, Dairy technology

1 Introduction

Flavored milk is a dairy product obtained by adding flavorings to milk, such as cocoa powder, essences, and fruit extracts. Optional ingredients can also be part of the formulation, including sweeteners, colorings, stabilizers, and thickeners [1, 2]. Flavored milk generally has an average of 5% fat, 4–6% sugar, 0.1–0.5% flavoring, 0.1–0.5% coloring, and 0.1–0.3% stabilizers [3]. Among its main characteristics, the simple and low-cost formulation, low acidity, viscosity, and fluid and homogeneous texture stand out, which results in good sensorial and market acceptance [2].

Flavored milk is estimated to represent 28.1% of total liquid milk consumption [4]. Consumption of these products is expected to reach 20.5 billion liters in 2023, with a growth of 4.1% in 2022. Currently, Asia leads its global production, with a share of around 40%, and China represents the most significant global consumer,

Adriano Gomes da Cruz et al. (eds.), *Dairy Foods Processing*, Methods and Protocols in Food Science,
https://doi.org/10.1007/978-1-0716-4144-6_2,

followed by the United States and India [5]. The growth of the child and adolescent population (primary consumers of flavored milk), the increase in demand for ready-to-drink products, and the launch of new flavors and packaging contributed to the rise of this market.

The preference for flavored milk flavors may vary according to the age of consumers. Children and teenagers, for example, tend to prefer sweeter flavors, such as chocolate and strawberry, while adults prefer more complex flavors, such as fruit, coffee, and green tea. Predilection may also vary between regions; Chinese consumers highly accept the green tea flavor, while American consumers highly accept the vanilla flavor. The chocolate flavor is popular, representing around 50% of the global flavored milk market [6].

Regarding the nutritional contribution of flavored milk, this product is a source of essential nutrients for the diet. It can contribute up to 6% of total energy in a usual diet, in addition to 12% fat, 3% sodium, 4% sugar, 5% vitamin A, 11% vitamin D, 6% calcium, 4% potassium, 3% magnesium, and 4% phosphorus [7]. Some studies report controversial effects of its consumption about health benefits, as it can contribute to a greater intake of sugars and an increased incidence of chronic noncommunicable diseases such as diabetes mellitus and obesity [8–10]. However, as it is a highly versatile food matrix, flavored milk presents countless possibilities for innovation, such as incorporating functional ingredients, replacing sugars with sweeteners, and processing conditions that meet different market demands [1, 11].

In this scenario, several emerging technologies (ET) emerge, mainly aiming at minimal effects on this product's sensorial and nutritional aspects and lower environmental impacts. The new consumer market is increasingly aware of how food is produced, prioritizing brands that guarantee quality, reliability, sustainability, and ethics and using clean technologies from renewable energy sources. Ohmic heating to replace conventional pasteurization, for example, can help form peptides with bioactive properties and easy digestibility and reduce the production of unpleasant aromas and flavors [12] (**Note 6**).

Another nonthermal technique, such as high-intensity ultrasound, has also been studied in manufacturing dairy products, guaranteeing microbiological quality and viability and probiotic cultures and improving the homogenization of ingredients [13]. Finally, microencapsulation has also proven to be a promising technology in the manufacture of flavored milk, mainly in improving the stability, solubility, and bioavailability of functional compounds and increasing the shelf life of these products (**Note 3**). Therefore, this chapter will cover the processing stages of flavored milk and the main innovations and perspectives in this sector.

2 Flavored Milk Processing

2.1 Adding Ingredients

Adding ingredients adds flavor and texture and, depending on the ingredient, increases the nutritional quality of flavored milk, and consumers' choice depends on a series of factors, such as the type of product, target audience, and cost. The most commonly added ingredients include sweeteners, flavoring syrups, colorings, thickeners, and stabilizers.

In the conventional processing of flavored milk, the ingredients are added before heat treatment, which significantly contributes to the safety of the final product. However, in establishments where fresh milk and ingredients are prohibited in the same equipment, these can be added to pasteurized milk without subsequent heating. When this occurs, there is a risk of cross-contamination, especially in cases where the added components have not undergone any treatment that guarantees their safety [3].

2.1.1 Sweeteners

Sweeteners are added to the formulation of flavored milk to give the milk a sweet flavor, and they can be natural or artificial (**Note 1**). The industry's choice of sweetener type depends on a series of factors, including achieving the desired flavor and nutritional value and meeting consumer demand, which is increasingly concerned about the adverse effects of consuming this ingredient (**Note 2**).

In general, the natural sweeteners most used in the processing of flavored milk are sucrose, fructose, glucose, xylitol, and stevia. Among the artificial ones are saccharin, aspartame, sucralose, and acesulfame-K, with a maximum concentration of around 20 mg per 200 ml portion of the product [14].

2.1.2 Dyes

Dyes or color additives are substances that color foods and beverages and directly influence consumer acceptance, consumption, preference, and purchase intention. They are added to flavored milk to give it a more attractive visual appearance or even mask unwanted colors [15].

These ingredients can also be classified as natural or artificial, with artificial ones being the most used in the manufacture of flavored milk, with emphasis on tartrazine, twilight yellow, azorubine, and brilliant blue due to their more excellent stability, lower cost, and more intense color perception [16].

Despite the advantages of artificial colors, there is a high market demand for "clean label" foods, putting pressure on industries to replace synthetic ingredients with natural ones. However, some natural dyes may have limitations, such as high instability, low dye potential, and interaction with other ingredients [17]. Therefore, the most commonly used by the industry are generally carmine (in red fruit, strawberry, raspberry, and cherry-flavored milk) and curcumin and beta-carotene (in yellow fruit flavored milk, such as banana, mango, and passion fruit).

2.1.3 Flavoring Syrups

Syrups are used as sweetening, coloring, and flavoring agents. They also help to stabilize the product, as they can bind to proteins, fat, and water present in milk, increase its viscosity, and prevent the formation of lumps. Furthermore, they can reduce the product's water activity and microbial growth, helping maintain its useful life [14].

The choice of syrup for flavored milk depends mainly on the desired flavor. Syrups based on glucose, fructose, and corn are used most in products with simple flavors, such as strawberry, banana, or chocolate. Syrups such as maple and agave are used for more complex flavors, such as mint and coffee [18].

2.1.4 Stabilizers and Thickeners

Stabilizers and thickeners are added to flavored milk to improve its texture and stability. These ingredients help keep the milk components suspended evenly, preventing phase separation and the formation of lumps, resulting in a creamy and uniform texture, even after stirring. Like syrups, they also reduce the water activity of milk, making it difficult for bacteria to grow [19].

The most common thickeners used to improve the texture and stability of flavored milk are xanthan gum, carrageenan, gelatin, and guar gum. The change in texture results from the interaction between proteins, fat, and the thickening agent under heating through the formation of complexes via electrostatic interaction that can absorb a large amount of water, maintaining the system's structural integrity [20].

2.2 Heat Treatment

Heat treatment is mandatory in the manufacture of flavored milk, with the main objective of ensuring its microbiological safety by reducing the risk of multiplication of pathogenic and spoilage agents. However, this stage can be considered a critical processing point, as, in addition to being related to the product's safety, it can influence its sensory aspects. When milk is heated to high temperatures and for a prolonged period, the serum proteins can denature, and the flavored milk can acquire a more viscous texture, which can hurt the characteristics of this product. Lactose can also undergo several modifications, including caramelization, hydrolysis, and Maillard reactions. Finally, fat can coalesce and separate in milk, and some water-soluble vitamins, such as vitamin C, can oxidize, influencing the sensorial and nutritional quality of the final product [21].

Heating milk can also promote the formation of flavor compounds such as lactone gamma-butyrolactone, which gives the product a butter flavor and is desirable in manufacturing some flavored milk flavors [22]. Furthermore, aldehydes and ketones may form, which give a sweet, fruity, and caramelized flavor, as well as free fatty acids, which contribute negatively to the "meat flavor" of the milk. The amount and type of flavor compounds formed depend on several factors, including temperature, heating time, and the presence of other ingredients in the milk [23]. When processing flavored milk, heat treatment can occur through pasteurization, UHT (ultrahigh temperature) treatment, and sterilization, preferably carried out after adding the ingredients.

2.2.1 Pasteurization

Pasteurization is carried out in the dairy industry with the main objective of guaranteeing its microbiological quality and can occur quickly or slowly. Rapid pasteurization occurs in a continuous flow system in plate heat exchangers, where the milk is heated at a binomial time and temperature of 72 °C to 75 °C for 15 to 20 s. In slow pasteurization, milk is heated by heat transfer in the range of 63 °C to 65 °C for 30 min inside double-jacketed tanks in a batch system [24].

In general, rapid pasteurization is the most used in the manufacture of flavored milk, as it is more efficient in terms of time, energy, and microbiological and enzymatic inactivation and has fewer adverse effects on the flavor and texture of the milk. On the other hand, slow pasteurization is more used in industries that work with smaller volumes of milk and do not have a plate heat exchanger, which presents a higher cost for installation and maintenance [24].

2.2.2 UHT (Ultrahigh Temperature) Treatment

UHT treatment can also be carried out on flavored milk to ensure greater commercial validity and storage for extended periods. The flavored milk is subjected to more intense heating, from 130 °C to 150 °C, for 2 to 4 s, then cooling to 32 °C, and homogenization. During this stage, the addition of stabilizers to the milk is permitted, which reduces the risk of protein precipitation and the increase in milk viscosity during the storage period.

The homogenization step is crucial in flavored milk subjected to UHT treatment, as it prevents the separation of fat and the formation of a surface layer of cream in the flavored milk packaging, which is considered a technological defect. This flavored milk must be packaged under aseptic conditions in sterilized, hermetically sealed packaging [24]. All of these steps, associated with selecting quality raw materials, are essential to guarantee a longer useful life for this product.

When compared to conventional pasteurization, UHT treatment can cause greater changes in the natural composition of the milk, resulting, for example, in a thinner and more fluid texture and the possible formation of a cooked flavor. The cooked taste in UHT milk is related to protein denaturation and the decomposition of amino acids into compounds with a bitter and pungent taste, such as acrolein, derived from glycine [25]. However, adding some ingredients such as sugar and other flavoring and flavoring agents to flavored milk helps mask this undesirable sensory change. The commercial validity of UHT flavored milk is approximately 4 months from the date of manufacture when kept in closed packaging and at room temperature. After the packaging is opened, UHT flavored milk must be consumed within 3 days and kept under refrigeration.

2.2.3 Sterilization

Unlike pasteurization and UHT treatment, sterilization is carried out with the flavored milk already in its packaging after adding ingredients. Thermal treatment occurs in autoclaves at 100 °C to 121 °C, ranging from 15 to 60 mins, and in a vacuum environment. This condition removes free oxygen, preventing the oxidation of milk components and minimizing the release of undesirable aromatic compounds derived from protein denaturation and enzymatic degradation [26].

Therefore, vacuum sterilization has some advantages over UHT treatment, such as fewer effects on flavor and texture, lower risk of formation of volatile compounds, and greater stability of flavored milk. This process can eliminate all pathogenic and spoilage microorganisms, guaranteeing the safety and commercial validity of the product for around 6 months without refrigeration [27]. Despite these advantages, it is a slower and more expensive alternative, requiring specialized equipment and high energy consumption.

2.3 Homogenization

After heat treatment, the milk is homogenized to distribute its components and added ingredients evenly. The process is usually carried out in a homogenizer, which consists of two chambers separated by a plate with small holes. The milk is pumped under pressure through the plate, which reduces fats and other ingredients into small particles, preventing phase separation and the formation of lumps [19].

The amount of pressure used in the homogenization process depends on the type of flavored milk. Flavored whole milk, for example, requires more pressure than flavored skimmed milk to break down fat globules efficiently. Furthermore, the pressure can also be changed according to the desired consistency and flavor, as higher pressure results in creamier milk with a more intensified flavor. In strawberry-flavored milk, for example, higher pressure may result in a product with a more uniform texture and a more concentrated flavor. In comparison, lower pressure may promote a grainier texture and milder flavor [28].

2.4 Packaging

Packaging is an important process to guarantee the quality and safety of the product. It must be carried out to avoid milk contamination and the loss of nutrients and flavor. Flavored milk is packaged below 4 °C in multi-carton packaging made of plastic or glass, which varies according to the heat treatment subjected.

Aseptic packaging of UHT flavored milk is generally carried out in multi-carton packaging, sterilized with 5% hydrogen peroxide and hot air at 270 °C. This step is carried out without oxygen and in a sterile environment, preventing the entry of microorganisms, humidity, and dirt. Most multi-carton packaging has layers of low-density polyethylene, aluminum, and paper, guaranteeing product safety through an impermeable structure [24]. On the other hand, the packaging of sterilized flavored milk is generally carried out in glass packaging, as they are highly resistant to the heat and pressure of the autoclave. In contrast, pasteurized flavored milk can be packaged in plastic or glass.

Factors that must be considered when choosing packaging include resistance, as it must remain intact after impacts and different storage conditions; safety to avoid the product's physical, chemical, and microbiological contamination; and sustainability, aiming to reduce environmental damage [29].

2.5 Optional Steps

In addition to the basic steps, flavored milk processing may include evaporation or drying steps after adding ingredients. Evaporation is a process that removes part of the water from milk to concentrate its flavor and nutrients and obtain a more consistent texture. This process generally takes place in boilers and can be carried out under vacuum at temperatures between 50 °C and 60 °C or under pressure at temperatures between 80 °C and 90 °C. The evaporation time and temperature binomial choice depend on factors such as the type of flavored milk to be produced and the available equipment [30].

On the other hand, drying is a process that removes all the water from milk, which results in a powdered product that can be reconstituted with water. The two main methods are spray drying and atomization. Spray drying is a batch process in which flavored milk is spread thinly on a hot surface, usually a drum or turntable. In contrast, spray drying consists of a continuous process in which flavored milk is sprayed in a fine jet through a spray nozzle and then dried by hot air. A spray dryer is slower than spray drying but guarantees greater product quality parameters maintenance [30].

In short, the processing of flavored milk involves a few steps, each of which is of great importance in guaranteeing the quality of this product. In other words, various factors and manufacturing conditions can influence its texture, viscosity, taste, color, and safety parameters, from adding ingredients to packaging.

In addition, because it is a highly versatile dairy matrix, flavored milk allows for various technological innovations that meet new consumer demands, from reducing or replacing sugars by adding functional ingredients to using emerging technologies, demonstrating greater efficiency in processes with less environmental impact. Therefore, the market and scientific research into flavored milks have a high potential for growth in the coming years, especially with new products with a healthier and more sustainable appeal (**Notes 4** and **5**).

3 Notes

1. To help reduce the incidence of chronic noncommunicable diseases, such as obesity and diabetes, the World Health Organization (WHO) has encouraged small changes in product formulation, such as replacing sugars with low-calorie sweeteners [31]. In this scenario, consumers still have a general predilection for natural sweeteners, given that, despite the maximum recommended doses of artificial sweeteners being safe for consumption, some studies have already investigated the relationship between the consumption of these ingredients and an increased risk of cancer [32], metabolic diseases, and insulin resistance [33].
2. Replacing artificial sweeteners with natural ones can challenge the flavored milk industry, as it can modify important sensory characteristics such as color, flavor, aroma, and texture. An alternative to this problem is the technique of microencapsulating compounds, which can improve the viability, stability, and solubility of sweeteners in the product [34].
3. The microencapsulation technique has been widely studied in dairy products, mainly in the microencapsulation of probiotics [34]. However, studies on applying this technique to flavored milk still need to be made available in the literature, as the heat treatment subjected to this product hinders the viability of probiotic strains [35]. On the other hand, other functional compounds, such as prebiotics, can be easily incorporated into these products. As they are substrates used selectively by beneficial host microorganisms resistant to stages of the digestive process, these compounds modify the composition and activity of the intestinal microbiota and generate health benefits. The most used ones consist of soluble fibers, gums, resistant starches, pectin, and oligosaccharides, commonly applied in products such as dairy drinks, yogurts, and ice creams.
4. Emerging technologies are inserted in the processing of flavored milk mainly in the context of replacing conventional heat treatments with greater efficiency, sustainability, and lower cost. Several emerging technologies have been applied to dairy products, and in particular, ohmic heating (OH) has demonstrated great efficiency in the processing of flavored milk [36], especially with a high protein content [11]. This technology converts electrical energy into thermal energy and is highly efficient in enzymatic and microbiological

inactivation, nutritional maintenance, the formation of bioactive compounds, and the reduction of operating costs [37, 38].

5. High-intensity ultrasound (US) to replace conventional heat treatments has also guaranteed the quality and safety of dairy products [39], resulting in improvements in their sensory and textural characteristics. The inactivation of microorganisms by this method, including sporulated ones, is highly efficient due to the phenomenon of cavitation, which can occur at frequencies <2.5 MHz [40].
6. ET can have positive rheological effects on products, such as the formation of gel networks [40], physicochemical and functional modifications, and the formation of bioactive compounds [41], which makes this technology potentially viable in the processing of flavored milk.

Acknowledgments

Ramon S. Rocha wants to thank the grant 2022/14824-4, São Paulo Research Foundation (FAPESP).

References

1. Arab M, Razavi SH, Hosseini SM, Nayebzadeh K, Meybodi NM, Khanniri E, Mardi P, Mortazavian AM (2019) Production and characterization of functional flavored milk and flavored fermented milk using microencapsulated canthaxanthin. LWT 114(2):108373. https://doi.org/10.1016/j.lwt.2019.108373
2. Filho ER, Prestes AA, Canella MH, Prudencio ES, Freitas MQ, Pimentel TC, Esmerion EA, da Cruz AG (2023) Sensory Aettributes of liquid milk products. In: Sensory profiling of dairy products, pp 81–102. https://doi.org/10.1002/9781119619383.ch6
3. Holkar CR, Jadhav AJ, Pinjari DV (2019) A critical review of possible remediation of sediment in cocoa/coffee flavored milk. Trends Food Sci Technol 86:199–208. https://doi.org/10.1016/j.tifs.2019.02.035
4. Chen B, Wang X, Zhang Y, Zhang W, Pang X, Zhang S, Lu J, Lv J (2023) Determination and risk assessment of flavor components in flavored milk. Food Secur 12(11):2151. https://doi.org/10.3390/foods12112151
5. Tetra Pak (2023) Tetra Pak dairy index. https://www.tetrapak.com/pt-pt/insights/tetra-pak-index/tetra-pak-index-2023. Accessed 11 Oct 2023
6. Mintel Flavored Milk: Global Trends and Opportunities. Mintel (2023) https://store.mintel.com/report/milk-and-flavoured-milk-china-2023. Accessed 11 Oct 2023
7. Nicklas TA, O'Neil CE, Fulgoni VL III (2013) The nutritional role of flavored and white milk in the diets of children. J Sch Health 83(10):728–733. https://doi.org/10.1111/josh.12087
8. Fayet-Moore F (2016) Effect of flavored milk vs plain milk on total milk intake and nutrient provision in children. Nutr Rev 74(1):1–17. https://doi.org/10.1093/nutri/nuv031
9. Mahato DK, Oliver P, Keast R, Liem DG, Russell CG, Mohebbi M, Cicerale S, Mahmud SC, Gamlath S (2021) Identifying ideal product composition of chocolate-flavored milk using preference mapping. J Food Sci 86(7):3205–3218. https://doi.org/10.1111/1750-3841.15817
10. Ricklefs-Johnson K, Pikosky MA (2023) Perspective: The Benefits of Including Flavored Milk in Healthy Dietary Patterns. Adv Nutr 14(15):959–972. https://doi.org/10.1016/j.advnut.2023.06.002
11. Rocha RS, Oliveira LB, Morais STB, Soutelino MEM, Silva MC, Freitas MQ, Junior SB, Pimentel TC, Esmerino EA, Cruz AG (2023) Effect of ohmic heating in bioactive peptides, volatile compounds, and fatty acid profile in a high-protein vanilla flavoured milk drink. Int

Dairy J 146:105733. https://doi.org/10.1016/j.idairyj.2023.105733

12. Rocha RS, Cavalcanti RN, Silva R et al (2020) Consumer acceptance and sensory drivers of liking of Minas Frescal Minas cheese manufactured using milk subjected to ohmic heating: Performance of machine learning methods. LWT 126:109342. https://doi.org/10.1016/j.lwt.2020.109342
13. Guimarães J, Silva EK, de Freitas MQ, de Almeida Meireles MA, Cruz AG (2018) Non-thermal emerging technologies and their effects on the functional properties of dairy products. Curr Opin Food Sci 22:62–66. https://doi.org/10.1016/j.cofs.2018.01.015
14. Ferreira AS, Oliveira ML (2023) Nutrition: science and practice. Manole, São Paulo, SP
15. Durazzo A, Carocho M, Heleno S, Barros L, Souto EB, Santini A, Lucarini M (2022) Food dyes and health: literature quantitative research analysis. Meas: Food (1):7, 100050. https://doi.org/10.1016/j.meafoo.2022.100050
16. Luzardo-Ocampo I, Ramírez-Jimenez AK, Yañez J, Mojica L, Luana-Vital DA (2021) Technological applications of natural colorants in food systems: a review. Food Secur 10(3): 634. https://doi.org/10.3390/foods10030634
17. Sigurdson GT, Tang P, Giusti M (2017) Natural colorants: food colorants from natural sources. Annu Rev Food Sci Technol 8:261–280. https://doi.org/10.1146/annurev-food-030216-025923
18. Ozuna C, Franco-Robles E (2022) Agave syrup: an alternative to conventional sweeteners? A review of its current technological applications and health effects. Lwt 162: 113434. https://doi.org/10.1016/j.lwt.2022.113434
19. Cunha JAR (2021) Food science and technology: fundamentals and applications. Editora Artmed, Porto Alegre
20. Le XT, Rioux LE, Turgeon SL (2017) Formation and functional properties of protein– polysaccharide electrostatic hydrogels in comparison to protein or polysaccharide hydrogels. Adv Colloid Interf Sci 239:127–135. https://doi.org/10.1016/j.cis.2016.04.006
21. Fox PF, Uniacke-Lowe T, Mcsweeney PLH, O'mahony JA (2015) Dairy chemistry and biochemistry, 2nd edn. Springer, p 584. Food Safety Brasil
22. Eisner MD (2021) Direct and indirect heating of milk–a technological perspective beyond time–temperature profiles. Int Dairy J 122: 105145. https://doi.org/10.1016/j.idairyj.2021.105145
23. Cheng Z, O'Sullivan MG, Miao S, Kerry JP, Kilcawley KN (2022) Sensorial, cultural and volatile properties of milk, dairy powders, yoghurt and butter: a review. Int J Dairy Technol 75(4):761–790. https://doi.org/10.1111/1471-0307.12898
24. Silva ACDO, Cortez MAS (2021) Tecnologia de leite e derivados lácteos. Niterói. Eduff
25. Rocha BH, da Silva AP, da Silva LC, de Oliveira VM, de Oliveira JC (2023) Flavor compounds formed by protein denaturation in milk. J Dairy Sci 106(5):5314–5324. https://doi.org/10.3168/jds.2022-19494
26. Alves LF, da Silva LC, de Oliveira VM, de Oliveira JC (2022) Effect of different heating methods on volatile compounds in milk. J Food Sci 87(12):4261–4269. https://doi.org/10.1111/1750-3841.16007
27. Cadwallader DC, Gerard PD, Drake MA (2023) The role of packaging on the flavor of fluid milk. J Dairy Sci 106(1):151–167. https://doi.org/10.3168/jds.2022-22060
28. Heymann HW (2012) Food science: principles and applications, New York, USA
29. ABIM (2023) Manual de Envasamento de Leites. Associação Brasileira da Indústria de Laticínios, São Paulo, Brazil, 300 p
30. Hart RW (2016) Milk processing technology, 3rd edn. John Wiley & Sons, New York
31. Ibrahim O (2018) Sweeteners in our diets and World health organization guidelines on free sugars intake. Int J Clin Nutr Diet 4(129): 10–15344. https://doi.org/10.15344/2456-8171/2018/129
32. Yan S, Yan F, Liu L, Li B, Liu S, Cui W (2022) Can artificial sweeteners increase the risk of cancer incidence and mortality: evidence from prospective studies. Nutrients 14(18):3742. https://doi.org/10.3390/nu14183742
33. Pearlman M, Obert J, Casey L (2017) The association between artificial sweeteners and obesity. Curr Gastroenterol Rep 19:1–8. https://doi.org/10.1007/s11894-017-0602-9
34. Andrade DP, Bastos SC, Ramos CL, Simoes LA, Fernandes NAT, Botrel DA, Magnani M, Schwan RF, Dias DR (2023) Microencapsulation of presumptive probiotic bacteria Lactiplantibacillus plantarum CCMA 0359: technology and potential application in cream cheese. Int. Dairy J. In Press 143:105669. https://doi.org/10.1016/j.idairyj.2023.105669
35. Oliveira D, Antúnez L, Giménez A, Castura JC, Deliza R, Ares G (2015) Sugar reduction

in probiotic chocolate-flavored milk: impact on dynamic sensory profile and liking. Food Res Int 75:148–156. https://doi.org/10.1016/j.foodres.2015.05.050

36. Rocha RS, Mahieu B, Tavares Filho ER, Zacarchenco PB, Freitas MQ, Mársico ET et al (2023) Free comment as a valuable approach to characterize and identify the drivers of liking of high-protein flavored milk drink submitted to ohmic heating. Food Res Int 165:112517. https://doi.org/10.1016/j.foodres.2023.112517
37. Makroo HA, Rastogi NK, Srivastava B (2020) Ohmic heating assisted inactivation of enzymes and microorganisms in foods: a review. Trends Food Sci Technol 97:451–465. https://doi.org/10.1016/j.tifs.2020.01.015
38. Rocha RS, Silva R, Ramos GL, Cabral LA, Pimentel TC, Campelo PH, Zacarchenco PB, Freitas MQ, Esmerino EA, Silva MC, Cruz AG (2022) Ohmic heating treatment in high-protein vanilla flavored milk: quality, processing factors, and biological activity. Food Res Int 161:111827. https://doi.org/10.1016/j.foodres.2022.111827
39. Sfakianakis P, Topakas E, Tzia C (2015) Comparative study on high-intensity ultrasound and pressure milk homogenization: effect on the kinetics of yogurt fermentation process. Food Bioprocess Technol 8(3):548–557. https://doi.org/10.1007/s11947-014-1383-2
40. Gregersen SB, Wiking L, Hammershøj M (2019) Acceleration of acid gel formation by high intensity ultrasound is linked to whey protein denaturation and formation of functional milk fat globule-protein complexes. J Food Eng 254:17–24. https://doi.org/10.1016/j.jfoodeng.2019.03.004
41. Zhang R, Pang X, Lu J, Liu L, Zhang S, Lv J (2018) Effect of high intensity ultrasound pretreatment on functional and structural properties of micellar casein concentrates. Ultrason Sonochem 47:10–16. https://doi.org/10.1016/j.ultsonch.2018.07.001

Chapter 3

Mozzarella Cheese

Patrizio Tremonte, Filomena Nazzaro, and Raffaele Coppola

Abstract

Mozzarella cheese is one of the most well-known food products. Due to the high variability in technological processes, it is possible to find various products. The heterogeneity condition is further enriched by the variability inherent in traditional approaches (e.g., natural whey starter or manual stretching and brine management) that often characterize specific processes. Therefore, the ready availability of control methods that are simple, inexpensive, and easily applicable to different cheesemaking conditions appears crucial. This chapter summarizes the raw ingredients and technological operations and tools for quality control during production and storage of the mozzarella cheese.

Key words Fresh cheese, Mozzarella cheese, Dairy microflora, Water buffalo milk, Cow milk

1 Introduction

Mozzarella is a fresh soft cheese belonging to the "pasta filata" category. Therefore, its manufacturing process involves a weak or moderate syneresis of the curd, its subsequent acidification, and the distinctive stretching operation. Since it is a fresh cheese, it does not have a ripening period and has a short shelf life under conditions that maintain its native characteristics and freshness.

Specifically, it appears porcelain white, glossy on the surface, and is characterized by a fibrous and slightly melting texture. The texture is elastic, slightly yielding, and particularly milky. Finally, the flavor is pleasantly acidic and has a distinctive musky aroma. Achieving these characteristics requires careful selection of raw materials, conscientious control of the cheesemaking, and constant monitoring of process and product parameters.

The choice of buffalo, cow, or other species milk influences the cheesemaking operations, the management of process parameters, and consequently, the product's final characteristics. The most famous type of mozzarella is the buffalo mozzarella "Mozzarella di Bufala Campana," which was the first cheese to obtain Protected Designation of Origin (PDO) recognition under the June 1996

Adriano Gomes da Cruz et al. (eds.), *Dairy Foods Processing*, Methods and Protocols in Food Science,
https://doi.org/10.1007/978-1-0716-4144-6_3,

European Union regulation [1]. This PDO cheese is produced in a specific area of Southern Italy, including two provinces in the Campania region and a small portion of the Lazio and Molise regions. Cheesemaking can involve raw or pasteurized milk, inoculation with natural whey culture (NWC), from the previous day's manufacture), and the use of calf rennet. When the curd has reached proper ripeness, it is stretched under boiling water, becoming elastic, and then molded to achieve the typical round final shape [2, 3]. However, the market offers also other types of mozzarella cheese, including those obtained by cow milk, which are increasingly popular. In addition, regardless of the origin of the milk, the choice of acidifying agent (directly using organic acid or biological via the action of natural or selected starter cultures) conditions the stretching operation and influences the final structure and the flavor profile of the final product [4–6]. Finally, the choice and management of certain operations, such as the intensity of syneresis, influence the moisture content of the product and its functional characteristics.

A proper monitoring parameter is essential to obtain high-quality products. The scientific literature is teeming with studies aimed at defining analytical methods for monitoring process and product parameters [7–10]. However, while highly performing for research purposes, some methods are often not validated for routine use suitable for process management. This chapter aims to offer through the following paragraphs a description of the process variants among the different types of mozzarella and the actions and methods required for quality control and assurance.

2 Materials

2.1 Milk

Milk is a colloidal dispersion of proteins distributed in water in a dynamic equilibrium with fat and lactose. Two kinds of milk are used for the mozzarella production: buffalo milk, for the water buffalo mozzarella cheese, in specific areas of the Central and Southern Italy (*see* **Note 1**), and cow milk, for widespread worldwide products. With reference to their composition, according to Garau et al. [11], the two milks highlight the main differences in the fat level and, consequently, in total dry matter, which are higher for buffalo milk (Table 1). No significant differences in quantitative protein levels are reported between the two milks. Both kinds of milk contain between 35 and 40 g of protein per liter, with two distinct groups of proteins: caseins and whey proteins.

Caseins, about 80% of the total protein content, consist of four main proteins: αs1-, αs2-, β-, and κ-casein. Casein binds calcium through phosphoserine residues in its structure. The number of these residues and, consequently, the ability to bind calcium vary among different caseins: αs1-casein contains 8, αs2-casein includes

Table 1
Dry matter and nutrients in cow, buffalo, and goat milk

	Total dry matter (g/L)	Lactose (g/L)	Proteins (g/L)	Fat (g/L)	Ash (g/L)
Cow	133 (±9)	48 (±6)	33 (±5)	43 (±10)	9 (±1)
Buffalo	161 (±11)	41 (±7)	37 (±11)	75 (±12)	8 (±1)
Goat	140 (±15)	40 (±9)	42 (±10)	51 (±16)	7 (±0.1)

10–13, β-casein contains 5, and κ-casein contains 1 residue per mole. The primary structure of αs1-casein appears to be homologous when comparing buffalo and cow milk, consisting of 199 amino acids. The same is true for β-casein, which consists of 209 amino acids. However, the literature reports a different distribution of αs1-, αs2-, β-, and κ-casein in buffalo milk proteins compared with cow milk [12]. There is a higher β-casein content (up to twice as high) in buffalo milk than in cow milk. Caseins are organized as micelles, spherical colloidal particles between 50 and 500 nm in diameter. Amorphous calcium phosphate binds to caseins at the ester phosphate groups, determining their characteristics. The αs1-, αs2-, and β-caseins are all insoluble at the calcium concentration present in milk, so they arrange themselves in such a way that the calcium-soluble κ-casein stabilizes the micelles. Therefore, the micelles have a surface layer rich in κ-casein that protrudes into the surrounding medium.

Although the casein micelles of buffalo milk are less hydrated and more mineralized than those of cow milk, both show similar behavior toward different physicochemical conditions. The internal structure of casein micelles has been much debated, and several models have been proposed: the sub-micelle model, the double bond model, and the nanocluster model [13]. These models have evolved due to advanced techniques that can measure the structure and composition of micelles. The oldest of these models, and the most frequently cited, is the sub-micelle model. With the appearance of the nanocluster model, several versions have been proposed, including an interlocking lattice and water channels within the structure. During the mozzarella production process, micelles undergo several changes. However, a certain level of substructure remains even after cheese production. As previously reported, the most substantial difference between buffalo and cow milk is related to fat level. The fat globules are dispersed in the milk and vary in size between 0.1 and 10 μm, with about 90% between 1 and 8 μm. These fat globules are surrounded by a lipoprotein membrane that stabilizes them from coalescence and protects them from lipases in the milk.

Table 2
Fatty acid profile in cow and buffalo milk

	Cow	Buffalo
SFA	69.6	70.1
MUFA	24.8	26.0
PUFA	5.5	3.9
SFA/UFA	2.3	2.3

The milk of the two species appears to have comparable average fatty acid profiles [14]. Although some differences have been described in several studies, they are not significant, and therefore, it can be asserted that the SFA and UFA (MUFA + PUFA) ratio is substantially similar in buffalo and cow milk (Table 2). Some of the differences are related to the acid profile: cow milk contains higher levels of linoleic acid (18:2 cis-9 and cis-12) and linolenic acid (18:3 cis-9, cis-12, and cis-15) than buffalo milk [11].

Other differences could be due to the size of the membrane surrounding the fat globule. Buffalo milk has larger fat globules than cow milk, so the relative proportion of polar lipids in the membrane is lower.

On the other hand, buffalo milk contains more beneficial compounds than cow milk, such as gangliosides (*see* **Note 2**), which can provide antioxidant and neuronal protection and improve bone, heart, and gastrointestinal health in humans [15].

In addition, besides cow and buffalo milk, there is growing interest in producing mozzarella cheese from the milk of small ruminants, particularly goat milk, which is gaining more and more interest due to its nutritional values and health benefits. The production of mozzarella from goat milk is difficult, mainly due to the high moisture content (HMM about 60% moisture); the resulting acidified curd has poor extensibility, and the cheese tends to harden after stretching.

New protocols are being developed to use goat milk in mozzarella production. Techniques based on novel binomials of pasteurization time and temperature and pre-acidification operations offer interesting prospects.

2.2 Coagulating Agent (Rennet)

The coagulating agent is one of the most important ingredients in the cheesemaking processes of predominantly presamic coagulation products such as mozzarella. Different types of coagulating enzymes (rennet) are available, ranging from those of animal (derived from the abomasum of calf or annutol) to vegetable (extracted from a wide variety of plants) to microbial (i.e., fungal rennet) origin (Fig. 1). The conventional and most widely used coagulating agent in the preparation of mozzarella is calf rennet (*see*

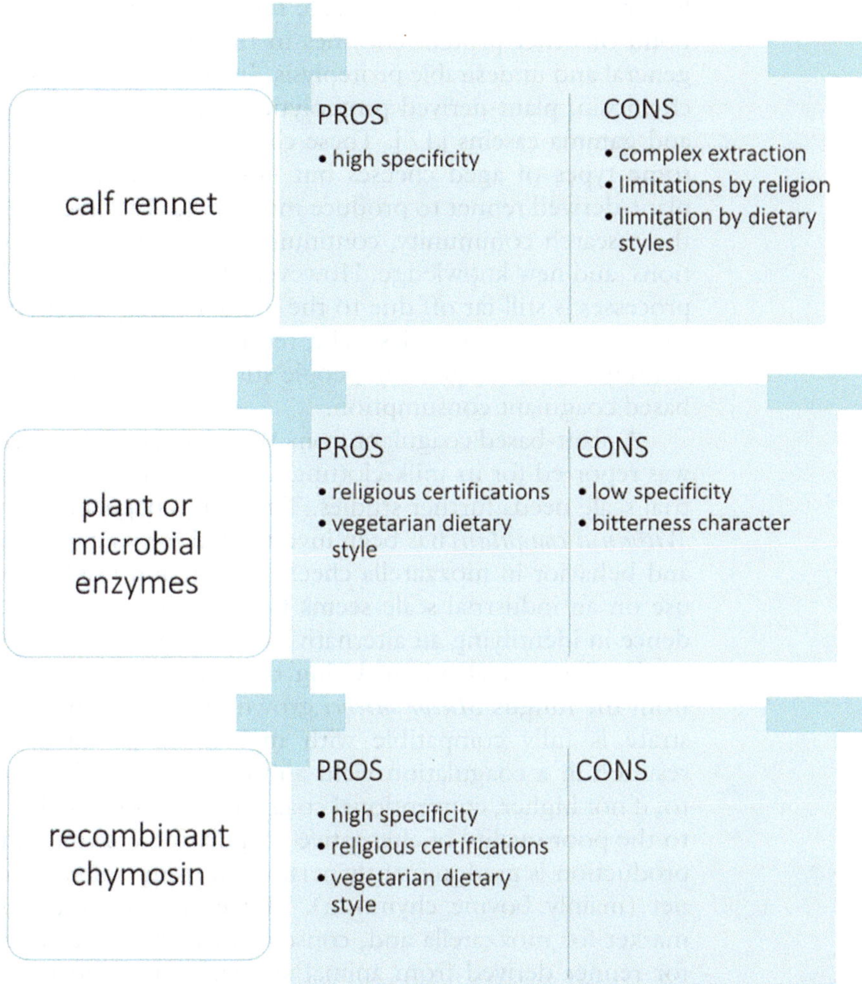

Fig. 1 Advantages and disadvantages associated with the use of different types of clotting enzymes

Note 3), predominantly characterized by the presence of chymosin, an enzyme that acts in a specific and limited manner on the Phe105-Met106 peptide bond of κ-casein. If, on the one hand, calf or annutol rennet is strongly linked to the cheesemaking of Mozzarella, concerning the PDO production segment, on the other hand, the calf rennet replacement is gaining strength for several reasons. Calf rennet highlights cost-related disadvantages due to the complex extraction process and subsequent purification. In addition, its use finds limitations by religion, such as Islam and Judaism, or dietary styles, such as vegetarianism. In the late 1900s, scientific research made significant efforts to identify milk-coagulating enzymes from plants, microorganisms, and fungi to reduce reliance on calf rennet [16]. Alternative coagulants often increase, albeit slightly, the bitterness character that cannot be matched to the sensory profile of mozzarella. The bitterness is usually due to the lower specificity of the vegetable rennet for the Phe105-Met-106 bond or other protein enzymes in the extract that cause a more general and undesirable proteolysis. In fact, unlike κ-casein-specific chymosin, plant-derived proteolytic enzymes cleave alpha-, beta-, and gamma-caseins [17]. These characteristics may be helpful for some types of aged cheeses but not for mozzarella. Identifying plant-derived rennet to produce mozzarella is a topic that intrigues the research community, continuing to inspire studies, investigations, and new knowledge. However, their application in industrial processes is still far off due to the many limitations of sensory and rheological characteristics. The rennet storage and the religious matters of some vegetarian people stimulated the interest in plant-based coagulant consumption.

A plant-based coagulant from *Withania coagulans* (*see* **Note 4**) was reported for its milk-clotting activity, but its use on an industrial scale needs further studies. The plant coagulant derived from *Withania coagulans* has been investigated for its coagulant activity and behavior in mozzarella cheese production [18]. However, its use on an industrial scale seems far away. More encouraging evidence in identifying an alternative rennet has been obtained using curds of microbial origin. Using the coagulating enzyme derived from the fungus *Mucor miehei* grown in a rice bran molasses substrate is fully compatible with mozzarella production. Its use resulted in a coagulation yield and curd extensibility comparable to, if not higher, conventional coagulation processes [19, 20]. Due to the poor quality of alternative coagulating enzymes, mozzarella production is predominantly performed using animal-derived rennet (mainly bovine chymosin). However, the expansion of the market for mozzarella and, consequently, the increase in demand for rennet derived from animal stomachs have led to a flurry of research into recombinant chymosin (r-chymosin). The recombinant enzyme (*see* **Note 5**), produced through modern microbial biotechnology, is particularly promising because it meets both the advantages of animal rennet and those of plant or microbial-derived rennet [21, 22]. Like animal rennet, recombinant chymosin has a high specificity in proteolytic action, and, therefore, it does not impart abnormal flavors; in addition, because microorganisms biotechnologically produce it without the intervention of animal components, it is compatible with religious certifications and vegetarian dietary style; in addition, it meets the favor of animal rights activists. The advancement of knowledge in this topic is remarkable and promising; expertise for the production and scale-up of r-chymosin from calf, lamb, and kid is available to date [23]. Based on these considerations, r-chymosin could precisely represent the future of milk coagulation to produce mozzarella.

2.3 Acidifying Agent (Biological or Chemical)

The acidifying agent is the crucial ingredient in the production of mozzarella since proper and appropriate acidification is essential for demineralization and, consequently, the stretchability of the curd. The choice of acidifying agent strongly affects the evolution of the cheesemaking process, shelf life, and quality of the finished product. Several agents are available to promote curd acidification. An early differentiation may be established between chemical and biological agents. The first ones, used in the cheesemaking process with direct acidification, are represented by citric acid, or less frequently by lactic acid, added to the milk immediately before rennet. The citric or lactic acid added must be labeled with the appropriate E number: E330 for citric acid and E270 for lactic acid. Although the appearance of mozzarella cheese is independent of the acidification method used, and it is very difficult for consumers to distinguish products made with chemical acidifier versus organic acidifier, the odor part of the volatile fraction containing different volatile organic compounds is significantly different between the two types of products. It is widely recognized that using the chemical acidifier results in a relatively poor finished product from a sensory point of view. On the other hand, chemical acidification allows for high repeatability and standardization of the cheesemaking process, ensuring a curd ready for stretching soon after coagulation. These conditions result in a more economical and less time-consuming cheesemaking process [24]. A more complex and laborious process is associated with using biological acidifying agents, represented by undefined natural whey cultures (NWCs) or selected microbial cultures (SMC). The NWC and SMC have the primary function of producing an adequate amount of lactic acid to ensure rapid acidification of the curd determining its demineralization. This condition allows it to be stretched in hot water and obtain pleasant sensory attributes. Despite advances in knowledge in dairy science and biotechnology, undefined microbial cultures are still widely used. Natural whey cultures (NWCs), consisting of whey rich in lactic acid bacteria (LAB-rich) from previous processing, are required by the standards (*see* **Note 6**) of Protected Designation of Origin water buffalo mozzarella cheese or Traditional Speciality Guaranteed cow milk mozzarella cheese. The study of NWC microbiota has attracted the attention of many researchers applying culture-dependent and culture-independent techniques to describe diversity at the species level (Fig. 2).

Although biomolecular methods can provide an accurate picture of the microbial composition of a natural whey starter, the inevitable variability attributable to various factors (season, temperature, cheesemaking operations) should be highlighted. Nowadays, NWCs cannot be considered undefined because available biomolecular methods can investigate each microbial actor inhabiting the ecosystem, providing a complete picture of the complexity and composition of the microbial community. NWCs have been

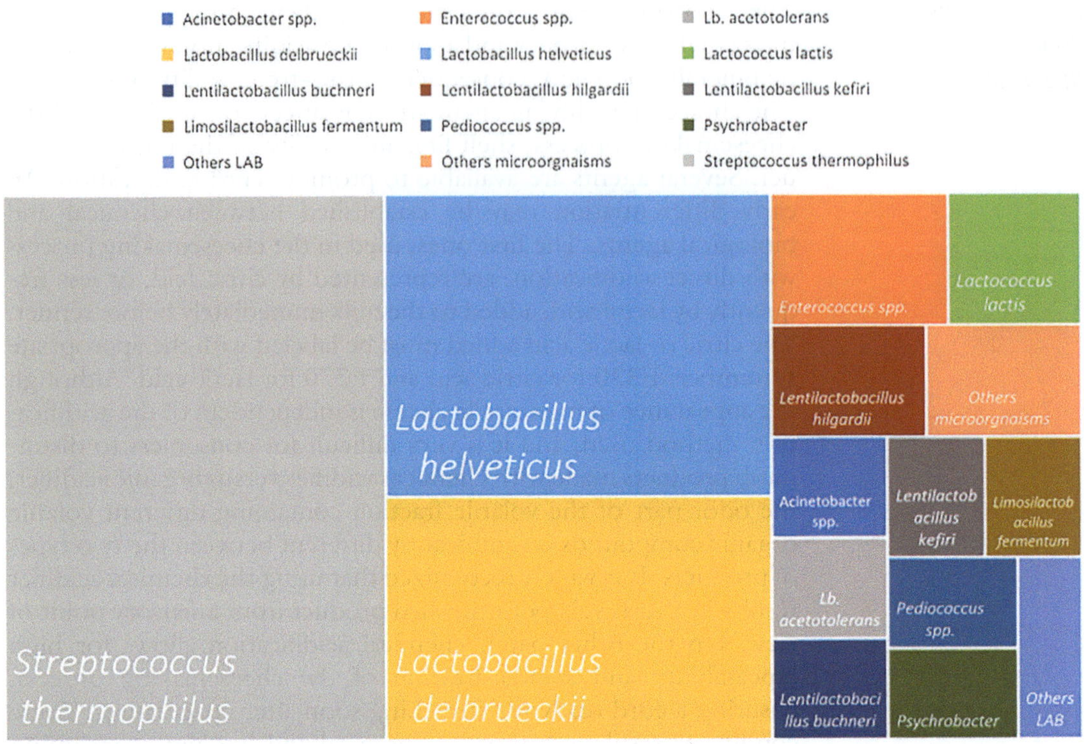

Fig. 2 Relative abundance of species or genera found in natural whey starter cultures investigated in recent decades [2, 3, 28, 29]

extensively studied using culture-dependent and culture-independent approaches as well as next-generation sequencing (NGS) and high-throughput sequencing (HTS) technologies [5, 25–27]. These natural cultures are usually dominated by thermophilic lactic acid bacteria; in addition, subdominant species, including mesophilic LAB (*Lactococcus*, *Leuconostoc*, *Limosilactobacillus*, *Pediococcus*, *Enterococcus*) and yeasts, may also be present in some cases. The various studies, even those accompanied by better-performing surveys and analytical techniques, highlight that the dominant species are *S. thermophilus*, followed by *L. delbrueckii* and *L. helveticus*.

Despite the small size of the mozzarella ecosystem, in terms of species able to guide the fermentative process, it has been extensively reported that the complexity and microbial variability of NWC are strictly connected to the ecosystem, cheesemaking technology, and production area. The presence and predominance of *S. thermophilus* in NWC used for cow and buffalo milk-based mozzarella production are well documented. However, a certain degree of diversity within isolates of this species has been demonstrated. In addition, the ratio of cocci to bacilli sometimes varies in favor of lactobacilli, with a predominance of *L. helveticus* species, phenomena attributable to both process and NWC management. NWCs, with their microbial biodiversity at both species and strain levels, significantly influence the metabolome of curd and, thus, of mozzarella cheese. Natural cultures are important in forming volatile organic compounds (VOCs). Indeed, products obtained by natural whey fermentation are always characterized by more complex profiles than products obtained by direct acidification or using selected cultures consisting of individual strains.

The crucial odor compounds that contribute to the flavor of mozzarella are 3-methylbutanal, ethyl acetate, and 2,3-butanedione, which are responsible for the ethereal, fruity, and buttery odor [30]. Unfortunately, the composition of natural starter cultures is complex, variable, and subjected to unpredictable performance. Therefore, in some cases, NWCs have been replaced by commercial starters, usually consisting of *S. thermophilus* alone or mixtures of *S. thermophilus* and *L. helveticus*. The effect of this substitution has been extensively studied by investigating the physical, rheological, aromatic, and functional characteristics of mozzarella. It is well-known that thermophilic lactic acid bacteria are essential in determining physical properties and a balanced ratio of rods to cocci positively influences the proteolysis, elasticity, and viscosity of mozzarella. However, the use of defined cultures consisting of one or two strains is responsible for a minor broad flavor. The use of cultures consisting only of *S. thermophilus* results in the partial utilization of lactose; concurrently, the accumulation of galactose gives rise to the formation of new organic compounds. Supplementation of a thermophilic lactobacillus only partially improves the aromatic picture but remains significantly poorer than that characterizing product obtained with NWCs. Therefore, in the last decade, developing indigenous multi-strain starter cultures, representative of the microbial composition of natural whey cultures, can be the way to ensure high mozzarella quality. Several studies have selected AMSs and established protocols to improve texture, sensory properties, and shelf life of traditional high-moisture mozzarella.

2.4 Governing Liquid

Governing liquid consists of tap water, sodium chloride, diluted whey, and sometimes citric acid and other salts. Its principal function is to preserve the freshness and shape of the mozzarella during storage by preventing its collapse and avoiding alteration processes. Although the storage period is relatively short (*see* **Note** 7), the composition of the liquid plays an important role in the texture and shelf life of mozzarella. The different compositions can influence the transfer of organic material between the mozzarella and the storage liquid. In this regard, scientific research has moved numerous studies. The general objective is to delay the alteration of mozzarella quality and obtain an extension of shelf life. The adoption of optimal salt concentrations, also balanced with respect to the weight and shape of the product, and the addition of other compounds, such as calcium salts and natural polysaccharide gels, have shown promising results [31].

Dairy industries have a great interest in finding new strategies to improve the management of liquids and, consequently, extend the shelf life of mozzarella. Adding natural bioactive compounds to the preserving liquid would extend the shelf life of mozzarella. Calcium milk alone or in combination with lactic acid or even bergamot extract significantly improved shelf life by delaying alterations caused by *Pseudomonas* spp. In addition, the use of lemon, sage, and thyme extracts in the preserving liquid retarded the proliferation of coliforms and *Pseudomonas* spp. From this perspective, great attention has also been paid to other strategies, such as using protective cultures in the governing liquid. The use of lactobacilli in the governing liquid seems to intensify the levels of free fatty acids and volatile organic compounds [32]. Although an appropriate validation on an industrial scale and a complete scale-up are still being completed, antagonistic strains as cultures added to the governing liquid are the future.

3 Methods

Mozzarella cheesemaking (Fig. 3) plays an important role in defining the quality of the final product, conditioning its yield, structure, and aromatic framework. Indeed, texture and sensory

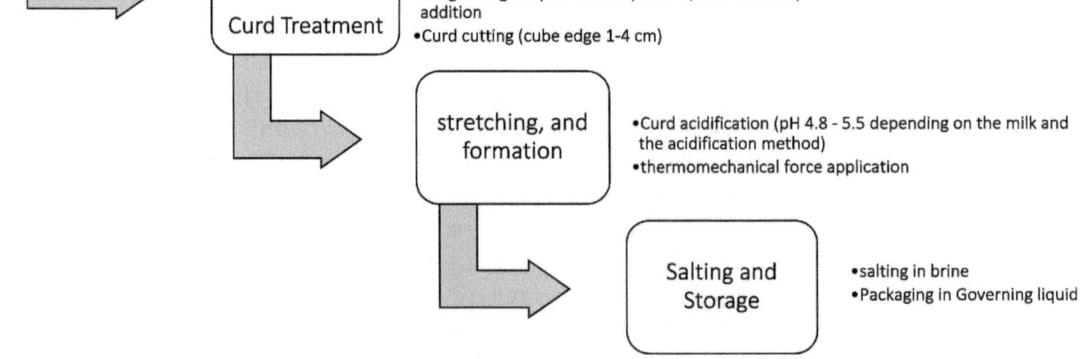

Fig. 3 Main operations of mozzarella cheesemaking

characteristics are the most critical attributes in consumer choice and in determining mozzarella's functional quality and use.

As explained in the following paragraphs, each stage of processing, starting from milk pretreatment to product packaging, can affect the structure and sensory profile of the product.

3.1 Milk Treatment

Milk management and pretreatment procedures, aiming to ensure the quality of the product with a particular focus on its microbiological safety, have been well established in recent decades. From this perspective, procedures such as refrigeration after milking and thermization (*see* **Note 8**) of milk are well established and represent fundamental steps in the cheesemaking process.

Precisely to ensure the hygienic safety of milk, EU [33] states that farm-fresh raw milk "shall be cooled immediately to not more than 8 °C in case of daily collection, or to not more than 6 °C if collection is not daily. During transport, the cold chain must be maintained, and upon arrival at the destination establishment, the temperature of the milk must not exceed 10 °C." However, while ensuring the hygienic quality of milk, these procedures could affect other quality attributes of the milk and milk product [34]. In fact, in addition to reduce the number of pathogens and spoilage microorganisms, refrigeration also decreases the native dairy microbiota and, if prolonged, promotes the growth of psychrotrophic bacteria, which can produce thermostable proteases that hydrolyze proteins, causing various problems, from decreased dairy yield to the formation of off-flavors. In defining milk quality, protein or fat content significantly influences consistency and rheological characteristics. Variations in milk composition can occur for various reasons, and standardization is used to ensure that the starting point of the mozzarella cheesemaking process is consistent across batches. Typically, this process is achieved by centrifugal separation of raw milk into skim milk and cream or even by ultrafiltration and subsequent adjustment of the chemical composition of the milk. In addition, the milk is subjected to the homogenization process with the formation of a new fat/water interface containing mainly caseins. The size of the fat globule and its dispersion in the casein matrix influence the meltability and fat loss of the resulting cheese [35, 36]. Milk homogenization at high pressures (6700 kPa) tends to reduce the elongation and melting properties; on the other hand, homogenization at low pressures (400 kPa) does not affect these properties of the resulting mozzarella cheese.

Based on the above considerations, monitoring specific quality parameters at the reception of milk is crucial (Table 3). Specifically, on each batch of milk, physical (temperature, cryoscopy), chemical (presence of inhibitors and antibiotics, protein and fat levels), and chemical–physical (pH and acidity) parameters are evaluated daily. No less than twice a month, total microbial load levels, presence of

Table 3
Quality parameters of raw milk for mozzarella cheesemaking

Parameters	Monitoring frequency	Methods	Threshold/range	Notes
Inhibitors and antibiotics	On each batch		Absence in 0.1 ml	
pH	On each batch	ISO/TS 22113:2012	6.65–6.85	
Acidity	On each batch	ISO/TS 22113:2012	5.8–7.6 °SH	
Freezing point	On each batch	ISO 5764|IDF 108: 2009	−0.518 °C < X > − 0.530 °C	The freezing point can be used to estimate the proportion of extraneous water in milk
Protein level	On each batch	ISO 8968-1:2014	>3.20% (bovine milk) >4.20% (bufalin milk)	Kjeldahl principle and crude protein calculation
Fat level	On each batch	ISO 19662:2018	>3.5% (bovine milk) >7.20% (bufalin milk)	
Aflatoxin M1	Every 15 days	ISO 14501:2021	<0.05 ppb	Acido-butyrometric (Gerber method)
Total microbial counts	Every 15 days	ISO 4833: 2003	< 10.0E5 CFU/ml	Enumeration on plate count agar after 72 h of incubation at 30 °C
Somatic cells	Every 15 days	ISO 13366-2 148-2: 2006	< 400,000/mL	Using fluoro-opto-electronic somatic cell counters in which either a rotating disc technique or flow cytometry is applied in the counting section
Dioxin	Annually			Reg. CE 1881/2006

aflatoxin M1 [37], and somatic cells are determined. Finally, dioxin is determined annually.

Heat treatments also deserve special attention as their effect on mozzarella quality is debated. Recently, heat treatments of milk, partly in response to legislative regulations, have become a widely used practice on farms, including artisanal farms. Standard milk pasteurization techniques (*see* **Note 9**) comprise heating at either 63 °C for 30 min (low-temperature long-time, LTLT), 72 °C for

15 s (high-temperature short-time, HTST), or any other equivalent thermal treatment [38]. Their effect on milk and end-product quality has been extensively studied. The primary purpose of heat treatments is to remove potentially pathogenic microbial forms. In addition, besides positively affecting cheeses' microbiological quality, heat treatments can change the structure of specific molecules, positively impacting the texture of final products. Heat treatments by denaturing specific proteins can improve the structure of cheeses and the technological yield, as well as the taste and aroma of milk and dairy products. However, heat treatments can also adversely affect physicochemical, sensory, and microbiological characteristics by significantly reducing the non-starter bacterial component. The extent and type of effects depend on the heating method and the time–temperature factor applied. Some authors have highlighted that minimally invasive heat treatments conducted at 72 °C for a few seconds resulted in mozzarella with superior organoleptic and storage characteristics compared to cheese made from raw milk.

On the contrary, the use of more severe heat treatments appears to significantly reduce the fluidity and extensibility of melted mozzarella [39]. Due to the importance of thermal remediation of milk used for fresh products such as mozzarella, it is crucial to evaluate the effectiveness of the treatment, which is controlled by monitoring alkaline phosphatase activity using a specific fluorimetric method. In addition, the count level of certain microorganisms, generally total coliforms, is determined (Table 4).

Besides conventional methods, attention is also turning to alternative remediation methods based on microwaves or radio frequencies. The latter, despite their highlighted positive effects on milk quality and consequently on dairy products, remain at laboratory or pilot scale and need appropriate scale-up so that they can be adopted in industrial processes.

Table 4
Quality parameters of pasteurized milk

Parameters	Monitoring frequency	Methods	Threshold/ range	Notes
Alkaline phosphatase activity	Monthly (min)	ISO 11816-1: 2013	<100 CFU/ mL	Expressed as milliunits of enzyme activity per liter of sample (mU/l); measured by a continuous fluorimetric direct kinetic assay
pH	Biweekly (min)	ISO/TS 22113:2012	6.65–6.85	

3.2 Coagulation, Curd Breaking, and Dehydration

Milk coagulation is the step that leads to dramatic effects on the structure of the milk system by transforming it from a single liquid state with proteins in colloidal dispersion to a system characterized by two portions, one represented by a liquid solution (the whey) and the other by a gel obtained from the aggregation of casein micelles. The enzymatic step is due to the action of coagulation agents (enzymes, predominantly chymosin) that hydrolyze κ-casein at the Phe105-Met106 site, with the production of para-κ-casein and soluble glycomacropeptides. Glycomacropeptides are released into the serum, while para-κ-casein remains attached to the outside of the micelle. Caseins αs1, αs2, and β are not hydrolyzed during coagulation but can be hydrolyzed during mozzarella ripening [40].

The subsequent nonenzymatic step consists of the aggregation of casein micelles with subsequent destabilization. Loss of protruding peptides and release of glycomacropeptides result in reduced intercellular repulsions and z potential, allowing paracasein micelles to aggregate. It has been observed that micelle aggregation occurs when κ-casein hydrolysis has reached a level of about 85% and the formation of a three-dimensional gel appears when about 97% of κ-casein has been hydrolyzed. Both enzymatic and nonenzymatic stages are affected by several factors, including temperature, pH, and ionic strength [41]. Exposing milk to high temperatures could, for example, affect the effectiveness of rennet. Treatments above 90 °C producing bonds between β-lactoglobulin and κ-casein interfere with rennet action and aggregation. During coagulation, temperatures 10 °C higher than ambient temperatures significantly accelerate rennet action. The slightly lower pH than that proper to milk also promotes the action of micelle aggregation.

On the other hand, pH values of about 5.9 result in a progressive decrease in the hydration of casein particles. Finally, ionic strength reduction, attributable to calcium loss, decreases the coagulation time. It happens because the lower ionic strength causes less shielding of the positive charge of κ-casein and the negative charge of chymosin, thus favoring the action of rennet. However, we should consider that calcium, which contributes to ionic strength, plays an additional role in coagulation. Calcium, either in ionic form or as colloidal calcium phosphate, acts on the micelle lattice. Without free calcium ions, micelles do not aggregate, and coagulation does not occur. Similarly, if colloidal calcium phosphate falls below 20%, coagulation does not occur.

Based on this scientific evidence and drawing on the experiential heritage of the "pasta filata" dairy craft, precise cheesemaking protocols for mozzarella have been defined.

Milk is heat-treated, generally at a temperature of 71–72 °C for 15–18 s and cooled and then brought to a temperature between 33 °C and 37 °C to start the coagulation operation. The temperatures adopted, which differ slightly between companies and

Table 5
Quality parameters of curd

Parameters	Monitoring frequency	Methods	Threshold/range	Notes
pH	On each batch	ISO/TS 22113: 2012	4.85–4.90 (buffalo mozzarella) 5.00–5.20 (bovine mozzarella) 5.5–5.7 (direct acidification)	Determination is made on the mature curd
Relative humidity				

depending on the functional and technological characteristics desired for the finished product, contribute to the diversity that distinguishes the mozzarella product. The starter culture, either natural and undefined (2%) or selected and well-defined, is added to the milk and thus heated, and immediately afterward, the coagulating enzyme rennet (0.007% with a titer of about 1:13000) is added.

Once it reaches the desired gel structure, the curd is cut to favor syneresis, i.e., the whey release (*see* **Note 10**). The space between the blades used to cut the curd is inversely proportional to the level of breakage and syneresis. To obtain a product, such as mozzarella, with a moisture content between 50% and 60%, the curd is cut with spaced blades, giving rise to cubes with an edge of 1 to 4 cm. In addition to shear, temperature also influences syneresis; temperatures higher than ambient temperature, close to 40 °C, accelerate water loss. Quality parameters of the curd are constantly monitored (Table 5).

3.3 Curd Acidification, Stretching, and Formation

Compared to other cheeses, the operations following syneresis represent the distinctive element of *pasta filata* cheeses, such as mozzarella.

Once the curd has reached optimal maturity, stretching is carried out, resorting today to traditional and mechanized methods.

Traditionally, the application of thermomechanical force to the mature curd is performed manually (*see* **Note 11**): hot water (70–80 °C) is added to the curd and then processed with the aid of appropriate tools along the walls of the vessels.

More widely, stretching is carried out with mechanical mixers with single or double auger systems designed with different materials, geometries, and heating systems. Also, in this case, the curd is stretched in hot water at a controlled temperature (82–85 °C) by steam injection. Mechanical stretching (*see* **Note 12**), on the one hand, allows for better modeling, but on the other, it determines a heterogeneous distribution of humidity in the product and more significant fat loss.

The increase in the temperature of the curd determines an increase in the strength of the hydrophobic interactions within its protein matrix, resulting in their contraction and the passage of water into the free state in the interstitial spaces around the fat globules. Therefore, the stretching determines the parallel alignment of the casein fibers, interspersed with water and fat channels, resulting in the formation of free whey.

Stretching gives mozzarella a unique fibrous consistency; it is a thermomechanical treatment influenced by factors such as pH, acidity, stretching time, and temperature. The right combination of pH and calcium content affects the extensibility of the curd under thermomechanical treatment.

Specifically, the plasticization of the curd is conditioned by the calcium content associated with the casein, which in turn is modulated by the total calcium content and the pH of the curd; precisely for this reason, the curd after syneresis is subjected to a further phase called maturation which allows its optimal acidification [36]. As previously reported, the acidifying agent can be represented by an organic acid, such as citric or lactic acid, or by a natural and undefined or appropriately selected microbial culture.

The type of acidifying agent affects the optimal insoluble/soluble calcium ratio. When the agent is biological, represented by microbial culture, the proper ratio is reached when the pH of curd is between 4.9 (for mozzarella produced from buffalo milk) and 5.2 (for mozzarella made from cow milk). At these pH values, dicalcium para-caseinate is converted to monocalcium para-caseinate, which favors fiber formation (*see* **Note 13**).

Since acidification is so important in mozzarella cheesemaking, it is crucial to predict and control the action of the acidifying agent. This task is easily accomplished when the acidifying agent is represented by an organic acid or even by a selected starter culture. At the same time, more attention is required if acidification is entrusted to a natural starter culture such as whey starter. The natural whey starter must undergo stringent control to ascertain its microbiological quality and to predict its efficacy. Precisely, as shown in Table 6, daily checks aimed at ascertaining titratable acidity and monthly microbiological assessments are planned.

3.4 Salting and Storage

The stretched and formed product is cooled (*see* **Note 14**) in the firming tanks and then subjected to salting. Mozzarella is generally subjected to salting in brine, which takes place after the cooling of the mozzarella in water. The brine, in which the cheese is immersed, must have the pH and calcium content adjusted to prevent the escape of calcium and lactic acid from the cheese, which results in defects in the cheese. Microbial quality of brine is monthly monitored, as reported in Table 7. Salt is present mainly in the aqueous phase and is taken up according to a gradient higher on the surface and lower in the center.

Table 6
Quality parameters of natural whey starter

Parameters	Monitoring frequency	Methods	Threshold/ range	Notes
Titratable acidity	On each batch	International standard IDF 81: 1977 ISO/TS 22113:2012	20–22 °SH	Determined by titration of a known amount of reconstituted milk with 0.1 N NaOH using phenolphthalein as indicator
Eumycetes	Monthly (min)	ISO 21527-1:2008	<5E3 CFU/mL	By means of the colony count technique at 25 °C ± 1 °C
Total coliforms	Monthly (min)	ISO 4832:2006	<1E3 CFU/ml	Technique of colony counting after incubation on a solid medium at 30 °C or at 37 °C
Escherichia coli	Monthly (min)	ISO 16649-2:2001	<1E2 CFU/mL	For the enumeration of β-glucuronidase-positive *Escherichia coli* using a colony count technique at 44 °C on a solid medium containing a chromogenic ingredient for detection of the enzyme β-glucuronidase

Table 7
Quality parameters of brine

Parameters	Monitoring frequency	Methods	Threshold/ range	Notes
Total coliforms	Monthly (min)	ISO 4832: 2006	≤1.000	Parameter routinely adopted within dairy farms
Eumycetes	Monthly (min)	ISO 21527-1: 2008	≤5.000	Parameter routinely adopted within dairy farms

Alongside, the more widely used conventional brined curd, dry brining can also be used, which involves the addition of salt directly to the shredded curd before stretching. The dry operation causes more significant dehydration of the curd, influencing its rheological characteristics.

After formation, mozzarella undergoes a significant structural rearrangement, and the free water that occupies the interstitial area around the fat channels is absorbed into the protein matrix.

During the first days of storage, mozzarella undergoes notable structural and functional changes. Part of the β-casein partially dissociates from the casein matrix and passes to the serum phase.

The proteins become less aggregated and more hydrated, increasing the melting character of the cheese. In addition, during storage, the residual enzymes present in the cheese hydrolyze the casein, causing the breakdown of the protein matrix. In particular, plasmin, a heat-stable enzyme, remains active, and during storage, it starts a hydrolysis process mainly involving β-casein. Protein hydrolysis determines a dramatic change in structural characteristics with a loss of elasticity and increased softness and melting. Furthermore, considering its high moisture content and the fresh taste expected by the consumer, mozzarella is characterized by a poor shelf life, which varies from 1 day for high-quality niche products without technological conditioning to 30 days for immediately refrigerated products. In Table 8 we report the quality parameters of the final product.

Therefore, the way of preservation after processing affects the functionality of the cheese.

Refrigeration, although it reduces some freshness characteristics (especially for high-moisture mozzarella obtained from buffalo milk), is the most widely adopted tool to guarantee a reasonable shelf-life extension.

Attention has also been paid to the frozen storage of mozzarella immediately after its production due to the potential logistical–commercial interest. However, adverse effects on rheological and functional properties and proteolysis are also reported.

Decidedly more promising are the approaches that use modified atmosphere packaging and/or calcium lactate accompanied by refrigeration [42–44]. They allow a significant slowdown of proteolytic (*see* **Note 15**) events and the growth of undesirable microorganisms. Based on what has been said in this chapter, it can be concluded that the mozzarella category includes a wide variety of products that differ in terms of types of raw materials, milk processing technologies, shape, size, and storage methods. Consequently, product quality is significantly influenced by the cheesemaking protocols adopted and by appropriate process and product monitoring tools. This chapter has, therefore, been devoted to an overview of the salient aspects of its production chain.

4 Notes

1. Buffalo mozzarella recognized with the PDO mark as “Mozzarella di Bufala Campana” is produced with milk from specific geographical areas in four different regions, such as Campania (entire provinces of Caserta and Salerno, some municipalities in the province of Naples and Benevento), Lazio (municipalities in the province of Frosinone and Rome), Molise (municipality of Venafro), and Puglia (municipalities in the province of Foggia).

Table 8
Quality parameters of mozzarella cheese

Parameters	Monitoring frequency	Methods	Threshold/range	Notes
Relative humidity	On each batch	As difference dry matter content ISO 5534: 2004	≤65%	
Fats	On each batch	ISO 23319: 2022	>52% for buffalo mozzarella	Gravimetric method
Aflatoxin M1	Semiannual	Aflatoxin M1	<0,05 μG/KG	Clean-up by immunoaffinity chromatography and determination by thin-layer chromatography
Eumycetes	Monthly (min)	ISO 21527-1:2008	<5E3 CFU/mL	By means of the colony count technique at 25 °C ± 1 °C
Total coliforms	Monthly (min)	ISO 4832: 2006	<1E3 CFU/ml	Technique of colony counting after incubation on a solid medium at 30 °C or at 37 °C
Escherichia coli	Monthly (min)	ISO 16649-2:2001	<1E2 CFU/mL	For the enumeration of β-glucuronidase-positive *Escherichia coli* using a colony count technique at 44 °C on a solid medium containing a chromogenic ingredient for detection of the enzyme β-glucuronidase
Salmonella	Monthly	ISO 6579: 2017	Absent/25 g	Microorganism which forms typical or less typical colonies on solid selective media and which displays the characteristics described when confirmation tests are carried out in accordance with ISO document

(continued)

Table 8 (continued)

Parameters	Monitoring frequency	Methods	Threshold/range	Notes
Pseudomonas spp.	Monthly	ISO/TS 11059: 2009	≤ 100.000 CFU/g	Species of bacteria of the genus *Pseudomonas* which form colonies in penicillin and pimaricin agar (PPA) at 25 °C displaying the biochemical characteristics described
Listeria monocytogenes	Monthly	ISO 11290-1:2005	Absent/25 g	
Coagulase-positive cocci	Monthly	ISO 6888-1: 2004	<100 CFU/g	Method using Baird–Parker agar medium
Heavy metals	Quarterly	MP 1288 rev.8 2011	Conforms to Reg. 1881/06	
Chemotherapeutics, phytopharmaceuticals, tetracycline residues, and amphenicides	Semiannual	UNI EN 15662: 2009	Conforms to Reg. 396/05 and Reg. 1881/06 and Reg. 37/10	

2. Gangliosides, glycosphingolipids containing sialic acid in turn bound to galactose units, are bioactive molecules of interest in biochemical functions in the human body, such as neural repair, calcium homeostasis, and immune system enhancement. There are different methodologies for their determination and quantification in foods, and there is no official method for their determination. Determination by HPLC is the most widely used approach.
3. Calf rennet, the main milk-coagulating enzyme used in cheese production, is a mixture of enzymes (mainly chymosin and pepsin) obtained from the fourth stomach (abomasum) of unweaned ruminants, such as calf, lamb, kid, and buffalo calf, while still nursing.
4. *Withania coagulans* belongs to the Solanaceae family commonly found in Middle Eastern areas and the Indo-Pakistani subcontinent. Commonly in these areas, the aqueous extract of the plant is used as a coagulant in the preparation of cheese made from milk of different species such as goat, sheep, camel, and buffalo.

5. R-chymosin is a heterologous enzyme, that is, an animal enzyme produced by microorganisms through genetic engineering actions. Conventionally, cloning a eukaryotic gene into a bacterial host involves the synthesis of complementary DNA (cDNA) by reverse transcription of mRNA. In the case of r-chymosin, RNA from the mucosal layer of the abomasum is isolated and the mRNA is fractionated by affinity chromatography. Then, the mRNA is transcribed into cDNA and subsequently cloned into a vector compatible with the microbial host. The chymosin gene was cloned and expressed using plasmid vectors and host organisms such as bacterial cells (*Escherichia coli*, *Bacillus subtilis*) or even fungi (*Saccharomyces cerevisiae*, *Kluyveromyces* spp., *Aspergillus* spp.).
6. Acidification of milk and curd is achieved by the addition of natural whey starter resulting from previous buffalo milk processing that took place on the same farm or on neighboring farms located in the same production area.
7. The shelf life of mozzarella varies from a few days to as long as 24–28 days. Mozzarella made with raw milk and use of natural whey starter generally has the shortest shelf life. In contrast, the application of pasteurization, the use of a selected culture together with a high specificity curd, and the application of protective approaches during storage allow the extension of shelf life to about 28 days.
8. Heat treatments of milk can be distinguished into thermization or pasteurization treatments. The former refers to a series of sub-pasteurization heat treatments of milk, generally in the temperature range of 60 °C to 68 °C with a holding time between 5 s and 15 min. These types of treatments can reduce microbial contamination by 3–4 logs. Pasteurization, on the other hand, is conducted at temperatures below 100 °C. It is used to reduce or eliminate pathogenic microorganisms and thus ensure milk safety.
9. Pasteurization is carried out by plate heat exchanger. The most used time–temperature combination is 72 °C for 18 s, followed by immediate cooling to 4 °C ± 2 for storage, or heating to about +35 °C/+37 °C for transfer to the polyvalent. The pasteurizer is generally equipped with a detour valve that can prevent the milk from being sent to the polyvalent if the temperature is less than or equal to 72 °C.
10. Curd breaking/cutting occurs in two stages: from coagulation and the second after 20 min.
11. The kneaded, melted, and finally stretched curd is portioned by breaking manually by two operators using the thumb and index finger of the hand. It is precisely this operation (defined in the

Southern Italian dialect as "mozzatura") that gives rise to the name of the product "mozzarella."

12. Mechanical stretching is carried out by continuous machine inside which the curd is finely minced by the chopper and then subjected to mechanical force by augers, starting to stretch when hot water +83/+84 °C is added.
13. When the agent is an organic acid, the acidification of milk leads to greater solubilization of colloidal calcium and a lower level of calcium associated with proteins in the curd. Such a situation allows elongation at higher pH, between 5.5 for buffalo milk and 5.7 for cow milk.
14. The newly formed mozzarella cheese is cooled in special firming tanks separated into two compartments, a pre-freezing tank, where the water is at a temperature of +18 °C, and a firming tank where the temperature is +13 °C.
15. Proteolysis is a complex biochemical event that affects cheese quality. Protein hydrolysis during storage results in a reduction of typical fibrousness and a decay of rheological characteristics.

References

1. Commission Regulation (EC) No. 1107/96
2. Coppola S, Blaiotta G, Ercolini D, Moschetti G (2001) Molecular evaluation of microbial diversity occurring in different types of mozzarella cheese. J Appl Microbiol 90(3):414–420
3. Fusco V, Chieffi D, De Angelis M (2022) Invited review: fresh pasta filata cheeses: composition, role, and evolution of the microbiota in their quality and safety. J Dairy Sci 105: 9347–9366
4. Coppola S, Villani F, Coppola R, Parente E (1990) Comparison of different starter systems for water-buffalo mozzarella cheese manufacture. Lait 70(5–6):411–423
5. Guidone A, Zotta T, Matera A, Ricciardi A, De Filippis F, Ercolini D, Parente E (2016) The microbiota of high-moisture mozzarella cheese produced with different acidification methods. Int J Food Microbiol 216:9–17
6. Gobbetti M, Neviani E, Fox P (2018) The cheeses of Italy: science and technology. Springer, Cham
7. Angeletti R, Gioacchini AM, Seraglia R, Piro R, Traldi P (1998) The potential of matrix-assisted laser desorption/ionization mass spectrometry in the quality control of water buffalo mozzarella cheese. J Mass Spectrom 33(6): 525–531
8. Manzo C, Pizzano R, Addeo F (2008) Detection of pH 4.6 insoluble β-lactoglobulin in heat-treated milk and mozzarella cheese. J Agric Food Chem 56(17):7929–7933
9. Rullo R, Caira S, Nicolae I, Marino F, Addeo F, Scaloni A (2023) A genotyping method for detecting foreign buffalo material in mozzarella di bufala campana cheese using allele-specific-and single-tube heminested-polymerase chain reaction. Food Secur 12(12):2399
10. Karaziack CB, Vidal C, Pasquini C, Barbin DF, Viotto WH (2024) Application of near-infrared hyperspectral imaging for determination of cheese chemical composition. J Food Compos Anal 127:105994
11. Garau V, Manis C, Scano P, Caboni P (2021) Compositional characteristics of Mediterranean buffalo milk and whey. Dairy 2(3): 469–488
12. Barłowska J, Szwajkowska Z, Litwinczuk M, KrÀol J (2011) Nutritional value and technological suitability of milk from various animal species used for dairy production. Compr Rev Food Sci Food Saf 10:291–302
13. Ferranti P, Scaloni A, Caira S, Chianese L, Malorni A, Addeo F (1998) The primary structure of water buffalo α s1-and β-casein: identification of phosphorylation sites and characterization of a novel β-casein variant. J Protein Chem 17:835–844
14. De Kruif CG, Huppertz T, Urban VS, Petukhov AV (2012) Casein micelles and their

internal structure. Adv Colloid Interf Sci 171: 36–52

15. Felice VD, Owens RA, Kennedy D, Hogan SA, Lane JA (2021) Comparative structural and compositional analyses of cow, buffalo, goat and sheep cream. Food Secur 10(11):2643
16. De Simone C, Ferranti P, Picariello G, Scognamiglio I, Dicitore A, Addeo F et al (2011) Peptides from water buffalo cheese whey induced senescence cell death via ceramide secretion in human colon adenocarcinoma cell line. Mol Nutr Food Res 55(2): 229–238
17. Garg SK, Johri BN (1994) Rennet: current trends and future research. Food Rev Intl 10(3):313–355
18. Manuelian CL, Boselli C, Vigolo V, Giangolini G, De Marchi M (2020) Effects of animal versus vegetal rennet on milk coagulation traits in Mediterranean buffalo bulk milk. J Dairy Sci 103(6):4958–4964
19. Nazish HA, Gulzar N, Nadeem M, Rafiq S, Sameen A, Ajmal M et al (2022) Efficacy of Withania coagulans fruit extract as a coagulant for mozzarella cheese at different coagulation temperatures from curd formation to pizza top. J Food Proc Preserv 46(12):e17167
20. Sheehan JJ, O'Sullivan K, Guinee TP (2004) Effect of coagulant type and storage temperature on the functionality of reduced-fat mozzarella cheese. Lait 84(6):551–566
21. Rusdan IH, Kusnadi J (2017, Apr) Production of mozzarella cheese using rennin enzyme from Mucor miehei grown at rice bran molasses medium. IOP Conf Ser Mater Sci Eng 193(1):012011
22. Mohanty AK, Mukhopadhyay UK, Grover S, Batish VK (1999) Bovine chymosin: production by rDNA technology and application in cheese manufacture. Biotechnol Adv 17(2–3): 205–217
23. Nawaz AF, Samra Zafar SLF, Shahzadi K, Fatima Z, Siddique I (2021) Use of nanomaterials for the immobilization of industrially important enzymes. J Nanotechnol Res 3(4): 45–57
24. Arain MA, Rasheed S, Jaweria A, Khaskheli GB, Barham GS, Ahmed S (2023) A review on processing opportunities for the development of camel dairy products. Food Sci Anim Resour 43(3):383
25. Natrella G, Faccia M, Lorenzo JM, De Palo P, Gambacorta G (2020) Sensory characteristics and volatile organic compound profile of high-moisture mozzarella made by traditional and direct acidification technology. J Dairy Sci 103(3):2089–2097
26. Ercolini D, Mauriello G, Blaiotta G, Moschetti G, Coppola S (2004) PCR–DGGE fingerprints of microbial succession during a manufacture of traditional water buffalo mozzarella cheese. J Appl Microbiol 96(2): 263–270
27. Zotta T, Ricciardi A, Condelli N, Parente E (2022) Metataxonomic and metagenomic approaches for the study of undefined strain starters for cheese manufacture. Crit Rev Food Sci Nutr 62(14):3898–3912
28. Levante A, Bertani G, Marrella M, Mucchetti G, Bernini V, Lazzi C, Neviani E (2023) The microbiota of mozzarella di bufala campana PDO cheese: a study across the manufacturing process. Front Microbiol 14: 1196879
29. Bonizzi I, Feligini M, Aleandri R, Enne G (2007) Genetic traceability of the geographical origin of typical Italian water buffalo mozzarella cheese: a preliminary approach. J Appl Microbiol 102(3):667–673
30. Ercolini D, De Filippis F, La Storia A, Iacono M (2012) "Remake" by high-throughput sequencing of the microbiota involved in the production of water buffalo mozzarella cheese. Appl Environ Microbiol 78:8142–8145
31. Natrella G, Gambacorta G, Faccia M (2022) Influence of the stretching temperature on the volatile compounds and odour intensity of high moisture mozzarella: a model study. Int Dairy J 130:105282
32. Faccia M, Gambacorta G, Natrella G, Caponio F (2019) Shelf life extension of Italian mozzarella by use of calcium lactate buffered brine. Food Control 100:287–291
33. Huang X, Nzekoue FK, Renzi S, Alesi A, Coman MM, Pucciarelli S et al (2022) Influence of modified governing liquid on shelf-life parameters of high-moisture mozzarella cheese. Food Res Int 159:111627
34. European Commission Regulation 853/2004 (EC, 2004)
35. Fox PF, McSweeney PLH (2004) Cheese: an overview. In: Fox PF, McSweeney PLH, Cogan TM, Guinee TP (eds) Cheese: chemistry, physics and microbiology, vol 1. Academic Press, Cambridge, pp 1–18
36. Rowney MK, Hickey MW, Roupas P, Everett DW (2003) The effect of homogenization and milk fat fractions on the functionality of mozzarella cheese. J Dairy Sci 86(3):712–718
37. Gonçalves MC, Cardarelli HR (2021) Mozzarella cheese stretching: a minireview. Food Technol Biotechnol 59(1):82–91
38. Reg. CE 1881/2006

39. Liu G, Carøe C, Qin Z, Munk DM, Crafack M, Petersen MA, Ahrné L (2020) Comparative study on quality of whole milk processed by high hydrostatic pressure or thermal pasteurization treatment. LWT 127:109370
40. Rynne NM, Beresford TP, Kelly AL, Guinee TP (2004) Effect of milk pasteurization temperature and in situ whey protein denaturation on the composition, texture and heat-induced functionality of half-fat Cheddar cheese. Int Dairy J 14(11):989–1001
41. Smith JR, Carr AJ, Golding M, Reid D (2018) Mozzarella cheese – a review of the structural development during processing. Food Biophys 13(1):1–10
42. Loi M, Quintieri L, De Angelis E, Monaci L, Logrieco AF, Caputo L, Mulè G (2020) Yield improvement of the Italian fresh giuncata cheese by laccase–induced protein crosslink. Int Dairy J 100:104555
43. Zappia A, Branca ML, Piscopo A, Poiana M (2020) Shelf-life extension of mozzarella cheese packed in preserving liquid with calcium lactate and bergamot juice concentrate. J Dairy Res 87(4):474–479
44. Singla M, Talwar G (2022) Effect of various packaging techniques on quality characteristics of mozzarella cheese. Industria Farmacéutica 9(11):523–528

Chapter 4

Prato Cheese

Renata Golin Bueno Costa, Denise Sobral, and Junio César Jacinto de Paula

Abstract

Prato cheese is a cheese of Brazilian origin and ranks second among the most produced cheeses in Brazil. Its production involves a process that includes semi-cooking the curd with partial removal of the whey and subsequent incorporation of hot water, pre-pressing, forming, pressing, salting, and the ripening period. The cheese-making process is presented in detail in this chapter.

Key words Semi-cooked pasta, Curd wash, Technology, Annatto coloring, Brazilian cheese, Ripening

1 Introduction

The typically Brazilian Prato cheese originated from Danbo or Gouda cheese [1], in which the cheese mass is semi-cooked with hot water before the whey is partially removed. Over the years, Prato cheese has undergone several changes over the years, with specific texture and flavor characteristics, both in manufacturing technology and how it is consumed [2].

Prato cheese can have a few small, well-distributed holes or no holes, a yellowish or straw-yellow color, a characteristic taste and smell, and no rind, or a thin, smooth rind with no cracks. It is classified as full-fat cheese (between 45.0% and 59.9% fat in the dry matter) and medium-moisture (between 36.0% and 45.9%), which must be ripened for at least 25 days before being marketed. It is presented semi-cooked, with the whey partially removed and heated with hot water. It can be shown in a cylindrical shape called Cobocó, a spherical shape called Spherical Prato, or a rectangular shape, the most traditional [3]. On average, after brining, it has a moisture content of 43–45%, fat of 26–29%, fat in dry matter of 46–53%, pH 5.1–5.2, and salt content of 1.2–1.3% [4].

In automated manufacturing processes, the cheese curd is not washed because the technique is continuous and avoids diluting the

Adriano Gomes da Cruz et al. (eds.), *Dairy Foods Processing*, Methods and Protocols in Food Science,
https://doi.org/10.1007/978-1-0716-4144-6_4,

whey with water, which damages the marketing value of the whey. In this form of production, the cheese becomes slightly more acidic and is sold as a culinary ingredient [4].

It is one of the most consumed ripened Brazilian cheeses [5] for direct and indirect consumption in sandwiches and culinary preparations [6]. Because it is an ingredient, Prato cheese must have functional properties such as sliceability, meltability, and oil release [7]. The ability to slice is one of the most important characteristics of Prato cheese due to its practicality and convenience at the time of purchase [8].

The traditional yellowish or straw-yellow color of Prato cheese is due to the annatto dye, a carotenoid. However, it is possible to replace the dye with lutein, another carotenoid that prevents diseases. Lutein can prevent age-related macular degeneration (AMD) and cataracts [9] and can slow down chronic diseases and some types of cancer [10]. Lutein was added as a liquid emulsion with polysorbate 80 in milk for the manufacturing of Prato cheese with retention of 95.25% lutein in the cheese, with no change in the physicochemical properties of cheese. However, the cheese with added lutein had a more intense yellow-orange color, impacting the lower sensory acceptability of cheese [11]. Two concentrations of lutein (16 and 32 mg/L) were added to Prato cheese without losing antioxidant activity over the ripening period. Prato cheese with the highest concentration of lutein showed similar coloration (yellow-orange color) to traditional [12].

In addition to the change in color, Prato cheese can be made with the addition of carbon dioxide (CO_2) under pressure (1.6×10^5 Pa at 8 °C) to pasteurized Prato cheese milk (pH 6.0). The addition of CO_2 decreased the curd formation time (30 min), the total manufacturing time (47 min), and the pH of Prato cheese, thus leading to reduced moisture content. However, there was no difference in cheese yield, microorganism counts, proteolysis, or sensory acceptance [13].

As for the addition of new cultures to Prato cheese, exopolysaccharide (EPS)-producing culture in a capsular form was used in the manufacturing of Prato cheese. This culture presents excellent properties to bind water and retain moisture [14]. The Prato cheese made with EPS-producing culture had a higher moisture content and an increase in yield of 0.25 kg/100 kg of milk compared to the traditional cheese made without EPS. There were also some changes in the textural characteristics, such as less hardness and chewiness in the cheeses made with EPS. However, there was no difference in the pH, proteolysis, and melting capacity of cheese, nor in the consumer acceptance test [15].

Prato cheese can be produced with probiotic cultures (*Lactobacillus acidophilus* La5 and *Bifidobacterium* Bb12) separately (1% v/v) or in combination (0.5% v/v of each culture) without altering the characteristics of the cheese. Probiotic counts were higher than

10^6 CFU/g and showed resistance to loss of viability during in vitro gastrointestinal simulation. However, there was an increase in proteolysis with a reduction in cheese hardness [16]. Also, to improve consumer health, Prato cheese is made with the addition of *Lactobacillus helveticus* LHB02 and salt reduction, together and separately. The probiotic culture favored the release of bioactive peptides during cheese ripening [5].

Another technology employed in Prato cheese was sodium reduction, with the use of sodium substitutes such as KCl, sub4-salt®, and Salona™, with no changes in physicochemical characteristics, pH, proteolysis indexes, melting capacity, and texture profile, with a significant reduction in sodium content [17]. The effect of partial substitution of NaCl with KCl and the flavor enhancer addition (arginine, yeast extract, and oregano extract) on probiotic Prato cheese processing with *Lactobacillus casei* 01 (7 log CFU/mL) was also investigated. Some changes occurred in Prato cheese, such as proteolysis and pH, due to adding flavor enhancers [18]. Prato cheese can be an adequate carrier of probiotics, as seen in previous studies.

In addition to cultures, Prato cheese can be made with a coagulant from *Thermomucor indicae-seudaticae* N31 to replace the commercial coagulant without causing changes in the proteolysis of the cheese [19].

Although there are several proposed technologies, this chapter presents the production technology for traditional and industrial Prato cheese.

2 Materials

To make Prato cheese, it needs a basic cheese production structure such as a stainless-steel cheese tank with knife blades or wires and a stirring fork, a stainless-steel milk agitator, pre-pressing plates, a pneumatic or weighted press, Prato cheese molds, brine (20% w/v salt), and a ripening chamber (10–12 °C). Also, a graduated bucket for volume measurement, thermometer, measuring cylinder for ingredients, acidimeter, Dornic solution, and phenolphthalein are used to measure milk and whey acidity. All utensils, such as knives, knife blades or wires, molds, and cloths, should be washed with soap and water and immersed in a chlorine solution (200 mg/L of active chlorine) for at least 20 min (*see* **Note 1**).

As for the ingredients, it will need:

1. Milk with an acidity of between 0.15% and 0.18% expressed as lactic acid.
2. Starter culture. Type O mesophilic culture (*Lactococcus lactis* subsp. *lactis* and *Lactococcus lactis* subsp. *cremoris*). Mixtures of type O mesophilic cultures (80–90%) and thermophilic cultures (*Streptococcus thermophilus*—10–20%) can be used to

manufacture in a continuous production system. A little *Lactobacillus helveticus* can be added to give more flavor to the cheese. This starter culture is sold ready-made with the addition of *L. helveticus*, plus mesophilic bacteria and *Streptococcus thermophilus*. The dosage used is that recommended by the manufacturer.

3. Calcium chloride solution (40% or 50% w/v). The dosage will be 40 mL of calcium chloride for every 100 L of milk if the solution used is 50% (w/v). If the solution is 40% (w/v), use 50 mL for every 100 L of milk.
4. Annatto dye solution. The amount used in this cheese is 5 to 10 mL per 100 L of milk. The dye can be diluted in 5 to 6 parts of water (*see* **Note 2**).
5. Sodium nitrate solution (40% or 50% w/v). If the solution used is 50% (w/v), the dosage will be 40 mL of sodium nitrate for every 100 L of milk, or 50 mL for every 100 L of milk if it is 40% (w/v).
6. Rennet or coagulant (liquid or powder) for cheese making (*see* **Note 3**).
7. Hot water (70 °C) for heating the coagulum. The volume will be 15–20% hot water relative to the milk volume.

3 Methods

3.1 Prato Cheese

1. Standardize the milk to 3.0–3.2% fat and pasteurize at 72 °C/15 s or 65 °C/30 min (*see* **Note 4**).
2. Cool the milk to 32–34 °C.
3. Add the recommended dose of lactic acid starter, mesophilic type O (*Lactococcus lactis* subsp. *lactis* and *Lactococcus lactis* subsp. *cremoris*). Mix well with a fork or stirrer.
4. Leave to ripen the milk at 32–34 °C for 30 min.
5. Add the calcium chloride to the milk. Mix well with a fork or stirrer.
6. Add the annatto dye to the milk. Mix well.
7. Add the sodium nitrate and mix well.
8. Measure the temperature of the milk to check that it is between 32 °C and 34 °C. If necessary, heat the milk until it reaches this temperature.
9. Dose the amount of rennet/coagulant as the manufacturer recommends. Dilute it in 10 to 15 parts of water and add it to the milk to cover the entire tank volume. Shake well to complete homogenization. Remove the agitator or fork from the tank.

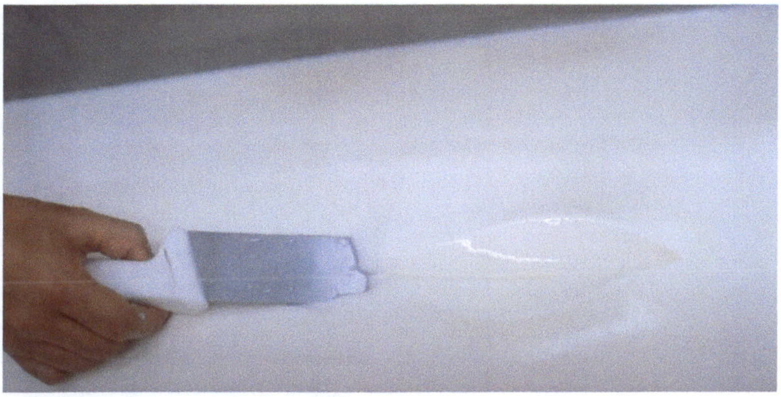

Fig. 1 Curd at the ideal cutting point

10. Wait around 40 min after adding the rennet to form the gel. The point of the curd is checked with a knife, as shown in Fig. 1. The ideal point should be when the curd opens a slit in the cut and the whey is greenish-yellow.
11. Cut the grains carefully and slowly, using knife blades or wires (horizontally and vertically). First, cut the curd lengthways with the horizontal knife blades or wires. Then, cut the curd again lengthways with the vertical knife blades and cut it crossways with the same knife blades, forming cubes. Keep cutting crossways until small pieces (grains between 0.3 and 0.5 cm in edge, approximately the size of popcorn grain, also called grain 03) are generated. Figure 2 shows the curd cut and the curd grain size.
12. Stir slowly with a suitable fork for 20 min (*see* **Notes 5** and **6**).
13. Remove 30–40% of the whey in relation to the milk volume.
14. Slowly add 15–20% hot water (temperature 70 °C) (in relation to the volume of milk) so that it heats up slowly (1 °C/ 2–3 min) until it reaches 41–42 °C. The water is added under continuous stirring of the curd grains. Under normal conditions, the point of the cheese occurs around 50 to 70 min after cutting the curd. At this point, the cheese is drier and softer.
15. Pre-pressing: Once the point has been reached, the curd must be pushed to the end of the tank with the pressing plates or placed in the drain press; then, all the whey must be drained off. Pre-pressing is carried out with double the weight relative to the curd amount (or 40–50 Lb/in^2 in a drain press) for 15–20 min.
16. Once the pre-pressing is complete, cut the block of curd, and place approximately 1 to 1.2 kg of curd in cylindrical molds (Cobocó cheese) or 2 kg molds (Prato Lanche cheese) with cloth for subsequent pressing.

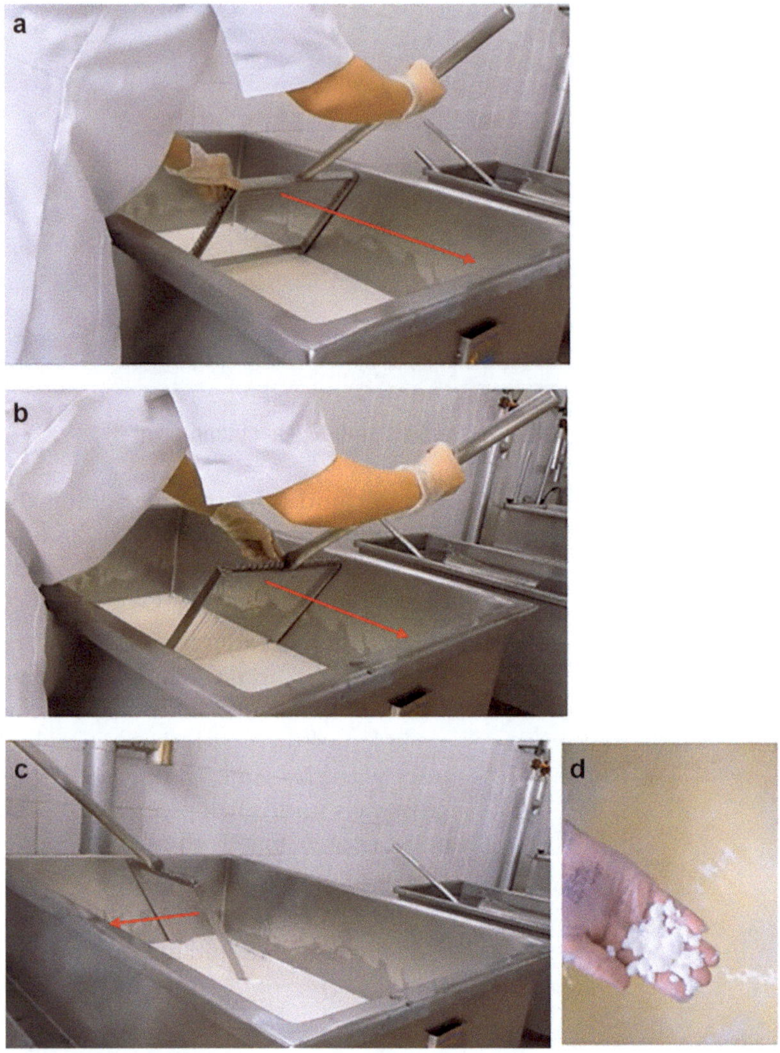

Fig. 2 (**a**) Cutting the curd lengthways with the horizontal knife blades or wires. The red arrow in the figure indicates the direction of the knife blades when cutting the curd. (**b**) Cutting the curd lengthways with the vertical knife blades. (**c**) Cutting it crossways with the same knife blades, forming cubes. (**d**) Curd grain size (grains between 0.3 and 0.5 cm in edge, approximately)

17. First, press the cheeses for 30 min with 20 Lb/in^2 or 10 times the cheese weight (20 kg for 2 kg cheese). Then, turn the cheeses over in the mold, and reverse the position of the cheeses inside the mold and in the press (move the cheeses that were on the top to the bottom), and press again for another 90 min with 10 times of the cheese weight (20 kg for 2 kg of cheese) or 30 Lb/in^2. Again, reverse the position of the cheeses inside the mold and in the press. Press for another 45 min with 10 times the cheese weight or 30 Lb/in^2. At the

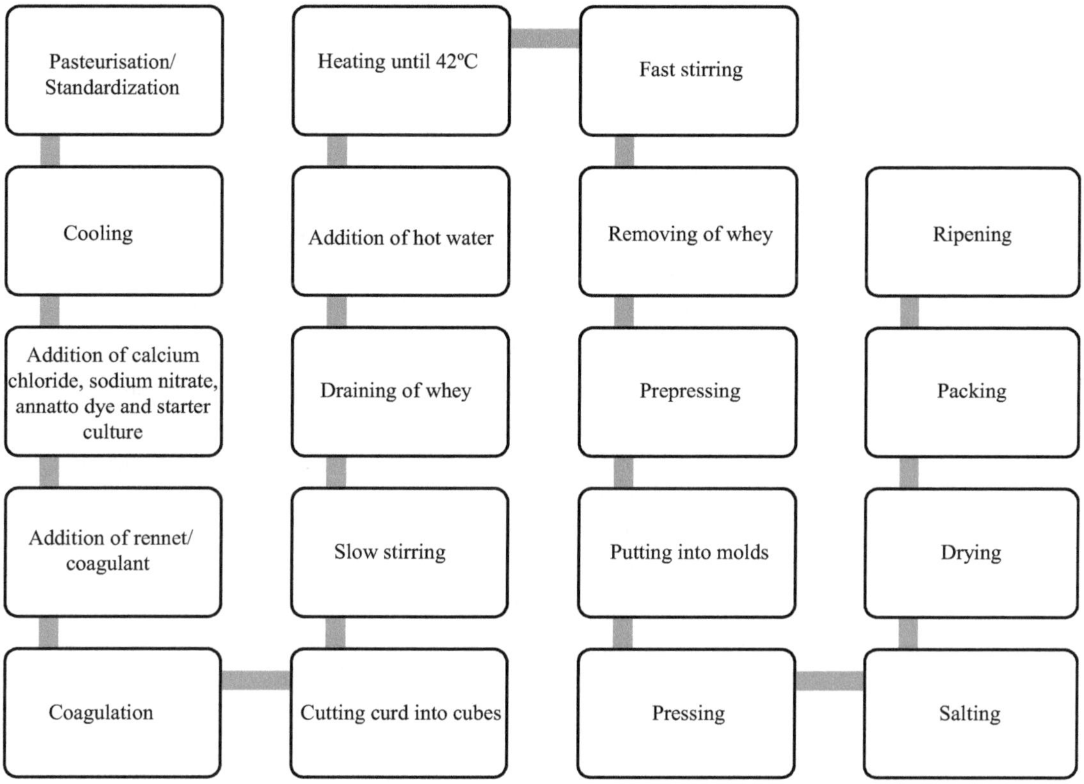

Fig. 3 Flowchart of traditional Prato cheese manufacture

end of the pressing time or as soon as the cheese reaches a pH of 5.2, cheese should be removed from the press and pressed again, without cloth, for about 15 min with 30 Lb/in^2 or 10 times their weight for a better finish. The cheese should then be salted.

18. Place the cheeses in the brine (20% w/v salt) at 10–12 °C for 10–12 h (1 kg of cheese) and 24 h for 2 kg of cheese.
19. After salting, remove the cheeses from the brine, and place them on the shelf in a chamber at 10–12 °C with relative humidity of approximately 70% for 24 to 48 h.
20. Pack the cheeses in a heat-shrinkable plastic bag and keep them in chambers at 10–12 °C for at least 25 days. At the end of this period, the cheeses can be sold or stored at 2–4 °C. Figure 3 presents the flowchart of traditional Prato cheese manufacture.

3.2 Prato Cheese Produced in a Continuous Production System [4]

1. Standardize the milk to about 3.4% fat and pasteurize at 72 °C/ 15 s (*see* **Note 4**).
2. Cool the milk to 36–38 °C.
3. Add the calcium chloride to the milk. Mix well with a stirrer.
4. Add the annatto dye to the milk and mix well.

5. Add the sodium nitrate and mix well. Instead of adding nitrate, a degerminating centrifuge or microfiltration is used.
6. Add the recommended dose of a mixture of mesophilic and thermophilic starter cultures. Mix well with a stirrer.
7. Leave to ripen the milk at 36–38 °C for 30 min. Generally, in large-volume tanks, the starter culture is added at the start of filling the tank, and the time taken to fill it is enough to ripen the milk.
8. Measure the temperature of the milk to check that it is between 36 and 38 °C (*see* **Note 7**).
9. Dose the amount of rennet/coagulant as the manufacturer recommends.
10. Wait around 40 min after adding the rennet to form the gel. The point of the curd is checked with a knife, as shown in Fig. 1. The ideal point should be when the curd opens a slit in the cut and the whey is greenish-yellow.
11. Cut the grains carefully and slowly, using knife blades or wires (horizontally and vertically) to generate small pieces (grains between 0.3 and 0.5 cm in edge, approximately the size of popcorn grain, also called grain 03).
12. Stir slowly (6–8 rpm) for 10–15 min.
13. Heat the curd with steam under rapid stirring (up to 12 rpm) until it reaches 41–42 °C.
14. The point of the cheese, under normal conditions, occurs around 60–70 min after cutting the curd. The whey is drained off.
15. The curd goes through a fermentation process (41–43 °C) for 60–90 min until the pH reaches 5.4–5.5.
16. After fermentation, the curd is shredded and dry-salted. The curd goes into a pressing tower to form rectangular blocks weighing up to 20 kg. The blocks are cooled in blast freezers, packed, and palletized for ripening (5–10 °C).
17. The blocks can be broken down into smaller portions before being sold. Figure 4 presents the flowchart of Prato cheese produced in continuous production system.

4 Notes

1. The chlorine solution (200 mg/L) is prepared by adding 200 mL of sodium hypochlorite (10% active chlorine) to 100 L of water.
2. Annatto dye precipitates in contact with calcium chloride and sodium nitrate. Avoid adding the dye with these two

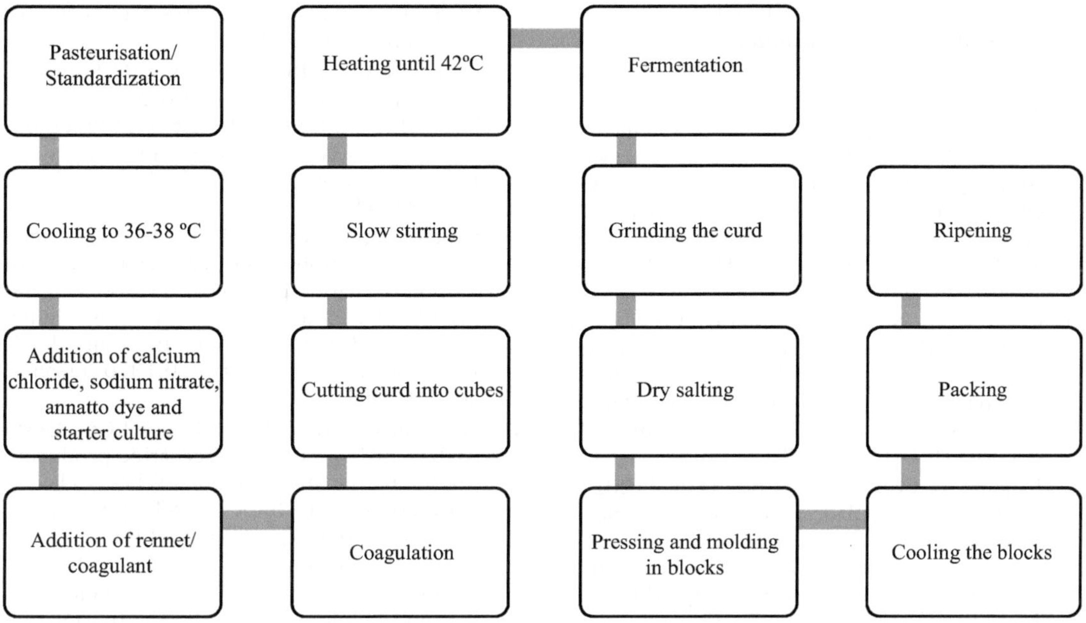

Fig. 4 Flowchart of Prato cheese produced in continuous production system. (Adapted from Furtado (2022) [4])

ingredients simultaneously to the production tank. Always wash the beaker before dosing the liquid ingredients.

3. The manufacturer recommends the dosage according to the instructions on the label. Dose only when adding to milk. It can be diluted in non-chlorinated water at room temperature.
4. The recommended casein-to-fat ratio for this cheese is 0.71–0.74 [20]. The casein-to-fat ratio of 0.72 is recommended for producing Prato cheese with good functional properties, such as intermediate melting and chewiness, and less release of free oil [8].
5. The whey should have an acidity equivalent to 2/3 of the milk initial acidity, i.e., for milk with 0.18%, the whey must have 0.12% expressed as lactic acid.
6. It can rest regularly if the grains break too much.
7. If necessary, heat the milk until it reaches this temperature.

References

1. Silva HLA, Balthazar CF, Esmerino EA et al (2018) Partial substitution of NaCl by KCl and addition of flavor enhancers on probiotic Prato cheese: a study covering manufacturing, ripening and storage time. Food Chem 248:192–200. https://doi.org/10.1016/j.foodchem.2017.12.064
2. Ribeiro MCE, Baptista DP, Queirós MS et al (2020) Brazilian traditional cheeses. In: Oluk CA, Karaca OB (eds) Traditional cheeses from selected regions in Asia, Europe, and South America. Bentham Science, Singapore
3. Brasil (1997) Portaria n° 358, de 04 de setembro de 1997. Regulamento Técnico para

Fixação de Identidade e Qualidade do queijo Prato. http://www.agricultura.gov.br
4. Furtado MM (2022) Receituário brasileiro de queijos, 1st edn. Autor, Valinhos
5. Baptista DP, Negrão F, Eberlin MN, Gigante ML (2020) Peptide profile and angiotensin-converting enzyme inhibitory activity of Prato cheese with salt reduction and lactobacillus helveticus as an adjunct culture. Food Res Int 133: 109190. https://doi.org/10.1016/j.foodres.2020.109190
6. Costa RGB, Sobral D, Teodoro VAM et al (2018) Sodium substitutes in Prato cheese: impact on the physicochemical parameters, rheology aspects and sensory acceptance. LWT Food Sci Technol 90:643–649. https://doi.org/10.1016/j.lwt.2017.12.051
7. Landin TB, Costa RGB, Sobral D et al (2022) Principais propriedades funcionais do queijo muçarela, queijo prato e requeijão culinário—uma revisão. Revista do Instituto de Laticínios Cândido Tostes 76:118–130. https://doi.org/10.14295/2238-6416.v76i2.848
8. Landin TB, Sobral D, Costa Júnior LCG et al (2022) Effect of the casein to fat ratio on the functional properties of Prato cheese. Res Soc Dev 11:e8811527812. https://doi.org/10.33448/rsd-v11i5.27812
9. Seddon JM, Ajani UA, Sperduto RD (1994) Dietary carotenoids, vitamins A, C, and E, and advanced age-related macular degeneration. JAMA J Am Med Assoc 272:1413–1420. https://doi.org/10.1001/jama.1994.03520180037032
10. Alves-Rodrigues A, Shao A (2004) The science behind lutein. Toxicol Lett 150:57–83. https://doi.org/10.1016/j.toxlet.2003.10.031
11. Kubo MTK, Maus D, Xavier AAO, Mercadante AZ, Viotto WH (2013) Transference of lutein during cheese making, color stability, and sensory acceptance of Prato cheese. Cienc Tecnol Aliment 33(Supl. 1):81–88. https://doi.org/10.1590/S0101-20612013000500013
12. Sobral D, Costa RGB, Machado GM, Paula JCP, Teodoro VAM, Nunes NM, Pires ACS, Pinto MS (2016) Can lutein replace annatto in the manufacture of Prato cheese? LWT Food Sci Technol 68:349–355. https://doi.org/10.1016/j.lwt.2015.12.051
13. Paula JCJ, Costa RGB, Carvalho EC, Sobral D, Paiva PHC, Pires ACS, Pinto MS, Teodoro VAM (2019) Effect of the use of carbon dioxide on Prato cheese making. Int Dairy J 91: 172–177. https://doi.org/10.1016/j.idairyj.2018.09.009
14. Lynch KM, McSweeney PLH, Arendt EK et al (2014) Isolation and characterisation of exopolysaccharide-producing Weissella and Lactobacillus and their application as adjunct cultures in Cheddar cheese. Int Dairy J 34: 125–134. https://doi.org/10.1016/j.idairyj.2013.07.013
15. Nepomuceno RSC, Costa Junior LCG, Costa RGB (2016) Exopolysaccharide-producing culture in the manufacture of Prato cheese. LWT Food Sci Technol 72:383–389. https://doi.org/10.1016/j.lwt.2016.04.053
16. Chaves KS, Gigante ML (2016) Prato cheese as suitable carrier for Lactobacillus acidophilus La5 and Bifidobacterium Bb12. Int Dairy J 52:10–18. https://doi.org/10.1016/j.idairyj.2015.08.009
17. Costa RGB, Sobral D, Teodoro VAM et al (2018) Sodium substitutes in Prato cheese: impact on the physicochemical parameters, rheology aspects and sensory acceptance. LWT Food Sci Technol 90:643–664
18. Silva HLA, Balthazar CF, Esmerino EA et al (2017) Effect of sodium reduction and flavor enhancer addition on probiotic Prato cheese processing. Food Res Int 99(Pt 1):247–255. https://doi.org/10.1016/j.foodres.2017.05.018
19. Merheb-Dini C, Garcia GAC, Penna ALB et al (2012) Use of a new milk-clotting protease from Thermomucor indicae-seudaticae N31as coagulant and changes during ripening of Prato cheese. Food Chem 130:859–865. https://doi.org/10.1016/j.foodchem.2011.07.105
20. Costa Júnior LCG, Pinheiro AJR (1999) Influência da relação caseína/gordura nas características sensoriais do queijo Prato. Revista do Instituto de Laticínios Cândido Tostes 54:40–46

Chapter 5

Processed Cheese

Patrícia Blumer Zacarchenco, Leila Maria Spadoti, Adriana Torres Silva e Alves, Adriano Gomes da Cruz, Natália Regina Coldebella Ferreira, and Silvani Verruck

Abstract

Processed cheese consumption has surged in recent decades due to its versatility. Besides being eaten on its own or with bread and toast, it is commonly used in home cooking, restaurant meals, and ready-to-eat frozen or chilled foods. Traditionally, this category of products is prepared with cheese, dairy fat, common salt, emulsifying salts, and sometimes spices. Processed cheeses are versatile and innovative, made with fiber, prebiotics, probiotics, and other functional ingredients. There are also variations prepared with dairy and vegetable protein concentrates, vegetable fats, and thickeners. Due to the flexibility in the formulation and process steps, processed cheeses may have low or no added fat and are reduced in sodium and lactose. In addition to this characteristic of versatility, processed cheeses are generally microbiologically safe due to the heating steps, emulsifying salts with bactericidal or bacteriostatic action, and hot filling packaging. The largest production volume is industrial, although there is also artisanal production of processed cheeses, even with geographical identification of origin. This chapter presents processed cheeses' process steps and formulations to serve as practical guidance for their manufacture.

Key words Processed cheese, Emulsifying salt, Whey protein concentrate, Milk protein concentrate, Lactic acid bacteria, Prebiotic fiber, Probiotic

1 Introduction

The basic ingredients of processed cheeses are cheese, cream or butter, sodium chloride, and melting salts. This category of products is very versatile, allowing the addition of different types of dairy and nondairy ingredients. Some studies highlighted this versatile characteristic of processed cheeses [1–3]. They are quite safe from a microbiological point of view and have good physical–chemical and sensorial stability over several months of storage [2]. The origin of the preparation of processed cheeses dates back centuries and is linked to the need to increase the shelf life of the cheeses with which they are produced. In the first processing of these products, different types of cheese were mixed with some

Adriano Gomes da Cruz et al. (eds.), *Dairy Foods Processing*, Methods and Protocols in Food Science,
https://doi.org/10.1007/978-1-0716-4144-6_5,

dairy fat (sour cream or butter) and heated under stirring to obtain a melted mass that could be preserved for longer than the original cheese [3, 4].

According to Ramos et al. [4], the microbiological characteristics of processed cheeses depend on the number of microorganisms present in the ingredients used, the temperature and duration of the heat treatment and cooling, the packaging system, and the storage temperature (*see* **Note 1**). However, when comparing the counts of spoilage microorganisms present in processed cheeses and other categories of cheeses with the same moisture content, the former have considerably lower values, which leads to longer storage times [1]. One of the ingredients in processed cheeses that control the multiplication of microorganisms is some melting salts (or emulsifying salts), in addition to their action in emulsifying the system by acting on the binding of calcium in caseins [2, 5].

Emulsifying salts are considered the most important ingredient in processed cheese blends. The amount, type, and composition of emulsifying salts can significantly affect processed cheese's pH and its textural, microstructural, melting, functional, and sensorial properties. The most commonly used groups of emulsifying salts for the manufacture of processed cheeses are citrate-based and phosphate-based salts [6, 7].

Due to the term "processed" associated with processed cheeses and the misinformation regarding so-called ultra-processed foods, in some countries, the term "spreadable cheese" or "cheese dip" is used on the label of some products that contain ingredients such as cheese, dairy fat, salt, and melting salts, as well as other additions such as condiments and spices [5]. This market movement may relate to the attempt to dissociate the processed cheese category from ultra-processed foods. Unfortunately, in the NOVA food classification system, processed cheeses are classified as ultra-processed, while alcoholic beverages, such as beer and wine, are in the processed category [8]. Processed cheeses can be produced with few ingredients, without the addition of preservative substances such as sorbate or benzoate, have a long shelf life, are microbiologically safe, and can be source foods or rich in proteins, with important levels of bioavailable minerals and vitamins [5, 6]. In this sense, evaluating the ingredients present in each type of processed cheese is important before classifying it inappropriately. Additionally, due to the versatility of this category of products, there are formulations and process steps already established for manufacturing versions with reduced fat, sodium, lactose, and added fiber content, as seen throughout the chapter [2, 8]. Thus, the aim of this chapter is to describe processed cheese manufacturing and possible steps to addition of different ingredients.

2 Materials

Materials used for processed cheese production can be divided into several groups: raw or pasteurized milk, starter culture, rennet (or coagulant agent), lactic acid 85%, different types of cheese (fresh or aged), butter, milk cream, butter oil, vegetable oils or fats, dairy protein concentrates (including milk protein concentrate, whey protein concentrate, caseinates), flavoring, spices, condiments, fibers or prebiotics (inulin, polydextrose), probiotics, salt (NaCl and/or KCl), and emulsifying salts (with regular sodium level or reduced). During the production of the spreadable processed cheese without fat addition, the use of whitening (titan dioxide) and anti-foaming agent (mono- or diglycerides) could be necessary. The processed cheeses undergo heating at high temperatures (above 70 °C). They are filled hot and kept under refrigeration, except for those treated by the Ultra High Temperature (UHT) system, which remain at room temperature. Despite these stages of destroying microorganisms and controlling their multiplication, it is essential to use ingredients with adequate microbiological characteristics as spores and some types of thermostable toxins may not be destroyed and cause food poisoning.

2.1 Milk and Dairy Proteins

The milk used to produce the curd can be full fat, partially skimmed or skimmed. The processed cheeses can be produced with milk of different species (cow, buffalo, sheep, goat). The milk is used to produce the curd applying the methods explained below. It also can be concentrated using ultrafiltration combined or not with other techniques like diafiltration, nanofiltration, or reverse osmosis. The whey produced in cheese making can also be concentrated by these membranes and generate the whey protein concentrates. In these cases, milk and whey protein concentrates with varying levels of protein, lactose, and minerals are obtained and can be added in the formulation of processed cheese (*see* Table 2).

2.2 Starter Cultures and Probiotic Strains

The mesophilic starter culture is added with the rennet (or coagulating agent) to produce curd by enzymatic coagulation for use in processed cheese. In acid coagulation by fermentation, the acidification of milk occurs only with the addition of these mesophilic lactic bacteria, which generally contain *Lactococcus lactis* ssp. *lactis* and *Lactococcus lactis* ssp. *cremoris*. Some strains of probiotic *Bacillus* were evaluated in processed cheese. *Lactobacillus acidophilus* and *Bifidobacterium animalis* subsp. *lactis* were added to processed cheese after the cooling step when the melting had already been done.

2.3 Emulsifying Salts

The melting salts bind to the calcium and the insoluble paracasein is transformed into a more soluble form. Thus, the cheese goes from the semisolid, coagulated, or flaked state to the soluble one. The amount of emulsifying salt to be added depends on the type of cheese (fresh or aged) and the texture, pH, and moisture content of the final processed cheese. The most used emulsifying salts are citrates and phosphates. The phosphates can be monophosphates or orthophosphates, polyphosphates of different chain sizes, metaphosphates, and cross-linked ultraphosphates. The polyphosphates are generally used in spreadable cheeses and citrates in block cheeses. There are emulsifying salts with greater bacteriostatic action (polyphosphates) and those are applied to adjust the final pH of the product. Originally, melting salts were sodium-based. Currently, there are salts where sodium has been replaced by potassium, calcium, or magnesium. There are studies to determine alternatives to the use of melting salts, such as the application of milk protein concentrates and hydrocolloids [9].

2.4 Milk Cream, Butter, and Other Sources of Fat

Traditionally, cream, butter, or butter oil is used to make processed cheeses. However, since a few decades ago, to reduce formulation costs, dairy fat has been replaced in some types of processed cheeses. This substitution reclassifies the product within a category called analogous cheese or product based on processed cheese or other denominations in accordance with the legislation of each country. Currently, replacing dairy fat with vegetable fats and/or oils makes it possible to obtain processed cheeses with higher levels of unsaturated fatty acids and even added bioactive compounds present in these oils.

3 Methods

Many processed cheeses are made from different cheeses such as cheddar, Gouda, Gorgonzola, Emmental, or fresh cheeses, using melting salts to make them spreadable, in blocks or slices. These different varieties of processed cheeses can also be added with combinations of flavoring, spices, condiments, fibers (prebiotic or not), vegetable oils or fats, and many beneficial microorganisms (such as probiotics) or with functional health effects. Products with different sensory characteristics are also obtained depending on the equipment used to manufacture processed cheeses.

Figure 1 presents the general stages of manufacturing processed cheese.

To illustrate the equipment that is used in the production of processed cheeses, Fig. 2 is an example of a jacketed tank with steam heating and manual agitation used in the production of the basic mass by hot acid precipitation; a hydraulic press for removing residual whey to obtain mass with lower moisture content; the

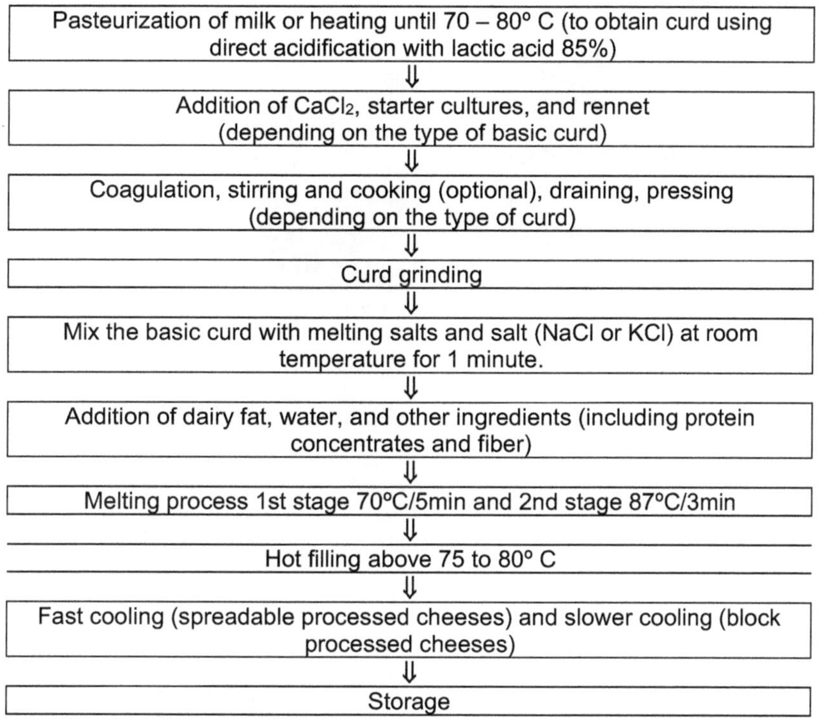

Fig. 1 Steps of the manufacture of processed cheese

homogenizer with direct (sanitary steam), indirect heating, two types of agitators, and two agitation levels (rpm); and a dosing machine and a seaming machine for glass cups with a metal lid.

3.1 Preparation of Curds for Use in Processed Cheeses

The so-called basic curd used in producing processed cheeses can be matured or unripened cheeses or curd recently obtained by coagulating milk after separating the whey and pressing it (fresh cheese) (*see* **Note 2**). It is common to use a single or several types of cheese, whether recently produced or highly matured cheeses. You should always observe the percentages of cheese in the mixture (if any) to standardize the texture and flavor of the final product and the cost of the formulation. Aged or ripened cheeses that present mechanical defects (or visual characteristics) but are microbiologically and sensorially adequate can be used in melting. When cheeses with a high degree of maturity are used in producing processed cheeses, they have an intense flavor and aroma and a more fluid texture. If this degree of maturation is too high, flavor and aroma defects, emulsion instability, granules, and liquid consistency may occur. This can be resolved by adjusting the types and concentrations of melting salts [10].

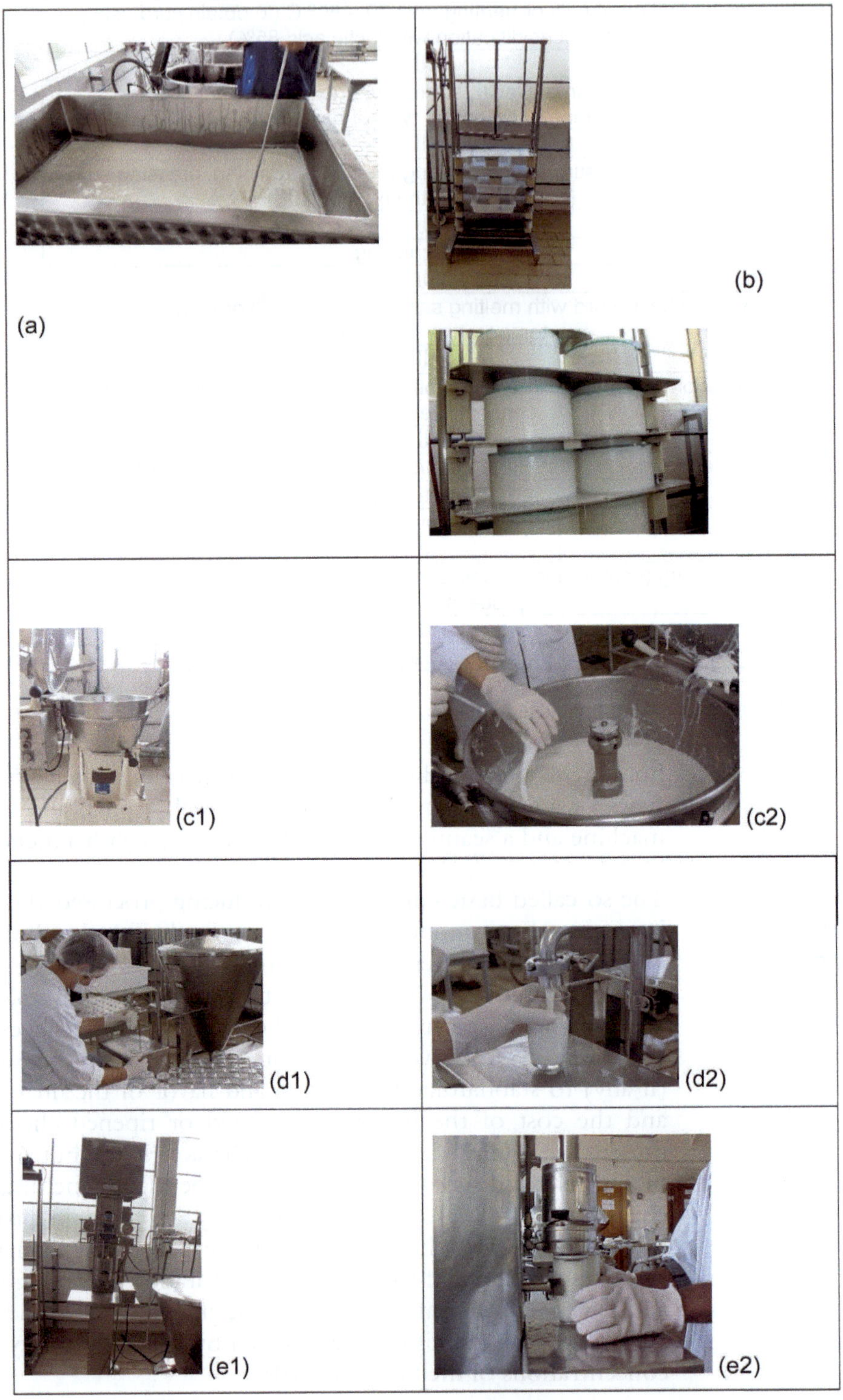

Fig. 2 Equipment used to manufacture spreadable processed cheese (**a**, tank to prepare cheese or curd; **b**, presses; **c1**, equipment for grinding, stirring, melting, or heating; **c2**, addition of whey protein previously diluted in water; **d1**, dosing the hot processed cheese; **d2**, filling the package with hot processed cheese before the sealing step; **e1**, a view of the sealing equipment; **e2**, sealing a glass container with metal lid (easy open system))

For the production of processed cheese, curd produced by the so-called co-precipitate containing casein and whey proteins precipitated using heat and acid can also be used. Milk contains two main groups of proteins: caseins and whey proteins. In the curd obtained by this method, called hot acid precipitation, albumin is in greater quantity, but it is not recommended to be greater than 10% relative to the total mass. The production of the well-known Latin American white cheese also uses heating and acidification in its manufacture and can also be used in producing processed cheeses. However, in this type of production, acid whey is generated, which, although it can be concentrated and dehydrated, has fewer opportunities for applications than the sweet whey resulting from enzymatic coagulation.

Portions of processed cheese or pre-melted cheese can be used to prepare processed cheese. This pre-melted cheese is strongly subjected to second melting creams and transfers its physical–chemical conversion of structure to the young processed cheese to which it was added. Control is needed to prevent the processed cheese from becoming over-creamed [5].

In the production of processed cheese, in addition to matured (or aged) cheeses, fresh curd can be produced by (1) traditional or lactic fermentation method (Fig. 3), (2) hot acid precipitation by direct acidification with 85% diluted lactic acid or acetic acid method (Fig. 4), and (3) enzymatic coagulation with the addition of lactic acid bacteria method (Fig. 5). The curd to be melted can also be obtained from ultrafiltered milk [11, 12].

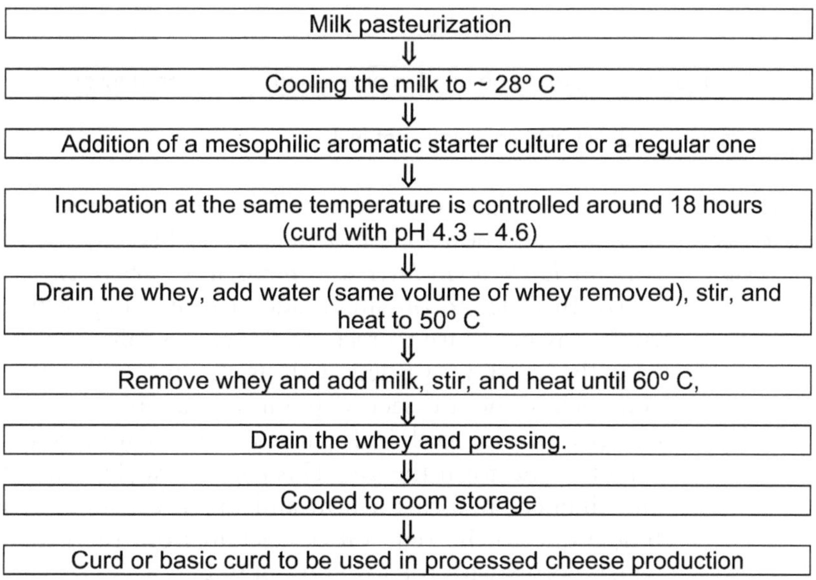

Fig. 3 Steps to prepare curd by fermentation (acid coagulation) of milk

Milk pasteurization

⇓

Cooling the milk to ~ 32° C

⇓

Addition of mesophilic starter culture, $CaCl_2$, and rennet

⇓

Coagulation

⇓

Cutting, stirring, and heating
(time and temperature will depend on the type of cheese)

⇓

Draining and pressing

⇓

Cooled to room storage and maturation

⇓

Cheese or basic curd to be used in processed cheese production

Fig. 4 Steps to prepare curd by enzymatic coagulation of milk

Whole or skimmed milk

⇓

Heating to 70° C (skimmed milk) or 82° C (whole milk)

⇓

Addition of 0.28% of lactic acid 85% diluted 1:10

⇓

Coagulation

⇓

Drainage

⇓

Pressing (optional) (80 min, 2 to 4 kg/cm^2)

⇓

Cooled to room storage

⇓

Curd (or basic curd) to be used in processed cheese production

Fig. 5 Steps to prepare curd by direct acid coagulation of milk

3.1.1 Production of Curd by Lactic Fermentation for Use in Processed Cheese

In acid coagulation by fermentation, the acidification of milk occurs with the addition of mesophilic lactic bacteria, which generally contain *Lactococcus lactis* ssp. *lactis* and *Lactococcus lactis* ssp. *cremoris*. In order to obtain an adequate melting of the acid casein curd, it is necessary to recover the curd's saline balance, which is why the curd is washed with water and milk. When the pH approaches the isoelectric point of casein (4.6), the electrical charges of the colloidal particles are neutralized; they come together and form the gel. This curd is fragile, but if fractionated and drained, it forms a firm and dry precipitate. However, it is fragile due to the removal of calcium from the protein network, which occurs with the lowering of pH.

Acid coagulation produces a curd with characteristics different from enzymatic coagulation. In addition to acidity, there is a difference in the content and state of minerals, which influences the texture. The coagulation endpoint is identified visually or by acidity (60 to 70°D, pH 4.5). The acidity of the curd (pH 4.3 to 4.5) must be reduced for fusion (pH 5.3 to 5.7); therefore, it is important to stop fermentation as soon as it coagulates so that the acidity does not intensify. Therefore, the bacterial percentage and fermentation temperature must be standardized. Once coagulation occurs, the curd is cut, stirred, and heated to 50 °C. The draining occurs, followed by resting and final draining. The whey should be clear and greenish. On average, two washes of the curd are carried out with filtered water in the same quantity as the extracted whey at 50 ° C. Deacidification is then carried out with the addition of 25 to 30% of the amount of whey used for coagulation and heating and stirring up to 60 °C. Skimmed or whole milk can be used. Draining is done and then pressing [10]. These steps are summarized in Fig. 3.

3.1.2 Production of Curd by Enzymatic Coagulation

In the production of basic curd by enzymatic coagulation, rennet (or renin) is used, which can be extracted from the abomasum of suckling or adult cattle or synthesized by microorganisms [13, 14]. Rennet obtained from calves consists mainly of chymosin. This coagulation occurs in two stages, the first being enzymatic, with the action of renin on the Phe105-Met106 peptide bond of the k-casein fraction (one of the four fractions of casein micelles). The second stage is nonenzymatic, and paracasein is formed, which, reacting with soluble calcium, results in calcium paracaseinate, the main constituent of the hydrated, malleable (elastic), and shiny curd. The melting salts are added to melt the curd, which complex with the calcium bound to casein, making the paracaseinate more soluble and leading to the entry of monovalent sodium into the protein complex or another ion such as potassium (processed cheeses with a low content of sodium). Figure 4 summarizes the basic steps for producing curd by enzymatic milk coagulation.

3.1.3 Production of Curd by Direct Hot Acidification for Use in Processed Cheese

In curd production, by adding diluted lactic acid to heated milk, also called direct hot acidification, coagulation occurs at a pH greater than 4.6 (casein's isoelectric point) due to the action of heating, in addition to the acid. The decrease in pH is sufficient to accelerate precipitation as the apparent isoelectric point is reached. There is the formation of a soft curd that contains casein and denatured whey proteins with the formation of grains, unlike the homogeneous one resulting from lactic fermentation or enzymatic coagulation by renin [10, 13]. The advantages of producing and using this type of curd are reduced costs with starter cultures, shorter process time, greater curd uniformity, better conservation, and automated steps. Figure 5 summarizes the basic steps for producing curd by direct hot acidification of milk.

3.2 Formulations and Manufacturing Steps for Different Types of Processed Cheese

There are many types of processed cheeses, and their number differs in different countries depending on the ingredients used in their production. In the United States (USA), the document "Code of Federal Regulations (CFR)" specified five categories of PCP (processed cheese products): (1) pasteurized blended cheese, (2) pasteurized processed cheese, (3) pasteurized processed cheese foods, (4) pasteurized processed cheese spread, and (5) pasteurized cheese spread. This classification is based on allowed ingredients and accepted moisture, fat, and pH levels. On the other hand, in the United Kingdom, the Cheese and Cream Regulations defined two categories of PCP: (1) processed cheese and (2) cheese spreads. Cheese spreads only differ from processed cheese in that milk products other than natural cheese can be used, whereas only natural cheese can be used in processed cheese [15]. In Brazil, the legal standards determine the main ingredients used to produce processed cheese (spreadable or creamy *requeijão*, slides, and block) and the maximum moisture and minimal fat they must contain. Table 1 contains suggested formulations for manufacturing block

Table 1
Ingredients and their amounts in the formulation of block processed cheese and fondue

Ingredients (kg)	Block processed cheese	Fondue
Basic curd[a]	13.25	–
Milk cream[b]	9	6.5
Water	2.38	0.5
Emulsifying salts[c]	0.18	0.3
Salt (NaCl)	0.2	
Sodium sorbate	0.008	0.008
Young Emmental cheese[d]	–	5
Aged Emmental cheese[d]	–	7.5
Dry white wine (L)[e]	–	5.25
Ground nutmeg	–	0.5
Ground garlic	–	0.08
Ground dried tomato	–	0.65
Annatto dye (mL)	–	20

[a]Curd with 38% of total dry matter and 12% of fat prepared by enzymatic or acid coagulation (*Lactococcus lactis* ssp. *cremoris* and subsp. *lactis*)
[b]Milk creams containing 50% of fat
[c]Emulsifying salts containing polyphosphate and citrates or only citrates (add a maximum of 3% considering the basic curd) (*see* **Note 3**)
[d]The mixture of young and aged cheese must contain 62% of total dry matter and 26% fat
[e]Depending on the acidity of the wine, it may need to add emulsifying salt to correct the acidity

Table 2
Ingredients and their amounts in the formulations of spreadable processed cheese—regular (in fat content), fat reduced, with fiber, and no fat added

Ingredients (kg)	Regulate SPC[a]	Reduced fat SPC with WPC	Reduced fat SPC with fiber	No fat added SPC
Basic curd[b]	4	15	8	5
Milk cream[c]	–	5	3.2	–
Butter[d]	3.5			
WPC 34%[e]	–	0.3	–	0.2
Calcium caseinate	–	–	–	0.11
Water	5.3	12.4	8.1	4.7
Melting salt (Joha S9)[f]	0.072	0.27	0.14	0.09
Salt (NaCl)	0.06	0.3	0.16	0.09
Fiber (HPX® inulin)—3.3% in relation to the final product	–	–	0.74	–
Whitening agent (titanium dioxide)	–	–	–	0.013
Anti-foaming agent	–	–	–	0.00041

[a]*SPC* spreadable processed cheese or *requeijão*
[b]Basic curd containing 0.5–2% of fat and 53–58% of moisture
[c]Milk creams containing 66% of fat and 32% of moisture
[d]Butter with 82% of fat
[e]This WPC contains 34% protein. If the protein content of the concentrates is higher, the quantity must be adjusted
[f]This melting salt contains a mixture of sodium phosphates with P_2O_5 ranging from 58.7 to 60.7 (%)

processed cheese (*requeijão em barra*) and fondue, according to Rodrigues [16].

In Table 2 the ingredients and quantities used to produce spreadable processed cheeses with regular fat content, with a reduced ingredient replaced by whey protein concentrate and inulin (fiber) and without added fat, are presented (*see* **Note 4**). The ingredients in spreadable processed cheeses with regular fat content were calculated (Table 2) by mass balance to contain 55–60% fat in the dry matter (or matter) and a maximum moisture content of 60–65%. The fat in the dry extract is the ratio between the fat content and the dry extract. The dry extract is 32–33% for cream cheeses with reduced fat content. Figure 6 shows details of the preparation steps for these products (*see* **Note 5**).

Figure 7 shows the steps for preparing spreadable processed cheese with reduced fat and lactose content, using inulin, as a fiber source intended to replace half of the fat in the original formulation (*see* **Note 6**). Replacing 50% of the dairy fat in the formulation can be achieved by adding small amounts of whey protein concentrate (Table 2), but this ingredient adds lactose. To prepare processed

Skimmed milk (to reduce fat SPC or no fat added) or whole milk

⇓

Hot acid coagulation with 85% lactic acid diluted 1:10

⇓

Resting, draining, and pressing the curd

⇓

Grinding the curd and weighing the ingredients

⇓

First heating (70°C) and addition of whey protein concentrate (WPC) or inulin - if fat reduced add whitening agent and anti-foaming agent—version "fat added".

⇓

Heating to 70°C /5min under vacuum and medium speed (~1500rmp)

⇓

Second heating until 87°C, then improve the speed to 3500rpm, also under vacuum and keep for 3min

⇓

Hot filling (greater than 75° C),

⇓

Sealing/crimping and cooling

Fig. 6 Steps to prepare spreadable processed cheese with no fat added or with reduction of fat added of whey protein concentrate (WPC) or inulin (fiber)

Skimmed milk, addition of lactase (0.07%), kept under refrigeration (4°C) for 18h

⇓

Hot acid coagulation with 85% lactic acid diluted 1:10

⇓

Resting, draining, and pressing the curd

⇓

Grinding the curd and weighing the ingredients (use butter)

⇓

First heating (70°C) and addition of inulin - if fat reduced

⇓

Heating to 70°C /5min under vacuum and medium speed (~1500rmp)

⇓

Second heating until 87°C, then improve the speed to 3500rpm, also under vacuum and keep for 3min

⇓

Hot filling (higher than 75° C)

⇓

Sealing/crimping and cooling

Fig. 7 Steps to prepare spreadable processed cheese include reducing fat and lactose and adding inulin (fiber)

cheese with reduced lactose content, the milk sugar must be hydrolyzed by adding lactase before being used to produce the basic curd (*see* **Note 7**). On the other hand, the use of butter adds less lactose than the traditional use of cream.

In the case of processed cheeses, one way to reduce the consumption of sodium by the consumer may be through the production of this food with a lower sodium chloride content. However, this decrease is a complex task due to the contribution of NaCl to the taste and preservation of food. This reduction must be balanced with preventing the development of pathogenic and deteriorating microorganisms and maintaining the proper flavor and texture. In addition to adding smaller amounts of NaCl, processed cheese makers can use other resources to adjust the sodium content in these cheeses, such as (1) applying common salt substitutes (e.g., KCl), (2) using melting salts with lower or sodium-free contents, and (3) using combinations of these practices with the addition of flavor modifying or enhancing ingredients. The use of salt substitutes is necessary, in general, when the amount of salt to be reduced is greater than 33%. The most popular salt substitute is potassium chloride (KCl) mixed with NaCl (*see* **Notes 3** and **8**).

4 Notes

1. There has been work on the microbiological safety of processed cheeses for decades, in particular, the bactericidal or bacteriostatic action of melting salts. This topic is still current with recent works published. To exemplify, Fusieger et al. [17] studied the performance of six mixtures of melting salts in controlling the multiplication of *Bacillus cereus* in processed cheeses and identified more suitable mixtures for this control along with low refrigeration temperatures. In another study by this group, three types of melting salts (ESSP, emulsifying salt composed of short polyP; BSLP, bacteriostatic salt composed of long polyP; and trisodium citrate) were evaluated for their efficiency in inhibiting the multiplication of *B. subtilis*, *B. thuringiensis*, *Clostridium perfringens*, *Enterococcus faecalis*, *Listeria monocytogenes*, *Pseudomonas fluorescens*, and *Staphylococcus aureus* in processed cheese. In this study, the sample considered the positive control was added with emulsifying salts without antimicrobial effect (ESSP 1.5%). The processed cheese sample added antimicrobial additives (nisin and potassium sorbate), established was the negative growth control. The sample T13 containing 1.5% of the BSLP (T13) performed similarly to the negative control for bacterial inhibition in the samples over 90 days of storage. In the sample T13 the reduction of the counts of *B. cereus* INV 10(3) and *B. thuringiensis* CFBP 3476 and inhibition of *C. perfringens*

ATCC 13124 and *L. monocytogenes* Scott A by the end of storage were observed. The negative control was more efficient better than T13 only against *S. aureus* ATCC 6538 ($p < 0.001$) [18].

2. The positive points of using newly produced unripened cheeses are their lower costs and the greater stability of the emulsion. However, using only these cheeses can generate processed cheese with a very mild flavor or a residual taste of melting salt, intense swelling, a very firm texture, and increased firmness during storage and incorporation of air due to high viscosity.
3. Research in Brazil found that a spreadable processed cheese (*requeijão*) produced without added fat, NaCl, and emulsifying salt Joha S9 contained 593 mg of sodium/100 g of the sample. If the same formulation was prepared by replacing 40% of NaCl with KCl, the sample, containing 476.86 mg sodium/100 g, met around 20% of reduction. If the formulation was produced by replacing 40% of NaCl with KCl and using the proportion of 0.8:1.2 until 1.2:0.8 of emulsifying salt Joha S9 to one of the emulsifying salts Joha B9, Joha S9K, Joha B50, and Joha SK75, the samples contained between 373.5 and 446.7 mg of sodium/100 g [9].
4. There are works such as those by Sangatash et al. [19] that produced samples of low-fat and low-sodium imitation pizza cheese using sodium caseinate (24.5%), canola oil (21%), inulin (0.19%), pregelatinized starch (0.4%), NaCl (1.3%), citric acid (0.5%), and disodium phosphate (0.5%). They concluded that inulin or starch could replace 3.6% of the fat in the formulation. They also replaced only part of the NaCl with KCl, the best combination being 0.35% NaCl and 0.50% KCl, and kept the same sodium-based melting salt.
5. Adding titanium dioxide to processed spreadable cheese without added fat could be necessary to give this product its characteristic white color, as it would be grayish without this ingredient [12]. In this processed cheese, an anti-foaming agent also could be necessary because the high protein content of the formulation without added fat led to intense foam formation that was incorporated into the final product.
6. According to Gibson et al. [20], prebiotics are a substrate that is selectively used by host microorganisms to confer health benefits. Inulin can be classified as a prebiotic and added to processed cheeses. Other prebiotics, such as galactooligosaccharides (GOS) and xylooligosaccharides, were added to processed cheeses and obtained good sensory results [21, 22]. In turn, adding probiotics (live microorganisms that could confer a health benefit on the host when administered in appropriate

amounts) to processed cheeses is not industrially common. However, it has already been studied in different forms, such as the use of strains resistant to characteristic process temperatures [23], the use of microencapsulated probiotics [24, 25], and modifications in process temperatures [25]. Postbiotics are an expanding field of knowledge, defined as the "preparation of inanimate microorganisms or their components that confer a health benefit on the host" [26]. In this context, processed cheeses can be vehicles for the consumption of postbiotics. However, as always in the case of ingredients with health benefits, these must be verified by in vivo *studies.*

7. The manufacturer determines the amount of lactase to be used. It is related to several factors, such as the activity of the enzyme extract, hydrolysis temperature, and time for hydrolysis that fits into the dairy production routine.
8. Research has shown that whey and milk permeate can contribute to giving a salty taste without significantly contributing to the sodium content. A study was carried out to explore the sensory characteristics and the composition of different permeates and their application in a product with low sodium content. The results demonstrate the potential of using whey permeate and lactose whey permeate as a salt substitute in product applications (Smith et al., 2016) [27].

References

1. Giha V, Ordoñez MJ, Villamil RA (2021 July) How does milk fat replacement influence cheese analogue microstructure, rheology, and texture profile? J Food Sci 86(7):2802–2815
2. Kamath R, Basak S, Gokhale J (2022 Feb) Recent trends in the development of healthy and functional cheese analogues – a review. LWT 155:112991
3. Lee SK, Klostermeyer H, Anema SG (2024 June) The denaturation of whey proteins in a processed cheese environment. Int Dairy J 153:105897
4. Ramos GLPA, Silva R, Spadoti LM, Alves ATS, Esmerino EA, Zacarchenco PB, Cruz AG (2022) Microbiology of processed cheese. In: El-Bakry M, Meahta BM (eds) Processed cheese science and technology: ingredients, manufacture, functionality, quality and regulations, 1st edn. Woodhead Publishing Elsevier Hardcover, Cambridge, pp 427–449
5. Muslow BB, Jaros D, Rohm H (2011) Processed cheese and analogues. In: Tamime AY (ed) Structure of dairy products, 1st edn. Society of Dairy Technology, Hyattsville, pp 210–235
6. El Dakhakhny E, Dabour N (2016 Nov/Dec) Processed cheese: basics and possibilities for the development of healthier products. Alexandria. J Food Sci Technol 13(2):45–62
7. Solowiej BG, Nastaj M, Szafrańska JO, Muszyński S, Gustaw W, Tomczynska-Mleko-M, Mleko S (2020 June) Effect of emulsifying salts replacement with polymerized whey protein isolate on textural, rheological and melting properties of acid casein model processed cheeses. Int Dairy J 105:104694
8. Monteiro CA, Cannon G, Moubarac JC, Levy RB, Louzada MLC, Jaime PC (2018 Jan) The UN Decade of Nutrition, the NOVA food classification and the trouble with ultra-processing. Public Health Nutr 21(1):5–17
9. Ramos GLDPA, Alves ATS, Spadoti LM, Zacarchenco PB, da Cruz AG (2022) Low salt and low sodium processed cheeses. In: El-Bakry M, Mehta BM (eds) Processed cheese science and technology: ingredients, manufacture, functionality, quality and regulations, 1st edn. Woodhead Publishing Elsevier Hardcover, Cambridge, pp 177–197

10. Van Dender AGF, Moreno I, Alves ATS, Spadoti LM (2014) Spreadable processed cheese ("requeijão cremoso") and other processed cheeses: basic manufacturing technology and process control parameters. In: Van Dender AGF (ed) Spreadable processed cheese and other processed cheeses—manufacturing technology, process control and market aspects, 2nd edn. Editora Ltd, São Paulo, pp 57–90
11. El-Aidie SA, Mabrouk AM, Abd-Elgawad AR, El-Garhi HEM (2023 Aug) Physicochemical, textural and organoleptic properties of functional processed cheese manufactured from ultrafiltered milk. Biocatal Agric Biotechnol 51:102798
12. Van Dender AGF, Spadoti LM, Bosi MG, Moreno I, Zacarchenco PB (2014) Spreadable processed cheese ("requeijão cremoso") and other processed cheeses: innovations and technological alternatives. In: Van Dender AGF (ed) Spreadable processed cheese and other processed cheeses—manufacturing technology, process control and market aspects, 2nd edn. Editora Ltda, São Paulo, pp 261–326
13. Salek RN, Buňka F, Černíková M (2022) The use of different cheese sources in processed cheese. In: El-BAKRY M, MEHTA BM (eds) Processed cheese science and technology, 1st edn. Woodhead Publishing Elsevier Hardcover, Cambridge, pp 79–113
14. Fox PF, Guinee TP, Cogan TM, Mcsweeney PL (2017) Fundamentals of cheese science, 2nd edn. Springer, New York
15. El-Bakry M, Mehta BM (2022) Overview of processed cheese and its product. In: El-Bakry M, Mehta BM (eds) Processed cheese science and technology, 1st edn. Woodhead Publishing Elsevier Hardcover, Cambridge, pp 1–27
16. Rodrigues F (2006) Cottage cheese, fondue, specialty, processed cheese, 1st edn. Templo, Juiz de Fora, 172p
17. Fusieger A, Russo N, Nero LA, Carvalho AFD, Randazzo C, Caggia C (2023 Nov) Emulsifying salts and P2O5 content as promising strategies to control Bacillus cereus growth during storage of processed cheese. Int J Dairy Technol 77(1):198–211
18. Fusieger A, de Jesus Silva SR, da Silva RR, Camargo AC, Honorato JA, Teixeira CG et al (2023 May) Bactericidal and bacteriostatic effects of sodium polyphosphate emulsifying salts on selected targets in processed cheese. Food Control 147:109580
19. Sangatash M, Dadras-Moghadam M, Mortazavi SA, Yarabbi H (2023 Jan/Feb) Evaluation of physicochemical and textural properties of low-fat and low-sodium imitation pizza cheese. Iran Food Sci Technol Res J 6(19):95–195
20. Gibson GR, Hutkins R, Sanders ME, Prescott SL, Reimer RA, Salminen SJ et al (2017 June) Expert consensus document: The International Scientific Association for Probiotics and Prebiotics (ISAPP) consensus statement on the definition and scope of prebiotics. Nat Rev Gastroenterol Hepatol [Internet] 14(8): 491–502
21. Belsito PC, Ferreira MVS, Cappato LP, Cavalcanti RN, Vidal VAS, Pimentel TC et al (2017 Oct) Manufacture of Creamy Requeijão processed cheese with galactooligosaccharide. Carbohydr Polym 174:869–875
22. Ferrão LL, Ferreira MVS, Cavalcanti RN, Carvalho AFA, Pimentel TC, Silva HL, Cruz AG (2018 May) The xylooligosaccharide addition and sodium reduction in requeijão creme processed cheese. Food Res Int 107:137–147
23. Soares MB, Almada CN, Martinez RC, Pereira EP, Balthazar CF, Sant'ana AS (2019 Oct) Behavior of different Bacillus strains with claimed probiotic properties throughout processed cheese manufacturing and storage. Int J Food Microbiol 307:108288
24. Pivetta FP, da Silva MN, Tagliapietra BL, Flores VA, Richards NSPS (2022 Apr) Probiotic viability of creamy processed cheese formulations. Res Soc Dev 11(6):e27211629027
25. Silva R, Pimentel TC, de Matos Junior FE, Esmerino EA, Freitas MQ, Fávaro-Trindade CS, Cruz AG (2022 Apr) Microencapsulation with spray-chilling as an innovative strategy for probiotic low sodium cream cheese processed cheese processing. Food Biosci 46:101517
26. Salminen S, Collado MC, Endo A, Hill C, Lebeer S, Quigley EM, Vinderola G (2021 May) The International Scientific Association of Probiotics and Prebiotics (ISAPP) consensus statement on the definition and scope of postbiotics. Nat Rev Gastroenterol Hepatol 18(9):649–667
27. Smith ST, Metzger L, Drake MA (2016 Nov) Evaluation of whey, milk, and delactosed permeates as salt substitutes. J Dairy Sci 99(11):8687–8698. https://doi.org/10.3168/jds.2016-10904. Epub 2016 Aug 31

Chapter 6

Fresh Cheeses

Ramon da Silva Rocha, Bianca Cristina Rocha de Oliveira, Maria Eduarda Marques Soutelino, and Adriana Cristina de Oliveira Silva

Abstract

Cheese, derived from the coagulation of milk, is a product consumed globally due to its diversity of flavors and shapes. Its origin dates back to 8000 BC and its appearance was accidental during the conservation of milk in ruminant stomachs. Over the years, studies have revealed the process of acidification and fermentation of milk, resulting in the formation of curds, the basis for the production of more than 2200 types of cheese, including fresh and matured ones. Fresh cheeses are characterized by high moisture and short shelf life. Despite this, its popularity is due to its mild flavor, slight acidity, and nutritional benefits. The global fresh cheese market is constantly expanding. Despite their short shelf life, these cheeses are appreciated in different cultures for their flavor, ease of production, and nutritional value.

Key words Fresh cheeses, Processing, Industrial technology, Traditional technology

1 Introduction

Cheese is a product obtained from the coagulation of raw or pasteurized milk and is a widely consumed product recognized for its various flavors and formats. It is a versatile food that contributes to health, as it is an excellent source of protein, calcium, and other essential nutrients; in addition, it is part of many culinary dishes and is present in different traditions and cultures worldwide [1].

Historically, it emerged around 5000 to 8000 BC, with the expansion of civilization to the east of the Mediterranean, between the Tigris and Euphrates rivers. It is believed that the discovery process occurred accidentally because, to maintain the conservation of milk, the ancients stored it in bags made from ruminant stomachs. After a specific time, the separation of the whey and formation of the cheese mass were observed [2].

Over the years, through accurate studies, it was observed that milk in contact with organic acids, specific enzymes, and lactic acid bacteria, under certain conditions of time and temperature, went

Adriano Gomes da Cruz et al. (eds.), *Dairy Foods Processing*, Methods and Protocols in Food Science,
https://doi.org/10.1007/978-1-0716-4144-6_6,

through the acidification and fermentation process and formed curd, a white mass made up of solid components of milk, serving as the basis for the development of a new product [3]. As a result, cheese became a popular consumer product, its manufacturing process was improved, and there are currently more than 2200 types worldwide, including fresh cheeses (mascarpone, ricotta, petit-suisse, and others) [4].

Fresh cheeses have as their main characteristic that they are not subjected to ripening, being ready for consumption immediately after manufacturing. They generally have high moisture, preserving in their composition a large amount of lactose milk sugar, an excellent nutrient for bacterial growth, which can facilitate the development of spoilage and pathogenic microorganisms, reducing the shelf life of these products [5].

The mild and slightly acidic flavor combined with its nutritional quality are factors that justify its popularity. The global fresh cheese market is constantly evolving. The demand for its use in baking, confectionery, desserts, sauces, seasonings, and condiments contributes to its expansion. In 2022, Italy had the highest import data for fresh cheeses worldwide [6].

Despite being a cheese that does not last long, it is well appreciated by the population due to its mild flavor and by those looking for a healthy diet, as some types have a lower fat content and a higher protein content of high biological value. In some countries, there are no specific regulations, and it can also be considered artisanal cheese. But at the same time, they can have advanced industrial techniques, such as ultrafiltration method that requires approximately 20% less milk. Furthermore, it is also important to highlight that regardless of the technology applied and the type of cheese, these products are commonly present in the population's diet, whether consumed directly or as an ingredient [4].

2 Material

2.1 Fresh Cheese Processing

Fresh cheeses encompass a variety of cheeses produced by the coagulation of milk, cream, or whey via acidification, acidification and the addition of specific enzymes, or a combination of acid and heat, being ready for consumption at the end of processing, without having to go through a ripening. Factors contributing to its popularity include smooth, creamy consistency, diet-conscious consumers' healthy perception of these products, their versatility in consistency and flavor, the possibility of adding sugar, fruits, and condiments, and the variety of attractive and convenient packaging [3]. Below, we describe some types of fresh cheese produced worldwide.

Fig. 1 Cottage cheese

2.1.1 Cottage Cheese

Cottage cheese originated in Great Britain and is characterized as fresh, soft, and unripened, which can be produced with whole or skimmed milk and is demonstrated in Fig. 1. It has a grainy consistency, with grains immersed in *dressing*, a mixture made up of standardized cream [7]. It is a very popular cheese in the United States of America (USA), where the annual per capita consumption is around 1 kg. The global production volume is 0.8 million tons, with an expected growth of 2–3% [8].

Historically, it was a food produced on farms in central Europe to use surplus milk production, hence the name *cottage cheese* (typical country cheese). In mid-1915, it began to be manufactured industrially in the USA [9]. There are also reports of its preparation in Pakistan, using Tatri (tartaric acid), which is currently banned in the country [10].

It is a cheese with a high moisture content, up to 80%, and an average fat content in the range of 4%, considered skimmed. However, cream can be added, making it tastier but with a higher fat content. According to the Codex [11], it has a specific granulation, must be individual and uniform, and measures approximately 3–12 mm.

According to the Food and Drug Administration (FDA), the creamy mixture accompanying the cheese is prepared from safe and suitable ingredients, but not limited to milk and dairy products. Ingredients not derived from milk must have a helpful function other than increasing the total solids content and be used in an amount no greater than that necessary to achieve the intended effect [12].

In its production process, the dough is obtained through acid or enzymatic coagulation, with acidifying fermenting cultures (BAL) and flavoring added. The cultures generally used are homofermentative *Lactococcus lactis* ssp. *cremoris* or *Lactococcus lactis* ssp. *lactis*. Other flavoring cultures can prepare cottage cheese, such as *Lactococcus lactis* ssp. *leuconostoc* and *Lactococcus lactis* ssp.

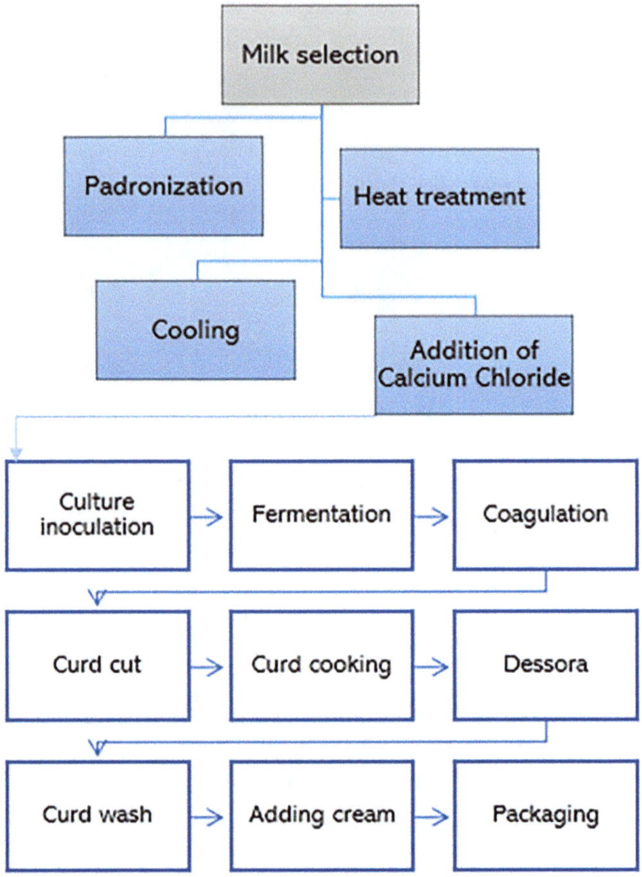

Fig. 2 Simple production flowchart for cottage cheese

diacetylactis [5]. However, there must be great control over this fermentation as problems may occur during the dough formation. For example, the agglutination of starter bacteria forms clusters deposited at the bottom of the vat during manufacturing [13]. The simple production flowchart for cottage cheese is presented in Fig. 2.

Brazilian legislation does not have a Technical Identity and Quality Regulation (RTIQ) for this product type. However, due to its percentage of moisture (>55%) and fat content in the dry extract (<10%), it qualifies as very high moisture skimmed cheese by cheese legislation [14]. Internationally, it is regulated by Codex [11] and FDA [12] specifications.

- **Manufacturing Steps**

Milk selection: Milk with good physical–chemical and microbiological quality; without antimicrobial or acid-neutralizing substances. Usually, titratable acidity must be 14 to 18°D.

Milk standardization: Before the milk pasteurization process, it must be centrifuged, clarified, and skimmed in the standardizing centrifuge.

Milk pasteurization: Fast pasteurization (High Temperature, Short Time (HTST)—72 °C/15–20 s) or slow pasteurization (Low Temperature, Long Time (LTLT)—63-65 °C/30 min) can be applied.

Milk cooling: After pasteurization, the milk will be cooled to between 30 and 36 °C. For the addition of thermophilic dairy yeast, the ideal temperature is 35–36 °C, while for mesophilic culture, this temperature will vary between 30 and 32 °C.

Add calcium chloride to the tank: Calcium chloride can be added to the milk in a 34% solution to make the curd firmer. Used to replace part of the soluble calcium that was insolubilized during the heat treatment of the milk and thus restore the firming action of the curd formed, the texture, and the yield during the manufacturing process.

Culture inoculation: A Direct Vat Set (DVS) lactic acid culture must be added at a rate of 5 U for every 100 L of milk. Flavoring cultures to produce diacetyl and acetaldehyde can also impart flavor and aroma to the cheese.

Fermentation and coagulation: The milk will remain in the fermentation tanks for a variable time, which can be 4.5 h and up to 18 h of fermentation (*see* **Note 1**).

Curd cutting point: After checking that the dough has reached the ideal pH (4.6 for cheese prepared without rennet and 4.8 for cheese with rennet added), the curd can be cut (*see* **Note 2**).

Cutting the curd: The dough will be cut slowly, just once, into large cubes with an edge of around 1.5 cm, carried out horizontally, with an interval of around 3 min before vertical cutting. Then, the grains will rest for 15 min.

Cooking the dough: After resting, the dough will be cooked until it reaches a temperature of 47–56 °C for approximately 2:30 h to expel the whey from the grains, making them firmer. The time may vary slightly depending on the yeast manufacturers' recommendations.

Curd test: Part of the curd will be collected and placed in ice water at 5 °C. A medium compression will be carried out on the grains, and the cooking will be finished if they separate.

Drainage: The whey must be clear without the presence of fines.

Washing: The grains will be washed to remove lactose and acid residue, which could shorten the product's shelf life (*see* **Note 3**). In general, two to three washes are carried out so that the flavoring substances are not lost, with water at temperatures of 30 °C in the first wash, 16 °C in the second, and 4 °C in the final wash. This step

Fig. 3 Ricotta cheese

will be completed after the washing water is clear or by measuring the solids content, which should reach the standard 0.8 to 1.2 brix.

Addition of cream: After removing the washing water, cream or fermented cream is pasteurized at 80–90 °C, cooled to 4 °C, with a fat content between 12% and 18%, and salted (0.5–1%); stabilizers must be added to the curd particles to form the product (*see* **Note 4**).

Packing and storage: Cottage cheese is packed in containers and stored at 4–5 °C before being distributed to retail stores.

2.1.2 Ricotta

Ricotta (Fig. 3) is a soft, unripened cheese of Italian origin, better known as whey or albumen cheese. It is produced by heating and acidifying serum, as the combination of acid and heat forms a mechanism capable of denaturing serum proteins and inducing their precipitation.

It is believed that it was first produced by farmers, who realized the possibility of using whey from cheese production by adding organic acids and heating. The word "Ricotta" has Italian origins and means "cooked" or "annealing," referring to the process of obtaining the cheese [15].

The average manufacturing yield is around 4–6%; it is a product that does not last long and is therefore considered a fresh cheese. Furthermore, it has a high moisture content (75–80%). It can be considered lean (4% fat) or not (22% fat) if the cream is added during the preparation process to allow for greater creaminess, softness of the dough, and increased flavor. It has around 6–9% highly digestible proteins [3].

According to the United States Department of Agriculture (USDA; 1981), in the USA, there are three types: whole Ricotta, with a fat content >11% and moisture <80%; partially skimmed Ricotta, with a fat content <11% and moisture <80%, and Ricotta made from whey, skimmed milk, or whey/skimmed milk, also

Table 1
Types of ricotta fresh cheese produced

Name	Ricotta type	Whey type	Country
Ricotta di Bufala Campana	Fresh	Buffalo	Italy (Campania)
Ricotta	Fresh	Cow	Brazil
Ricotta	Fresh	Cow; Buffalo	India
Ricotta	Fresh	Cow; sheep	Serbia
Ricotta	Fresh	Cow	United States
Ricotta	Fresh	Cow; Buffalo	Egypt
Ricotta di Capra	Fresh	Goat	Italy (Sicily)
Ricotta Romana	Fresh	Sheep	Italy (Lazio)

Reference: Adapted from [17]

known as Ricottone, with a fat content <1% and moisture <82.5% [16].

There is a wide variety of Ricotta (Table 1) due to regionalism and different manufacturing processes that include the way the cheese whey is obtained, the different milk-producing species, the different types of technology aids, such as the acid used, causing changes in the final characteristics of the product.

For example, Ricotta says Bufala Campana, which, in addition to cream, is produced with "sweet" buffalo whey and is richer in fat and total solids [18]. In Brazilian legislation [19] and the Codex [20], for example, Ricotta is classified as a cheese obtained from coagulating cheese whey. However, Italian legislation [21], which gave rise to Ricotta, considers it a by-product of cheese making, not cheese. The absence of microbiological and physical–chemical standards allows for a great diversity of Ricotta composition, making it difficult for the official inspection service to act.

In addition to cheese produced from whey being a product with excellent nutritional characteristics, whey prevents it from being inappropriately dumped as waste material in the environment, causing severe damage to fauna and flora. Fresh cheeses obtained by precipitation of whey have a higher protein content, and their constitution is composed of whey proteins (-β-lactoglobulin, α-lactalbumin, bovine serum albumin, and glycomacropeptides) [18]. The simple production flowchart for ricotta cheese is presented in Fig. 4.

- **Manufacturing Steps**

Obtaining whey: Whey can be obtained from cheese production through acid or enzymatic coagulation (*see* **Note 5**).

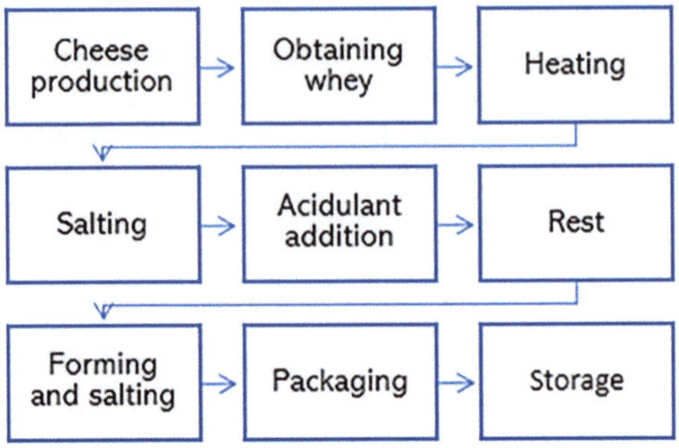

Fig. 4 Simple production flowchart for Ricotta cheese

Heating: The serum is heated indirectly in stainless steel tanks at temperatures around 75–90 °C or by injecting steam directly into the tanks or heating in a plate exchanger. Heating facilitates the denaturation of thermosensitive whey proteins, and the temperature used is directly related to the species from which the milk was derived (goat, sheep, cow, or buffalo). When using direct steam, coagulation happens much faster.

Salting: 0.5 to 1–2% NaCl can be added when the temperature reaches 55 °C. Salting must interfere with the product's coagulation, acceptability, and color.

Addition of acidifier (*see* Note 6):

Industrially, citric, lactic, or acetic acids are commonly used in a proportion of 1.5–2.5%, in addition to lactic acid bacteria (LAB) (starter culture of *Streptococcus thermophilus*) (500 mL/10 L of serum). White vinegar (20 mL/10 L of whey) or lemon juice (400 mL/10 L) can be used in artisanal productions. At this point, serum protein flocculation occurs when pH values reach around 5.8 to 6.0 under heating.

Resting: The serum and precipitate will rest for 10–30 min.

Forming and salting (optional): Ricotta is removed from the tanks with the help of a skimmer and added into perforated molds. If there has been no prior salting, salt can be added at this stage. Wait around 12 h for the draining to finish and to be unmolded and packaged.

Packaging and storage: Ricotta is then packaged in pots or polyethylene bags and must be kept refrigerated (temperature below 10 °C) for 7–15 days. The more artisanal the cheese, the shorter its durability.

The product will be considered low-fat cheese in the processing described above. In cases where a softer product is desired, with a pleasant flavor and greater yield, whole milk or cream is added at 55 °C in a proportion of 5–15% to the whey during heating, forming a mixed mixture before adding the acidulant.

Fig. 5 Minas Frescal cheese

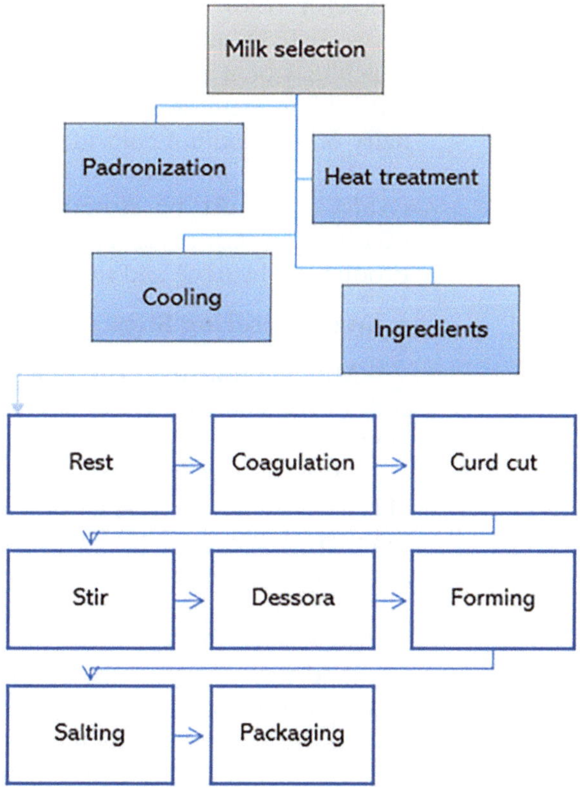

Fig. 6 Simple production flowchart for Minas Frescal cheese

2.1.3 Minas Frescal Cheese

Figure 5 shows the typical Minas frescal cheese, a typical Brazilian cheese originating in Minas Gerais, while Fig. 6 shows the simple production flowchart. It is widely produced and consumed in the country due to its high yield, low cost, simple manufacturing process, and high consumer acceptance. According to Embrapa, in 2018, it represented around 6% of the Brazilian consumer market, behind only mozzarella, dish, and curd cheese [22].

The origin of this cheese dates back to the colonial period. No wonder it is one of the most famous cheeses in Brazil. Its production co-occurred with the occupation of captaincies and gold mining. The Portuguese brought techniques and knowledge about cheese production, and the people of Minas Gerais adapted these techniques to the local conditions and ingredients of the region. Over time, production was industrialized, resulting in greater supply and availability throughout the country [23]. Minas Frescal cheese has its own identity in Brazil, protected by geographical identification seals, to guarantee its quality and authenticity and preserve its cultural legacy.

Minas Frescal cheese has a soft, moist, creamy texture, mild and slightly acidic and salty flavor. Generally, it has a white or slightly yellowish color due to the presence of milk fat, which varies between 0.5% and 10%. It is formed through milk coagulation, which may be with rennet and coagulating enzymes, with or without adding lactic acid bacteria, powdered milk, cream, solids of dairy origin, sodium chloride, and calcium chloride [24].

As it is an unripened cheese and has a high moisture content (>55%), it must be stored at a refrigerated temperature (not exceeding 8 °C). It has a short shelf life of 15 days, depending on production hygiene and milk quality [25].

- **Manufacturing Steps**

Milk standardization: Before the milk pasteurization process, it must be centrifuged, clarified, and skimmed in the standardizing centrifuge.

Milk pasteurization: Rapid pasteurization can be applied (HTST: 72 °C 15–20 s) or slowly occur in the cheese production tanks (LTLT: 63–65 °C/30 min).

Milk cooling: Pasteurized milk must be cooled to 36 °C. If it comes from refrigerated pasteurized milk tanks, it must be heated until it reaches 36 °C.

Addition of ingredients: 40% Calcium chloride (0.5 mL/L) is added to the milk to replace the calcium that was insolubilized during the heat treatment and **rennet** (enzymatic coagulant) in the volume recommended by the manufacturer. **Calcium** replacement is essential for the firmness of the curd and for greater cheese yield.

The **rennet** must be added little by little, being the last ingredient to be added. The operation generally takes 3 min. It can be of animal, microbial, or vegetable origin. Enzymes of animal origin are obtained from the stomach of ruminants (abomasum), while those of microbial origin are extracted from *Rhizomucor* fungi *miehei, Endothia parasitica, and Aspergillus niger.* There is also the use of enzymes extracted from vegetables, but they have little significance on the market [5].

In general, there is a preference for **chymosin** rennet taken from the abomasum of ruminants, as it is a more specific enzyme in breaking down the casein micelle, thus avoiding the breaking down of low molecular weight peptides that would cause flavor defects.

Some industries also add 1–1.5% **lactic acid bacteria** to the amount of milk so that there is greater acidification of the environment through lactose and transformation into lactic acid, increasing the cheese's shelf life and flavor. Generally, bacteria with good starter capacity, such as *Streptococcus thermophilus, are used.* Bacteria *Lactococcus lactis and Lactococcus cremoris* are also used.

Resting: The mixture is kept at rest for about 40–45 min until complete coagulation.

Cutting the curd: After checking the cutting point,[1] the curd must be cut into 2–3 cm cubes using horizontal and vertical lines.

Special care must be taken when cutting the lyres. First, pass the horizontal lyre, then the vertical lyre in the same direction as the horizontal lyre, and finally, the vertical lyre in a transverse position about the first two cuts, resulting in the formation of cubes, which should be very uniform.

Stirring: The curd should be stirred slowly for 1 min and then rested for 3 min. This operation must be repeated for 30 min to allow serum to escape the curd.

Draining and forming: The whey must be removed, and the curd is placed in molds to complete the draining. During this period, the first turning takes place 30 min after forming and every hour for four turns.

Salting: Salting can be carried out in molds with sea salt during the cheese-turning period. Another way of salting is to place the cheeses in brine with 30% NaCl for 2 h.

Packaging: After the cheeses have wholly drained at room temperature, they will be vacuum-packed in plastic bags and stored under refrigeration (5–10 °C).

2.1.4 Petit-Suisse

Petit-suisse (Fig. 7) is a cheese of French origin, developed by Charles Chervais in the nineteenth century. It is made from a base dough of quark cheese, a product originating in eastern and central Europe, traditionally obtained by the mixed coagulation of bovine milk through the addition of starter and coagulant yeast, with the possible addition of fruit pulp, sugar, and cream [26].

Despite being a cheese of French origin, its popularity has increased in several countries. It is generally consumed as a dessert

[1] There are correct ways to check the exact moment of the cut. One is the insertion of a sanitized knife or spatula into the inner wall of the manufacturing tank, between the tank and the formed mass. If there is detachment, it will be the right time to cut. The other way is to make a cut in the dough with a knife and insert the knife below the cut, forcing it upward. It will be the right time to cut if a straight crack appears. The serum will have a clear, light green color in both cases.

Fig. 7 Petit-suisse cheese

and is aimed at children, but consumers of all ages accept it, and it has the potential to be explored by adults. In Brazil, petit-suisse cheese is among the five most produced; between 2007 and 2011, it reached 54.5 thousand tons [27].

It has high moisture, greater than 55%, being classified as a very high moisture cheese. It has a high nutritional value due to its high milk protein content, at least 6%. It is also rich in fat-soluble vitamins, calcium, and phosphorus due to the desilting and concentration of milk components [26, 28]. Furthermore, its manufacturing technology is suitable for incorporating probiotic bacteria, as it is an unmatured cheese stored at refrigerated temperatures for a short period.

Mandatory ingredients for petit-suisse processing are specific BAL, rennet, and other appropriate coagulating enzymes. As optional ingredients, concentrated milk, cream, butter, anhydrous milk fat, food caseinates, dairy proteins, other solid ingredients of dairy origin, dairy whey, and whey concentrates can be used. To preserve its sensory characteristics, petit-suisse cheese must be kept at a temperature no higher than 10 °C [28].

- **Manufacturing Steps**

The making of petit-suisse cheese varies greatly depending on the country in which it is produced, and greater or lesser technology may be used. The production steps after coagulation will depend on this condition, where more modern or traditional equipment can be used. These methodologies will be described later. It is a type of cheese where coagulation is called mixed because it is obtained after the addition of fermentative dairy cultures in addition to the coagulating enzyme. Figure 8 shows three different methods to produce petit-suisse.

Milk selection: Milk of good physical–chemical and microbiological quality without antimicrobial, preservative, or acid-neutralizing substances. The titratable acidity must be between 14 and 18°D; the milk must be of good quality.

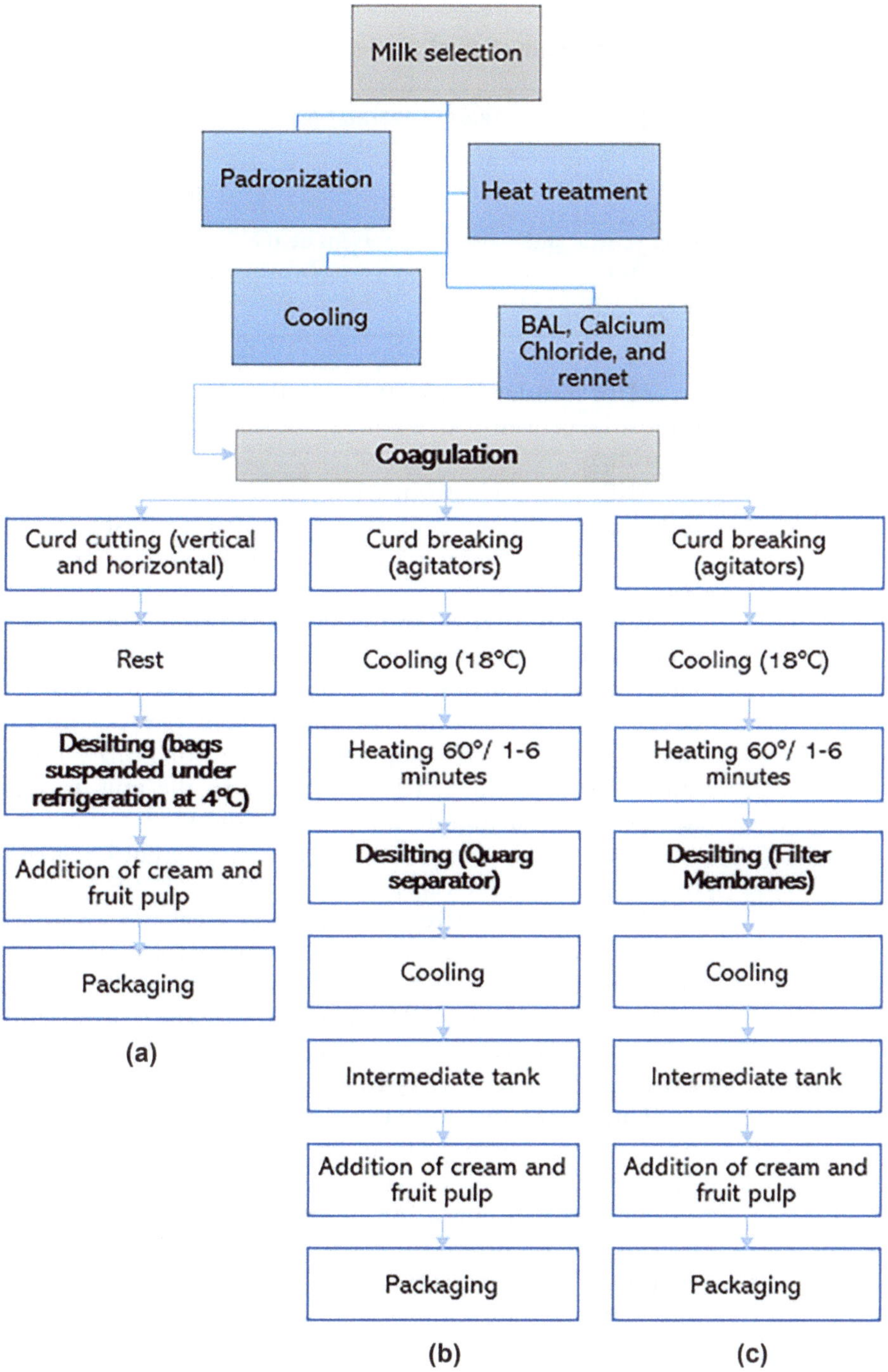

Fig. 8 Basic production flowchart of petit-suisse cheese. (**a**) traditional method; (**b**) whey draining using Quarg separator; (**c**) whey draining through membrane filtration

Table 2
Different technologies are used in the production of *petit-suisse cheese*

Technology	Traditional (A)	Quarg separator (B)	Ultrafiltration (C)
Fermentation/ coagulation	Open cheese production tanks	Ripening closed tanks tank	Closed tanks Ripening tank
Curd breaking	Manual using vertical and horizontal lyres	With agitators inside the coagulation tanks	With agitators inside the coagulation tanks
Whey removal	*Organza* bags, muslin	Centrifuge	Filter membranes

Milk standardization: Before the milk pasteurization process, it must be centrifuged, clarified, and skimmed in the standardizing centrifuge.

Milk pasteurization: Milk pasteurization generally occurs in plate equipment at 72–75 °C for 15–20 s (*see* **Note 7**).

Cooling: Pasteurized milk is cooled to 30–43 °C to add milk bacteria (*see* **Note 8**).

Addition of BAL and calcium chloride: Mesophilic or thermophilic BAL are added in a proportion of 0.3–4% so that the milk ferments and remains at rest until it reaches pH 6.3. The most commonly used cultures are *Streptococcus thermophilus*, *Lc. lactis subsp. lactis*, *Lc. lactis subsp. cremoris*, and/or *Leuconostoc* spp. Calcium chloride (2.5 g/10 L) is added to the milk to replace the insolubilized calcium during pasteurization.

The following steps can occur with different levels of technology, as presented in Table 2.

Coagulation: After reaching the ideal pH (6.3), rennet is added at a rate of 10% of what is used for typical cheeses, or approximately 2 ml of liquid rennet/100 kg of milk, keeping the milk between 30 and 43 °C, depending on the type of culture used, until the milk coagulates and the whey begins to separate, which generally happens after 7–16 h when the pH reaches 4.5–4.7.

Cutting or breaking the curd: After checking that the pH has been reached, different types of processing can occur, as shown in Table 1 and Fig. 9. (*see* **Notes 9–11**).

The **fruit preparation** generally consists of sugar fruit pulp, previously pasteurized (90 °C for 3 min), which can be added later. Xanthan gum, guar, and carrageenan can also be added. The average shelf life is around 28 days.

2.1.5 Mascarpone Cheese

Mascarpone is a type of creamy Italian cheese with a high fat content and mild flavor, which makes this cheese an ingredient used in various culinary dishes. It is recognized as a regional and traditional food product in the Lodi region, southwest of Milan, probably at the end of the sixteenth century [29].

Fig. 9 Mascarpone cheese

The global mascarpone cheese market size assessed in 2022 was US$1.3 billion and is expected to reach US$1.9 billion by 2032. The European region dominated the market in 2022 and is likely to remain dominant as it is a cheese popularly used in various culinary dishes in European countries [30].

It is appreciated for its high-fat content, which gives the product flavor, texture, and aroma characteristics that resemble butter. It is a cheese considered extra fat, according to Brazil [14] and the European Union [31], as it has a fat content greater than 60%. The product has a fat content of 45–55%, with the dry extract reaching 85%. Fresh cheeses can be classified according to the fat content in the dry extract (GES), as skimmed when it is less than 10%, lean when it is between 10 and 24.9%, and extra fat or double cream when it is greater than 60% [14]. The fat content of cheese can be controlled and normalized by adjusting the fat content of the milk used in making the cheese by adding cream or skimming the milk.

It is considered a noble cheese whose texture is smooth, dense, and homogeneous; its color varies from white to pale yellow. Unlike other cheeses, it is produced from the cream of milk and is, therefore, soft and spreadable [32]. It has a sweet flavor, with slight acidity and aroma of cream or butter; it also has high moisture and a melting sensation when tasting [33]. Mascarpone production technology is relatively simple and low-cost, making it suitable for artisanal or small industries.

- **Manufacturing Steps**

Obtaining cream: Cream with 19–24% fat is obtained by centrifuging and skimming the milk (*see* **Note 12**).

Pasteurization: The cream must be pasteurized at 80 °C for around 10 min due to the cream's greater thermal resistance (*see* **Note 13**).

Heating: The cream or the milk/sour cream mixture is heated to 95 °C to accelerate the denaturation of the serum proteins.

Addition of acid: To pasteurized cream or heated milk–cream mixture, acid is added, generally citric, acetic, or lactic, in a proportion of 10% (10 mL/kg).

Rest and coagulation: The acid is mixed, and the cream remains at rest for approximately 15 min until complete coagulation is verified. The final pH of the curd is around 5.5–6.5 (*see* **Note 14**).

Draining: After checking coagulation, the dough is placed in a double cream cheese separator for draining, which generally occurs after 24 h (*see* **Note 15**).

Packing and storage: After complete draining, the dough is homogenized, refrigerated, packaged, and storage at 4 °C.

3 Notes

1. This time will be determined by the type of culture used and its concentration. In general, the more concentrated, the faster the fermentation. When the pH of the milk reaches between 6.2 and 6.3, rennet can be added at a rate of 0.5 mL/3.700 L of milk. The enzyme is used to make the grains firmer. If the mass remains in the fermentation tanks for a short period, the grains formed will have a mild flavor. At the end of fermentation, the curd must have a titratable acidity of 65 to 70°D and a pH between 4.6 and 4.8.
2. At this point, the dough should be smooth, shiny, and flexible and not fall apart when touched.
3. The water used must be lightly chlorinated at 5–10 ppm, cold (7–12 °C), with pH adjusted to approximately 4.8 using phosphoric acid so that there is no change in the pH of the grains.
4. At the end of this process, grains of uniform size should be obtained, 3–12 mm, depending on the type of curd desired. The curd is creamy-white in color, shiny, firm in texture, and has a mild, slightly acidic flavor.
5. The whey obtained by enzymatic coagulation without acidification is considered a "sweet" whey. The Ricotta obtained from this whey will have a milder flavor, and the dough will be more elastic compared to the Ricotta produced with acidic whey, which will have a more granular structure. In artisanal productions, the serum obtained will generally be raw. In industrial productions, whey derives from cheese production, where the milk has generally been previously pasteurized.
6. The acidifier added may vary depending on the country, location, and type of production, artisanal or industrial. Furthermore, the choice, which will depend on many factors, will affect

the texture of the dough, the cheese yield, the moisture content, and the sensorial characteristics.

7. During heat treatment, sugar, protein concentrate, and stabilizers may be added to the milk, thus constituting a mixture. In this case, in general, a higher time-temperature binomial is used, around 85 °C/10–15 min. Slow pasteurization (63–65 °C/30 min) can be used inside cheese-making tanks in artisanal productions.
8. The temperature will depend on the type of bacteria used (mesophilic or thermophilic). If it comes from refrigerated pasteurized milk tanks, it must be heated until it reaches the appropriate temperature.
9. *In the traditional method,* **coagulation** occurs in open cheese-making tanks. When the pH reaches 4.65–5.0 and the cutting point is verified, **the curd must be cut** into 2–3 cm cubes using horizontal and vertical lines. This size is used for this high-moisture cheese. Then, the curd rests for approximately 15 min to separate the whey and partially precipitate the mass. Draining occurs next, and the dough is placed in suspended *organza bags and kept under refrigeration (4 °C)* until complete draining. After draining, the mass obtained is called *Quarg*. To make petit-suisse cheese, **cream and, in general, a fruit preparation are added.** The cream used was previously standardized to approximately 30% fat and pasteurized, mixed with the quarg cheese little by little until the cream formed without lumps. Then, the fruit mixture is added.
10. In the method in which *Quarg is used for* **draining** *the separator*, coagulation occurs inside **closed** coagulation tanks (ripening tank) and after **checking the pH (4.5–4.7), the curd** begins to cool to 18°–20 °C and the agitators inside the tank are turned on, causing **the curd to break**. This cooling favors the reduction of the metabolism of the added microorganisms. After this step, the Quarg is sent again, heated to 60 °C/1–6 min and sent to **the centrifugal separator** (operational temperature 32–44 °C) where the serum will be removed. It will then be **cooled** in a plate cooler at 10–12 °C temperatures. The cooled product is sent to an intermediate tank. To produce petit-suisse, **cream** was previously standardized to approximately 30% fat and pasteurized, and, in most cases, **fruit preparation is added to the tank**, before the product is packaged.
11. When using *Ultrafiltration Whey Separation Technology, the* **coagulation, cooling, and curd breaking** steps occur as described in method B. Then, the Quarg will be sent to filter membranes (operating temperature 40–45 °C), with a concentration of Quarg. The remaining steps are similar.

12. In some productions, for a small fermentation to occur, giving the product a more acidic flavor, the cream obtained may remain stored for around 24 h at 10–12 °C.
13. If pasteurized milk is mixed in mascarpone preparation, it must have a fat content of around 30%, pasteurized at approximately 75 °C/15–20 s (HTST).
14. This pH gives less firmness to soft cheeses.
15. Dessora can also be carried out traditionally using organza or muslin bags. In both cases, draining occurs at refrigeration temperatures.

Acknowledgments

Ramon S. Rocha wants to thank the grant 2022/14824-4, São Paulo Research Foundation (FAPESP).

References

1. Watson RR, Collier RJ, Preedy, VR (2017) Dairy in human health and disease across the lifespan. Elsevier Inc, London, 473p
2. Kindstedt, PS (2012) Cheese and culture: a history of cheese and its place in western civilization. Chelsea Green Publishing, White River Junction, 256p
3. Fox PF, Guinee TP, Cogan TM, McSweeney PL (2017) Fundamentals of cheese science. Springer, Boston
4. Tetra Pak International (2023) Produção de queijo fresco de alta qualidade: segredos e boas práticas. Grupo Tetra Pak. www.tetrapak.com
5. Silva ACDO, Cortez MAS (2021) Tecnologia de Leite e Derivados Lácteos. Eduff, Niterói
6. Market Research Reports (2023) Global fresh cheese market research report 2023. https://www.marketresearchreports.com/mrrpb5/global-fresh-cheese-market-research-report-2023
7. Khatun T, Rahman M, Pramanik AH, Wadud A (2019) Qualitative study on the cottage cheese prepared from partially skimmed milk. Int J Nat Soc 6(3):1–10
8. Halabiotec (2022) Cottage: um queijo simples com ingredientes naturais. www.halabiotec.com.br
9. Kosikowski FV (1996) Cheese and fermented milk foods, 2nd edn. Edward Bros, Inc, Ann Arbor
10. Ali MB, Murtaza MS, Shahbaz M et al (2022) Functional, textural, physicochemical, and sensorial evaluation of cottage cheese standardized with food grade coagulants. Food Sci Technol 42. https://doi.org/10.1590/fst.33420
11. Codex (1968) Codex international individual standard for cottage cheese including creamed cottage cheese. Codex stan C-16-1968. https://dairyconsultant.co.uk/pdf/codex_cottage_cheese_specification.pdf
12. FDA (2023) U.S. Food and Drug Administration. Code of Federal Regulations Title 21, volume 2, part 133 – Cheeses and related cheese products. https://www.accessdata.fda.gov/scripts/cdrh/cfdocs/cfcfr/CFRSearch.cfm?CFRPart=133
13. Emmons DB, Elliot JA, Beckkett DC (1966) Effect of lactic streptococci agglutinins on card formation and manufacture of cottage cheese. J Dairy Sci 49:1357
14. Brasil (1996) Ministério da Agricultura Pecuária e Abastecimento. Regulamento Técnico de Identidade e Qualidade dos Queijos, Brasília
15. Del Nobile MA, Conte A, Incoronato AL, Panza O (2009) Modified atmosphere packaging to improve the microbial stability of ricotta. Afr J Microbiol Res 3:137–142. https://doi.org/10.5897/AJMR.9000503
16. USDA (1981) Specifications for ricotta cheeses. United States Department of Agriculture, Agricultural Marketing Service, Dairy Division. https://www.ams.usda.gov/sites/default/files/media/ricottachees.pdf

17. Mangione G, Caccamo M, Natalello A, Licitra G (2023) Graduate student literature review: history, technologies of production, and characteristics of ricotta cheese. J Dairy Sci 106: 3807–3826. https://doi.org/10.3168/jds.2022-22460
18. Official Journal of the European Union (2009) Council Regulation n° 510/2006. 'ricotta di bufala campana'. Italy
19. Brasil (2020) Ministério da Agricultura Pecuária e Abastecimento. Regulamento Técnico de Identidade e Qualidade da Ricota, Brasília
20. Codex (2011) Standard for whey cheeses. Pages 284–1971 in Codex Alimentarius. Milk and milk products, 2nd ed. WHO and FAO. https://www.fao.org/3/i2085e/i2085e00.pdf
21. IRD (1925) Italian Regal Decree of 15/10/1925 n. 2033. Official Journal of Italian Reign 281. 4774.
22. Embrapa (2018) Empresa Brasileira de Pesquisa Agropecuária. Anuário leite 2018. embrapa.br/gado-de-leite
23. Netto, MM (2011) A geografia do queijo minas artesanal. Tese (Doutorado em Geografia)—Universidade Estadual paulista Júlio de Mesquita Filho, Rio Claro, SP.
24. Brasil (1997) Ministério da Agricultura, Pecuária e Abastecimento. Regulamento técnico para fixação de identidade e qualidade do queijo Minas Frescal. Brasília
25. Aragão SC, Oliveira MT, Abreu RA et al (2022) Análise qualitativa de laudos microbiológicos do queijo minas frescal observados no Laticínio Santa Heleine nos anos de 2020 e 2021 e comercializados no município de Naviraí/MS. Int J Dev Res 12(05): 55798–55806. https://doi.org/10.37118/ijdr.24500.05.2022
26. Cardarelli HR, Saad SMI, Gibson GR, Vulevic J (2007) Functional petit-suisse cheese: measure of the prebiotic effect. Anaerobe, London 13(5/6):200–207. https://doi.org/10.1016/j.anaerobe.2007.05.003
27. Lopes MMT, Rodrigues MCP, Araújo AMS (2018) Influence of expectation measure on the sensory acceptance of Petit Suisse product. J Food Sci 83(3):798–803. https://doi.org/10.1111/1750-3841.14067
28. Brasil (2000) Ministério da Agricultura e do Abastecimento. Aprova o Regulamento Técnico de Identidade e Qualidade de Queijo "Petit-suisse". Brasília
29. SDA (2018) Manual de Procedimentos de Inspeção Fiscalização de Leite e Derivados em Estabelecimentos Sob Inspeção Federal. Descrição dos Queijos Holandeses
30. Allied Market Research (2023) Mascarpone cheese market by type, by applications, by distribution channel: global opportunity analysis and industry forecast, 2023–2032. https://www.alliedmarketresearch.com/mascarpone-cheese-market-A15128
31. European Commission (2017) Guidance document describing the food categories in Part E of Annex II to Regulation, n° 1333/2008 on food additives. Off J Eur Union. https://food.ec.europa.eu/safety/food-improvement-agents/additives/eu-rules_en
32. Capozzi V, Lonzarich V, Khomenko I et al (2020) Unveiling the molecular basis of mascarpone cheese aroma: VOCs analysis by SPME-GC/MS and PTR-ToF-MS. Molecules 25(5):1–14. https://doi.org/10.3390/molecules25051242
33. Troiani C (2015) La solida nicchia del mascarpone. Il Latte. Retrieved from https://www.lattenews.it/la-solida-nicchia-del-mascarpone/

Chapter 7

Parmesan Cheese

S. C. Rangel-Ortega, L. G. Campos-Múzquiz, D. E. Cruz-Casas, and R. Rodríguez-Herrera

Abstract

The process of elaboration of parmesan cheese begins with skimming the milk collected the previous evening; fresh milk and fermented whey are added. This mixture is placed in a copper vat at 32–35 °C to form the curd, which is finely cut. Afterward, it is heated to 50 °C while being stirred until it forms a paste. The paste is then placed in cheesecloth for draining. Subsequently, the cheese is shaped in a mold; and a marking band is applied. It is then placed in a metal mold and added to a container with a saline solution. Afterward, it is aged for 12–30 months.

Key words Parmigiano-reggiano, Cheese elaboration, Milk coagulation, Cheese molding, Cheese salted, Cheese aging

1 Introduction

Parmigiano-reggiano is a hard cheese crafted from partially skimmed, raw cow's milk. Originating in a specific Italian geographical area, including provinces such as Parma, Reggio Emilia, and Bologna, it strictly adheres to rigorous production standards. Acknowledged as a Protected Designation of Origin (PDO) in the European Union since 1996, cheeses crafted outside that specific area of Italy are labeled as parmesan [1].

In the United States, "parmesan-type" cheeses are produced with pasteurized milk. These cheeses are characterized by their smaller size, lower cooking temperature (around 50 °C), heavier salting, and shorter maturation period, often used in grated form [2]. Meeting US standards, parmesan cheese must contain less than 32% moisture and age for a minimum of 10 months [3].

The production process involves utilizing morning and evening milk in traditional copper vats. Coagulation occurs by adding rennet and a whey starter collected from the cheese production of the previous day, followed by cooking at 55 °C. The resulting mass is

Adriano Gomes da Cruz et al. (eds.), *Dairy Foods Processing*, Methods and Protocols in Food Science,
https://doi.org/10.1007/978-1-0716-4144-6_7,

then divided into twin wheels, wrapped in linen cloth, and shaped in molds [4].

Various factors influence the final product, encompassing animal breed, milk composition, processing parameters, starter culture and nonstarter lactic acid bacteria (NSLAB) that come from the evening milk and lactic whey, and aging time. Typically, parmesan ages for 10 to 24 months, fostering flavor and aroma development [5].

This protocol aims to explain and describe the artisan process of elaborating parmesan cheese and the possible problems that may arise and how to solve them.

2 Materials

2.1 Ingredients

1. Cow's milk is collected the previous afternoon under hygienic conditions and left to rest at room temperature (*see* **Note 1**).
2. Cow's milk collected on the morning of the cheese-making day and refrigerated between 15 and 20 °C.
3. Whey collected from the cheese-making process of the previous day, allowed to rest between 30 and 35 °C, and adjusted to an acidity level of 32°D (*see* ref. 6).
4. Commercial enzymatic rennet, prepared according to the manufacturer's instructions based on the volume of milk intended for cheese-making (*see* **Note 2**).
5. Brine, prepared with water at 20% NaCl, also includes 0.2% CaCl, stabilizing the pH at 5.2 by adding 10 ml of white vinegar/10 L of brine (*see* **Note 3**). The total volume of brine to be prepared should be 5–10 L/kg of cheese to be salted (*see* **Note 4**).

For Acidity Determination in °D

6. Sodium hydroxide (NaOH) 0.1 M (prepared with distilled water).
7. 1% phenolphthalein indicator (prepared at 1% in 95% ethanol).

2.2 Utensils

In the production of parmesan cheese, various dairy utensils are needed:

1. Thermometer.
2. Buret.
3. Scale.
4. Stainless steel vats with a double bottom, 2000 L capacity.
5. Copper boilers in inverted bell form, 1200 L of capacity.
6. Shovel, can be made of wood or food-grade plastic.
7. Cheese harp "spino" for cutting the curd.

8. Cheesecloth, 1.5 m × 1 m sections.
9. Plastic molds (40 kg of capacity).
10. Wooden disks (60 cm in diameter and 6 cm of thickness).
11. Metal molds (40 kg of capacity).
12. Stainless steel transport carts.
13. Plastic bands for cheese marking should be the same size as the plastic molds.
14. Casein plate labels.

2.3 Facilities

1. Curding room.
2. Salting room.
3. Drying room.
4. Aging room.

3 Methods

1. It starts with skimming the milk collected from the evening milking (*see* **Note 5**), combined with the whole milk from the morning milking, using 550 L of each type of milk.
2. To the previous mixture of milks, fermented whey from the cheese production of the previous day is added in a proportion of 3% of the total volume of the milk.
3. Then, the mixture of milk and fermented whey is obteined, it is tranferred to a copper boiler and heated to a tempertaure of 32–35 °C (*see* **Note 6**).
4. Subsequently, considering the final volume of milk, a commercial rennet is prepared following the producer's instructions and added to the milk. The curd will form after approximately 30 min (*see* **Note 7**).
5. Once the milk has completely coagulated (*see* **Note 8**), using a manual curd cutter "spino," the curd is gently cut into small pieces the size of wheat grain (*see* **Note 9**).
6. After cutting the curd is heated to reach 50 °C (*see* **Note 10**). The curd should be stirred constantly for 10 min and then rest for 1 h. At this point, the curd will form a spongy mass at the bottom of the copper boiler (*see* **Note 11**).
7. The formed curd at the bottom is cut into two parts, and each part is transferred to a piece of cheesecloth. The cheesecloth containing the curd is hung to allow the whey to drain by gravity (*see* **Note 12**). The drained whey is collected in a container for the next day's cheese-making process.

8. The next step is to transfer the curd to the plastic molds that will shape the cheese. In this step, the cheesecloth also enters the mold to protect the curd and facilitate unmolding. In the molds, whey drainage continues with the help of a wooden disk placed on the surface to exert slight pressure. The cloth is removed from molds at night, and a marking band is placed around the cheese and inside the mold. On the surface, a casein plate is placed (*see* **Note 13**). This step may take several days (*see* **Note 14**) before unmolding.
9. Days later, the marking band and the mold are removed to place the cheese in a metal mold that will give it the traditional shape of this cheese (*see* **Note 15**).
10. Subsequently, the cheese is removed from the mold and immersed in a brine tank at a temperature of 15–17 °C. To facilitate the proper diffusion of salt, the cheeses should be turned in the brine once a day for 20–24 days.
11. At the end of the salting process, the cheeses should be dried for several hours at 28 °C before moving on to aging (*see* **Note 16**).
12. Once dry, the cheeses are taken to the aging room, maintained at 11 to 14 °C with a relative humidity of 80%. The cheeses are placed on shelves arranged vertically, with spaces between them. Cheese maturation is at least 12 months but can extend up to 30 months.
13. Every 10 days, the cheeses are inspected and cleaned on the surface with a brush to ensure the removal of any mold or other elements that may have accumulated. During this step, the cheese is also rotated.
14. At the end of the maturation period, each cheese that meets the quality standards is stamped with a fire seal, ensuring compliance with the PDO requirements. If not, all markings on the surface are removed.

4 Notes

1. The milk is left to ferment in stainless steel vats, ensuring that the milk's surface height in the vat is short. This exposes a larger amount of milk to the surface, allowing for greater recovery of the cream from the milk.
2. It is important to check the rennet strength before adding it to the milk, as it may reduce its coagulating effect due to various reasons. To do this, take the necessary amount to coagulate 1 L of milk and note the time it takes for the curd to form. If this time is longer than indicated in the rennet instructions, increase the volume of rennet to be used.

3. Brine can be reused several times. Before using it again, any remaining cheese residue should be removed, and the brine should be brought to a boil. It must be stored well-covered.
4. Brine volume should be generous and should allow sufficient space between one cheese and another.
5. The milk skimming process is carried out by collecting the milk that remains below the layer of fat formed on the surface. For this purpose, the stainless steel vat is equipped with a drainage valve that allows milk recovery at that level.
6. An increase or decrease in this temperature could lengthen or stop the milk curd process.
7. When the curd takes more than 30 min to form, it is necessary to monitor the milk temperature, ensuring that it is between 32 and 35 °C and that the room where the cheese is being made is at a warm temperature. If, despite this, the curd still does not appear, recheck the rennet strength and supplement by adding more rennet to the milk. Install temperature and humidity control systems in the different stages of processing parmesan cheese for proper curd formation, whey drainage, and ripening [7].
8. The curd should not be too soft, nor should be too firm; rather, it should have a smooth yet consistent texture.
9. It is important that the curd granules formed when cutting the curd are of a uniform size. This process should be done gently and with constant stirring using a skimmer. The cheese maker can frequently check the curd size by taking a sample in hand and assessing both the size and uniformity.
10. Once the temperature of 50 °C is reached, the heating of the curd is stopped.
11. The cheese maker can join the curd granules in one piece to form a single paste using a cheese shovel.
12. To facilitate hanging the cloth with the curd, it can be supported by a resting cheese paddle on the surface of the coagulation vat. The curd will remain in this step until most of the whey has drained. The remaining whey will be drained once it is in the mold.
13. In the labeling process, information such as the month and year of production, the dairy registration number, and dots inscribed around the circumference of the pieces is printed. The casein plate, shaped like a label, assigns a consecutive unique number to each cheese piece, providing it with identity.
14. The time the curd spends inside the mold will depend on the speed at which the whey drains, and it can take up to 3 days.

15. Metal molds are particularly used to give the characteristic shape to parmesan cheese, preserving its cylindrical form with rounded ends.
16. The cheese at this stage must be completely dry on the surface before entering the aging room, as failure to do so could lead to the rapid development of molds. The drying time may take up to a day.

References

1. Consorzio del Formaggio Parmigiano Reggiano (2017) Parmigiano Reggiano Guidelines. Rome, Ministero delle Politiche Agricole Alimentari, Forestali e del Turismo
2. Fox PF, Guinee TP, Cogan TM, McSweeney PL (2017) Fundamentals of cheese science, vol 1. Springer, Boston, p 271
3. Code of Federal Regulations (2023) § 133.165 Parmesan and reggiano cheese. https://www.ecfr.gov/current/title-21/section-133.165
4. Consorzio del Formaggio Parmigiano Reggiano (n.d.) The art of making – Parmigiano Reggiano. https://www.parmigianoreggiano.com/product-art-of-making/
5. Aubrey, MH (2020) The parmesan cheese model system: alleviation of browning defect in parmesan cheese. Electronic theses and dissertations 3684
6. Vázquez-Román S, Garcia-Lara NR, Escuder-Vieco D, Chaves-Sánchez F, De la Cruz-Bertolo J, Pallas-Alonso CR (2013) Determination of dornic acidity as a method to select donor milk in a milk bank. Breastfeed Med 8(1): 99–104. https://doi.org/10.1089/bfm.2011.0091
7. Mehta BM, El-Bakry M (2022) Processed cheese science and technology: ingredients, manufacture, functionality, quality, and regulations. Elsevier Science. isbn: 9780128214602, 0128214600

Chapter 8

Cheddar Cheese

Fabiane Picinin de Castro Cislaghi, Andréa Cátia Leal Badaró, and Fabiane Andresa de Oliveira

Abstract

Cheddar cheese is one of the most popular cheeses in the world. This cheese is typically made from pasteurized cow's milk and it can be classified as hard cheese. The manufacture of Cheddar cheese consists of preparing and standardizing milk, adding the ingredients, coagulating milk, cutting the coagulum into small cubes, heating and agitating the cubes, removing the whey, cheddaring, milling, dry-salting, molding, pressing, packaging, and ripening. This chapter is a practical guidance for Cheddar cheese manufacture.

Key words Cheddar, Cheddaring, Cheese, Ripening, Dairy

1 Introduction

Cheddar cheese originated from the village of Cheddar in Somerset county, England, in the twelfth century. Due to the strong colonial influence of England in the eighteenth and nineteenth centuries, its manufacturing expanded to countries under British influence at the time, such as the United States, Canada, New Zealand, and Australia. This cheese is one of the most popular cheeses in the world today [1, 2].

Although Cheddar originated in England, the name *Cheddar* can be used for cheese made anywhere because it is not protected. *Cheddar* is typically made from pasteurized cow's milk and is characterized by its rich, nutty flavor, and it ranges from white or ivory to light yellow or orange in color [3, 4].

Traditionally, the color was an indication of the quality of the cheese, with deeper shades suggesting higher levels of beta-carotene derived from grass-fed cows. The color came to be associated with Cheddar cheese. However, modern Cheddar cheese producers often use natural colorants to achieve the desired hue [1].

Adriano Gomes da Cruz et al. (eds.), *Dairy Foods Processing*, Methods and Protocols in Food Science,
https://doi.org/10.1007/978-1-0716-4144-6_8,

Some consumers are of the impression that colored cheeses taste better or are more intensely ripened than their equivalent white counterpart. A recent consumer study identified color as one of the key attributes influencing consumer choice, with higher attractiveness for orange color in comparison to white Cheddar. However, white cheeses were considered to be more "natural" because of the absence of annatto [3, 5].

Cheddar cheese generally has a minimum of 50% fat-on-dry basis (FDB) and a moisture on a fat-free basis (MFFB) of 55%, which means it can be classified as hard cheese, although it is on the verge of semihard types [6]. Typical moisture content is 35–39% [7]. This cheese varies extensively in consistency depending on how long it has been aged. Young Cheddar is mild flavored and slightly soft and pasty becoming more crumbly in texture and more strongly flavored as it matures during ripening [2]. There are typically four dominant flavor profiles for more mature variants: farmy/earthy, meaty/savory, sharp/tangy, and sweet/nutty [8]. The Cheddar cheese may be manufactured with or without rind which may be coated [4].

Cheddar cheese is a representative of high-fat food and has high protein contents and nutritional value. In addition, it contains all essential amino acids at various levels and small amounts of other nutrients such as vitamins A, B_2, B_6, and B_{12} [9].

The original Cheddar, which has been aged for a long time, has a typical pronounced, slightly acidic flavor. Due to its physical–chemical and sensorial characteristics and also its durability, Cheddar has become the main raw material for the production of processed cheeses. A large part of its production is not intended for direct consumption, being used as a base in mixtures of processed cheeses of different presentations and flavors [10–12].

In the United States, where it has been manufactured since 1795, Cheddar is the most consumed cheese and is often called American cheese, with several varieties. In a general classification, Cheddar is divided into two broad categories: the "long-hold" style, aged for 1 year or more, with a moisture content of less than 36% and a stronger flavor, and "short-hold" Cheddar, aged between 1 and 6 months, with moisture between 37 and 39%, softer and with a milder flavor. Due to the typical characteristics of cheese consumption in the United States, American Cheddar no longer has the same characteristics as the British original [10]. Furthermore, Cheddar can be commonly divided into four categories, according to the degree of ripening and sensory characteristics [13]:

- *Mild* or *mellow*: Aged for a shorter period, typically around 3 months. It has a milder flavor and slightly more elastic texture.
- *Medium*: Aged for a slightly longer period than mild Cheddar, usually for around 6 months. It has a more pronounced flavor

Table 8.1
Cheddar cheese composition

Moisture, %	37
Fat, %	33
Protein, %	25
Ash, %	4
Carbohydrate, %	1
pH	5.1

Source: Adapted from [14, 15]

and a slightly firmer texture than mild Cheddar but is still relatively mild compared with other aged Cheddar varieties.

- *Sharp*: Is normally aged for 9 months. It has a pronounced flavor.
- *Extra-sharp* (extra-mature): Aged for 12 months or more. It has sharper and tangier flavor and a firmer and crumbly texture.

Table 8.1 shows composition of a typical Cheddar cheese. Some chemical composition of Cheddar cheese might change according to ripening time. This chapter is directed toward the design of a Cheddar cheese protocol.

2 Materials and Methods

Cheddar is a cheese that requires progressive acidification in the tank to obtain the appropriate degree of demineralization, which gives it its typical consistency and traditional slightly acidic flavor. Therefore, the presence of mesophilic bacteria is essential, and the cooking temperature (38 °C) remains within a range at all times that allows fermentation within the curd grain [10]. Figure 8.1 shows the production of Cheddar cheese. To produce quality cheese, the steps must be carried out in a controlled manner, adapting the desired final characteristics. Each step is described separately in this protocol.

2.1 Pre-treatment

Pre-treatment includes raw milk standardization and heat treatment (*see* **Notes 1** and **2**). Firstly, the raw milk should be standardized to fulfill the requirement of production. It is recommended to use milk with around 3.5% fat, maintaining a casein/fat ratio of 0.70 (*see* **Note 3**). After standardization, milk should be heated for pasteurization. Usually, an appropriate heating condition should be 72 °C for 15 s. Hot milk should be cooled to ~30 °C for inoculation of the starter cultures (*see* **Note 4**).

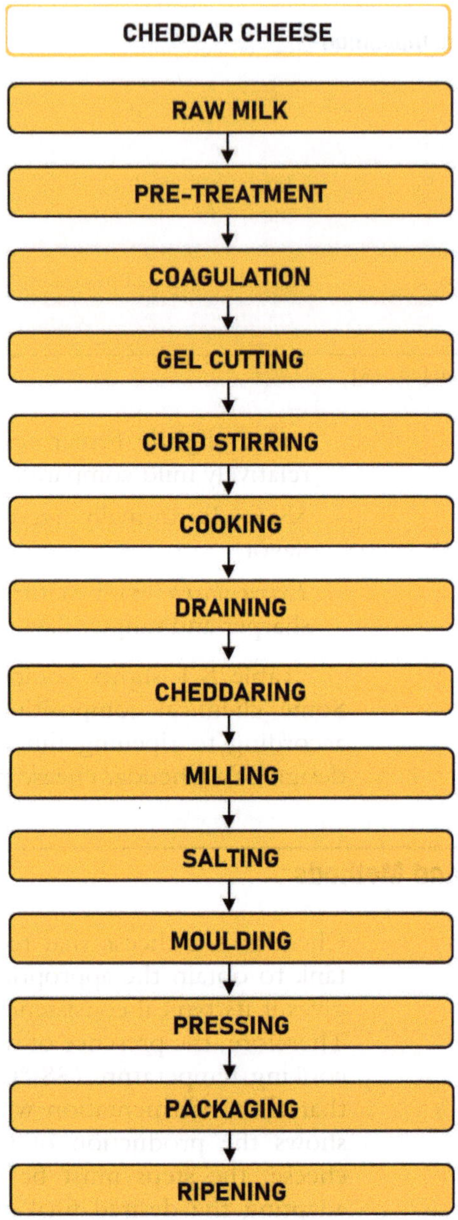

Fig. 8.1 Flowchart of Cheddar cheese production

It is recommended to use mesophilic homofermentative lactic culture (*Lactococcus lactis* subsp. *lactis* and *Lactococcus lactis* subsp. *cremoris*). These bacteria do not produce gas, which contributes to the closed texture of Cheddar cheese (*see* **Note 5**).

Calcium chloride (20 g 100 L^{-1}) and coagulant are added to the milk (*see* **Notes 6** and 7). The use of annatto (5 to 10 mL 100 L^{-1}) or similar colorant is optional and gives the cheese a yellow/orange color (*see* **Note 8**). The mixture is allowed to set and form curds at a temperature of around 30–32 °C for 30 to 40 min [6, 10, 16].

2.2 Coagulation

After coagulation action, a gel is formed from milk by enzymatic coagulation, in which coagulating enzymes are used (e.g., chymosin) to hydrolyze the κ-casein fraction of casein micelles (breaking of the Phe_{105}-Met_{106} peptide bond). The release of glycomacropeptide—a highly hydrophilic fragment—destabilizes casein micelles, which results in the aggregation and gelation of resulting casein aggregates [17].

2.3 Gel Cutting, Curd Stirring, Cooking, and Draining

Before the coagulum is cut, a simple test is normally carried out to establish its whey-eliminating quality. Typically, a knife is stuck into the clotted milk surface and then drawn slowly upward until proper breaking occurs. The curd may be considered ready for cutting as soon as a glass-like splitting flaw can be observed [6].

The curd is cut into small cubes (1 cm^2) (*see* **Note 9**). Immediately after cutting, the curd grains are very sensitive to mechanical treatment; therefore the stirring has to be gentle. It must, however, be fast enough to keep the grains suspended in the whey. Sedimentation of curd in the bottom of the tank causes formation of lumps. Lumps may influence the texture of the cheese, as well as causing loss of casein in the whey [6].

The curd is cooked to 37–39 °C. Cheddar cheese is cooked by steam in the jacket. Heating must be slow, at a rate of 1 °C/3 min (about 30 min) (*see* **Note 10**). The curd is trenched on the sides of the vat to facilitate further draining. As soon as the required acidity and firmness of the curd have been attained—and checked by the producer—the whey is drained off (*see* **Note 11**), and the curd is subjected to a special form of treatment called cheddaring [6, 10, 18, 19].

2.4 Cheddaring and Milling

Cheddaring is a unique step in Cheddar cheese making to give the typical Cheddar flavor. The curd is allowed to stick together (matting) to form loaves. Thus, loaves of curds are cut about 15 cm wide along each side of the vat. After 10 min, the loaves are turned and stacked every 10 min. During cheddaring, starter activity continues to reduce the pH, solubilizes colloidal calcium phosphate, and causes physicochemical changes to the curd that changes from being soft and friable to being tough and pliable. This process is complete when the pH is around 5.3–5.4. During this period, approximately 90 to 120 min, keep the temperature between 35 and 38 °C. The fully cheddared curd has a dry, firm, mellow, and meaty texture exhibiting chicken breast structure. When proper acidity level is attained, the Cheddar cheese loaves are ready for milling into small cuts. Then, a curd mill is used to cut the loaves into cubes. The milling process facilitates uniform salt distribution into the curds and promotes whey drainage from the curds (*see* **Notes 12**, **13**, and **14**) [10, 15, 17, 20].

2.5 Salting

Traditionally, Cheddar cheese is dry-salted (in the curd mass). It is recommended 2.5 to 3.0 kg of salt for every 1000 liters of coagulated milk, to provide between 1.6 and 1.8% salt in the finished cheese. For complete distribution, the curd may be stirred for 5–10 min. In Cheddar cheese about 60% added salt is retained in the curd, 35% lost in the whey from press, and 5% lost during salting and hooping. To avoid greater loss of salt and fat in the whey, which comes out of the mass during pressing, a "mellowing time" of 15 min is recommended (time between the end of salting and the beginning of pressing). During this period, the curd grain slowly absorbs the salt, forming a very thin shell without causing significant fat loss (*see* **Notes 15** and **16**) [6, 10, 21].

Salt plays an important role in the quality of Cheddar cheese. Dry-salting causes a rapid increase in the salt-in-moisture content of the curd and starter activity stops abruptly [20]. In addition, the inhibitory effect of salt on the proteolysis of β-casein in Cheddar cheese is particularly important in reducing the incidence of bitterness [15, 22–25].

2.6 Molding and Pressing

After salting, the curds are placed into molds (molds for Cheddar cheese are called hoops). The hooped curd is then placed in a cheese press where pressure is applied gradually at first and then increased as the curd cools, to form blocks of Cheddar cheese [6, 20, 21]. A longer period at high pressure is used for Cheddar, 100–170 kPa for up to 24 hours (*see* **Notes 17** and **18**) [18, 21]. Vacuum treatment can be applied before or after pressing to reduce or eliminate mechanical openings in the blocks [18].

2.7 Packaging

Cheddar cheese is normally covered with a protective coating at some stage of the ripening process. Coating varies from vegetable oils to special plastic films. The shape and characteristics of cheese and changes to be induced in cheese dictate the type of packaging material (*see* **Note 19**) [6, 18, 21].

Cheddar cheese can be provided with a plastic film, paraffin, or wax coating prior to ripening. The films are usually laminants of different plastic films and waxes with specific properties. The conventional materials used for coating production are petroleum based (e.g., paraffin wax and polyvinyl chloride), but nowadays with the environmental and sustainability issues as well as legislation restrictions, the use of edible film and coating materials is being exploited. Moreover, paraffining is now replaced by film packaging (cheap and easy-to-apply films) as it causes considerable loss of cheese while removing paraffin [6, 21, 26].

2.8 Ripening

Ripening takes place for different periods, from a few weeks up to 12 months or more, depending on the extent of ripening desired. The degree of ripening has a significant impact on the development of Cheddar cheese characteristics [20, 27]. Cheddar cheese is often ripened at low temperatures, between 4° and 8 °C, and a relative humidity lower than 80% as they are normally wrapped before ripening. Ripening of wrapped cheese can be done at lower humidity levels without excessive moisture losses from the cheese (*see* **Note 20**) [6, 20, 21, 28].

During ripening, cheese undergoes lipolysis, proteolysis, and glycolysis to produce the desired flavor and aroma [29]. The primary function of glycolysis is lactose conversion, generating lactic acid, which is removed together with whey during cheese making [30]. Casein peptides are degraded by the action of proteases and peptidases to produce free amino acids (FAAs) during cheese ripening, which contribute to the flavor and aroma of Cheddar cheese [30]. Free fatty acids (FFAs) released from milk fat by lipolysis, such as short- and medium-chain FFAs, also affect cheese flavor [29]. In addition, secondary biochemical events such as the metabolism of FFAs and free amino acids (FAAs) are also very essential for flavor development [9]. Proteolysis is the most extensive biochemical event occurring during ripening of Cheddar cheese (*see* **Notes 21** and **22**) [21].

3 Notes

1. To manufacture Cheddar cheese with quality, good quality milk that has relatively low microbial count is needed. Proteinases from psychrotrophic bacteria in milk such as *Pseudomonas fluorescens* and *P. putrefaciens* are heat stable and can hydrolyze C-terminal region of β-casein and α_{s1}-casein, resulting in bitter hydrophobic peptides that can cause bitter flavor to accumulate during cheese ripening [15].
2. Furthermore, during Cheddar cheese manufacture, if antibiotic residues are present, rate of acidification during draining and salting is reduced, which will result in longer manufacture time and may cause a higher pH in cheese. This resultant cheese may also have an uneven texture with pasty body and unusual off-flavors [15].
3. The manufacturing yield (cheeses with around 37% moisture, from milk with 3.5% fat) varies from 9 to 9.8% (i.e., from 10.2 to 11.0 L/kg) [10].
4. The starter is normally added to the milk at renneting temperature, while the cheese tank is being filled [6].

5. Other strains of *Lactobacilli* are used for improving the flavor and also to accelerate cheese ripening [31–33]. Furthermore, *Streptococcus thermophilus* produces acid rapidly during scalding and may confer an additional measure of bacteriophage (phage) protection [10, 34, 35].
6. Calcium chloride is used to obtain a constant coagulation time and results in sufficient firmness of the coagulum [6].
7. To facilitate distribution, the coagulant may be diluted with at least double the amount of water. After coagulant dosage, the milk is stirred carefully for not more than 5 min. It is important that the milk comes to a standstill within another 5–8 min in order to avoid disturbing the coagulation process and causing loss of casein in the whey [6].
8. Although Cheddar cheese containing annatto is usually orange in appearance, commercially it is referred to as red Cheddar [3].
9. Very important for the cheese yield is the moment of cutting the gel. If the gel is weak, the syneresis is not pronounced and fat and protein are lost with the way. If the gel is too hard, the curd is broken and also the fat and protein are lost.
10. The heat treatment promotes contraction of the curd accompanied by expulsion of whey (syneresis).
11. When the pH of the curd mass has reached about 6.0–6.1 (about 2 hours after renneting), the whey is drained off.
12. The cheddaring process affects the structure of the cheese by making the closed structure of the cheese more enhanced due to the removal of entrapped air from the cheese. The cheddaring also affects the final moisture and texture of the cheese [36].
13. Moreover, the pH is also a factor affecting the texture of Cheddar cheese. Hence, the higher the pH, the softer the texture of the cheese will be [36].
14. Traditionally, the milled-curd method is applied for Cheddar cheese manufacture. However, because of the mechanization of cheese making and time constraints, the stirred-curd method is the method of choice in modern and highly mechanized production plants, because of its relatively short manufacturing time. In the stirred-curd method of manufacture, the curds are continuously stirred after whey drainage until reaching the desired pH, hence eliminating the traditional cheddaring and milling process. Although the constant agitating does not allow knitting of curds, the use of "block former" hooping and vacuum pressing system yields Cheddar cheese with a close-texture characteristic [15, 36, 37].
15. After dry salt is distributed over the surface of cheese curd, some salt dissolves and slowly moves inward the curd. This

causes a counterflow of whey from the curd to the surface and dissolves the remaining salt crystals, creating a supersaturated brine solution around each curd particle. The rate of the salt uptake by the cheese curds highly depends on the initial moisture in the curd and the amount of the salt added. An increase in salting level results in an increase in the rate of both salt absorption by the cheese curd and whey drainage from the cheese curd [15, 22].

16. Acid production is insufficient if the cheese contains over 5 to 5.5% salt in water, whereas at less than 4.5% salt concentration, the lactic acid bacteria ferment too fast. In either case the flavor development is unsatisfactory or off-flavors can be produced [38].
17. Cheddar cheese was traditionally produced as large cylindrical-shaped loafs (weighing nearly 30 kg), but nowadays under modern industrial-scale process, the cheese is usually formed into rectangular blocks [36].
18. Traditionally, the salted curd is putted into molds and pressed. The inconvenience is that long pressing time and high pressure is needed. An alternative is pressing under vacuum, in columns or towers (block former system) [6, 10].
19. In the United Kingdom, many dairy industries still manufacture cheese in cylindrical format or even in blocks, but they ripen it without packaging for a period of time, to improve its flavor and texture [10].
20. Alternative ripening conditions (including the addition of ripening enhancing enzymes) may be used.
21. Some matured Cheddar cheeses may have small crystals of calcium lactate throughout or small crystals of tyrosine within the interior [1, 8, 39, 40].
22. As it has a higher acidity, it tends to retain more coagulant enzymes in the curd mass, undergoing strong proteolysis from the beginning of ripening, with the risk of forming a bitter taste. Therefore, the standardization of certain parameters is essential to obtain quality cheese [10].

References

1. Hayes G (2023) Cheddar cheese: a delicious journey into the world of savory delights. J Pharmacogn Nat Prod 9:241. https://doi.org/10.37421/2472-0992.2023.9.241
2. O'Sullivan MG (2023) Sensory characteristics of Cheddar and related cheeses varieties. In: Tuohy JJ (ed) Sensory profiling of dairy products. Wiley, Hoboken
3. Sharma P, Segat A, Kelly AL et al (2020) Colorants in cheese manufacture: production, chemistry, interactions, and regulation. Compr Rev Food Sci Food Saf 19:1220–1242. https://doi.org/10.1111/1541-4337.12519
4. Codex Alimentarius Commission (2022) Standards for Cheddar, CXS 263–1966. Codex Alimentarius: International Food Standards. Joint

FAO/WHO Food Standards Programme, Rome, Italy. https://www.fao.org/input/download/standards/186/CXS_263e.pdf. Accessed 27 Feb 2024

5. Speight KC, Schiano AN, Harwood WS et al (2019) Consumer insights on prepackaged Cheddar cheese shreds using focus groups, conjoint analysis, and qualitative multivariate analysis. J Dairy Sci 102:6971–6986. https://doi.org/10.3168/jds.2018-16209
6. Bylund G (2015) Dairy processing handbook. Tetra Pak Processing Systems, Lund, Sweden. https://dairyprocessinghandbook.tetrapak.com/chapter/cheese. Accessed 27 Feb 2024
7. Hill A, Ferrer MA (2021) Cheese making technology e-Book. https://books.lib.uoguelph.ca/cheesemakingtechnologyebook/chapter/8-3-firm-to-hard-cheese-low-temperature-provolone-cheddar/ Accessed 27 Feb 2024
8. Dairy UK (2018) Code of practice on compositional standards for UK named variety cheeses. https://www.dairyuk.org/wp-content/uploads/2018/12/FINAL-Compositional-Standards-Cheese.pdf. Accessed 27 Feb 2024
9. Murtaza MA, Ur-Rehman S, Anjum FM et al (2014) Cheddar cheese ripening and flavor characterization: a review. Crit Rev. Food Sci Nutr 54:1309–1321. https://doi.org/10.1080/10408398.2011.634531
10. Furtado MM (2013) Queijos especiais. Setembro Editora, São Paulo
11. Riebel B, Govindasamy-Lucey S, Jaeggi JJ et al (2024) Functionality of process cheese made from Cheddar cheese with various rennet levels and high-pressure processing treatments. J Dairy Sci 107:74–90. https://doi.org/10.3168/jds.2023-23825
12. Wang J, Yang ZJ, Xu LY et al (2021) Key aroma compounds identified in Cheddar cheese with different ripening times by aroma extract dilution analysis, odor activity value, aroma recombination, and omission. J Dairy Sci 104:1576–1590. https://doi.org/10.3168/jds.2020-18757
13. Pace N, Verma A, Mayank et al (2024) The utility of a slice defect score method in understanding factors impacting the sliceability of commercial Cheddar cheese blocks. Int Dairy J 151:105865. https://doi.org/10.1016/j.idairyj.2023.105865
14. Ayala-Bribiesca E, Lussier M, Chabot D et al (2016) Effect of calcium enrichment of Cheddar cheese on its structure, in vitro digestion and lipid bioaccessibility. Int Dairy J 53:1–9. https://doi.org/10.1016/j.idairyj.2015.09.002
15. Phadungath C (2019) Cheddar cheese review: I cheese manufacture. Koch ChaSarn J Sci 37: 1–17
16. Zhu D, Damodaranand S (2012) Short communication: Annatto in Cheddar cheese-derived whey protein concentrate is primarily associated with milk fat globule membrane. J Dairy Sci 95:614–617. https://doi.org/10.3168/jds.2011-4857
17. Cruz AG, Silva MC, Pimentel TC et al (eds) (2023) Probiotic foods and beverages: technologies and protocols. Humana, New York. https://doi.org/10.1007/978-1-0716-3187-4
18. Chandan RC (2014) Cheese in the marketplace. In: Batt CA, Tortorello ML (eds) Encyclopedia of food microbiology, 2nd edn. Academic Press, Amsterdam. https://doi.org/10.1016/B978-0-12-384730-0.00058-6
19. Jiménez-Flores R, Yee J (2007) Low temperature hard cheeses and semi-hard washed cheeses. In: Weimer BC (ed) Improving the flavour of cheese. Woodhead Publishing, England. https://doi.org/10.1533/9781845693053.4.444
20. McSweeney PLH (2007) Cheddar cheese. In: McSweeney PLH (ed) Cheese Problems Solved. Woodhead Publishing, England. https://doi.org/10.1533/9781845693534.214
21. Indira Gandhi National Open University (2017) Principle and method of manufacture of Cheddar cheese. http://egyankosh.ac.in//handle/123456789/9597. Accessed 27 Feb 2024
22. Guinee TP (2004) Salting and the role of salt in cheese. Int J Dairy Technol 57:99–109. https://doi.org/10.1111/j.1471-0307.2004.00145.x
23. Kuhfeld RF, Eshpari H, Kim BJ et al (2024) Identification of bitter peptides in aged Cheddar cheese by crossflow filtration-based Fractionation, Peptidomics, statistical screening and sensory analysis. Food Chem 439: 138111. https://doi.org/10.1016/j.foodchem.2023.138111
24. Bisig W (2017) The importance of salt in the manufacturing and ripening of cheese. Special Issue of the International Dairy Federation. https://fil-idf.org/wp-content/uploads/2017/03/Factsheet-001_2017-The-importance-of-salt-in-the-manufacturing-and-ripening-of-cheese.pdf. Accessed 27 Feb 2024
25. McSweeney PLH (2007) Salt in cheese. In: McSweeney PLH (ed) Cheese problems solved. Woodhead Publishing, England.

https://doi.org/10.1533/9781845693534.80
26. Costa MJ, Maciel LC, Teixeira JA et al (2018) Use of edible films and coatings in cheese preservation: opportunities and challenges. Food Res Int 107:84–92. https://doi.org/10.1016/j.foodres.2018.02.013
27. Song Z, Cao Y, Qiao H et al (2023) Analysis of the effect of *Tenebrio Molitor* rennet on the flavor formation of Cheddar cheese during ripening based on gas chromatography-ion mobility spectrometry (GC-IMS). Food Res Int 171:113074. https://doi.org/10.1016/j.foodres.2023.113074
28. Soodam K, Ong L, Powell IB et al (2017) Effect of elevated temperature on the microstructure of full fat Cheddar cheese during ripening. Food Struct 14:8–16. https://doi.org/10.1016/j.foostr.2017.05.003
29. Khattab AR, Guirguis HA, Tawfik SM et al (2019) Cheese ripening: a review on modern technologies towards flavor enhancement, process acceleration and improved quality assessment. Trends Food Sci Technol 88:343–360. https://doi.org/10.1016/j.tifs.2019.03.009
30. Tekin A, Güler Z (2019) Glycolysis, lipolysis and proteolysis in raw sheep milk Tulum cheese during production and ripening: effect of ripening materials. Food Chem 286:160–169. https://doi.org/10.1016/j.foodchem.2019.01.190
31. Duan C, Li S, Zhao Z et al (2019) Proteolytic activity of *Lactobacillus plantarum* strains in Cheddar cheese as adjunct cultures. J Food Prot 82:2108–2118. https://doi.org/10.4315/0362-028X.JFP-19-276
32. El Soda M, Madkor SA, Tong OS (2000) Adjunct cultures: recent developments and potential significance to the cheese industry. J Dairy Sci 83:609–619. https://doi.org/10.3168/jds.S0022-0302(00)74920-4
33. Stefanovic E, Kilcawley KN, Roces C et al (2018) Evaluation of the potential of *Lactobacillus paracasei* adjuncts for flavor compounds development and diversification in short-aged Cheddar cheese. Front Microbiol 9:1506. https://doi.org/10.3389/fmicb.2018.01506
34. Mullan WMA (2014) Starter cultures. Importance of selected genera. In: Batt CA, Tortorello ML (eds) Encyclopedia of food microbiology, 2nd edn. Academic Press, Amsterdam. https://doi.org/10.1016/B978-0-12-384730-0.00321-9
35. Michel V, Martley FG (2001) *Streptococcus thermophilus* in Cheddar cheese – production and fate of galactose. J Dairy Res 68:317–325. https://doi.org/10.1017/s0022029901004812
36. Salek RN, Buňka F, Černíková M (2022) The use of different cheese sources in processed cheese. In: El-Bakry M, Mehta BM (eds) Processed Cheese science and technology. Woodhead Publishing, England. https://doi.org/10.1016/B978-0-12-821445-9.00010-8
37. Shakeel-ur-Rehman DMA, Farkye NY (2008) Differences between Cheddar cheese manufactured by the milled-curd and stirred-curd methods using different commercial starters. J Dairy Sci 91:76–84. https://doi.org/10.3168/jds.2007-0003
38. Walstra P, Wouters JTM, Geurts TJ (2005) Dairy science and technology, 2nd edn. CRC Press, Boca Raton. https://doi.org/10.1201/9781420028010
39. Kindstedt PS, Polowsky PJ (2021) Invited review: Crystals in cheese: more than a curiosity. J Dairy Sci 104:1233–1250. https://doi.org/10.3168/jds.2020-19119
40. Tansman GF, Kindstedt PS, Hughes JM (2015) Crystal fingerprinting: elucidating the crystals of Cheddar, Parmigiano-Reggiano, Gouda, and soft washed-rind cheeses using powder x-ray diffractometry. Dairy Sci Technol 95:651–664. https://doi.org/10.1007/s13594-015-0225-6

Chapter 9

Blue Cheese

Esmaeel Mirzakhani, Masoud Aman Mohammadi, Fereshteh Ansari, and Hadi Pourjafar

Abstract

Blue cheese is a type of cheese that is grown with blue mold (*Penicillium roqueforti*). Blue cheese is a general name for types of cheese that are produced from cow, sheep, or goat milk with the help of *Penicillium* mold. This type of cheese is mostly used in salads or as a taste and has a special taste. According to experts, this cheese is served with pistachios. Depending on the type of cheese, bacteria can be added during rinding or after the cheese has hardened. This cheese is usually made in a temperature-controlled process, such as using a special machine. This product can be consumed alone or sprinkled on food or heated with food until it melts. One of the characteristics of blue cheese is the tendency of its aroma and taste to be sharp and salty. The smell of this cheese depends on the type of bacteria and the container and mold in which it grows. This chapter presents detailed explanations regarding the required materials and the different stages of production from milk preparation to packaging and storage of blue cheese.

Key words Blue cheese, *Penicillium roqueforti*, Culture, Cheese production, Cheese storage

1 Introduction

Blue cheese is a type of cheese that has blue, blue-green, or blue-gray spots or veins due to the presence and development of *Penicillium roqueforti* mold in it [1]. All over the world, in different countries, local types of blue cheese have been produced with different methods, each of which has its characteristics [2]. The types of blue cheese produced by *P. roqueforti* include Italian Gorgonzola, French Bleu and Roquefort, Spanish Cabrales, English Stilton, Picón Bejes-Tresviso and Valdeón, and many others from the United States and Denmark [3, 4]. Roquefort, Stilton, Gorgonzola, and Danablu are the most well-known varieties among them [2]. The fermentation of blue cheeses is usually performed by mesophilic bacteria such as *Streptococcus lactis* subsp. *diacetylactis*, *Streptococcus lactis*, and *Leuconostoc* spp. *P. roqueforti* then produces the compound responsible for the fragrance and the specific flavor of the blue cheese during the cheese ripening process,

Adriano Gomes da Cruz et al. (eds.), *Dairy Foods Processing*, Methods and Protocols in Food Science, https://doi.org/10.1007/978-1-0716-4144-6_9,

with lipolytic and proteolytic activity of its enzymatic system on the main components of milk, fat, and protein [1, 5]. Before blue cheeses and their production methods were recorded and written, they were probably produced for a long time, either accidentally or intentionally. In fact, initially, molds were not consciously inoculated in the blue cheese production process, but milk was accidentally contaminated with spores. Since the late eighteenth century, *P. roqueforti* spores were collected from naturally rotten bread and inoculated into the curd at the beginning of the cheese-making process. In the past decades, strains grown in laboratory conditions from monospore isolates have been used for inoculation to increase safety and also to make the cheese production process more basic and repeatable [6]. In this chapter, the general protocol for the production of blue cheese and some key points during the production process that lead to the formation of its special characteristics are described.

2 Materials

Milk: The foundation of blue cheese, commonly sourced from cows, sheep, or goats, each offering unique flavors and textures.

Starter cultures (fermentation agents): A carefully chosen blend of lactic acid bacteria, including:

Lactococcus lactis subsp. *lactis*

Lactococcus lactis subsp. *cremoris*

Lactococcus lactis subsp. *biovar diacetylactis*

Leuconostoc mesenteroides (optional)

Streptococcus thermophilus (optional in some recipes)

These bacteria are crucial for lowering the pH, developing initial flavors, and setting the stage for the cheese's texture.

P. roqueforti (mold spores): The essential mold for creating the characteristic veins and tangy flavor profile in blue cheese. Added after starter cultures; it's important for the distinct ripening stage.

Rennet (coagulant): An enzyme that curdles milk, transforming it into curds and whey. Chymosin is the primary enzyme in rennet that facilitates the creation of a firm curd base for cheese.

Salt (flavor and preservative): Cheese salt is applied after curds reach the target pH within the molds. It draws out residual whey, enhances flavor, extends shelf life, and prevents undesirable microbial growth.

Water (curd washing, optional): Used in some blue cheese variants to create a milder flavor profile by dilution of whey's lactose. It's mixed with the curds if a washing step is incorporated.

3 Methods

3.1 Milk Selection and Pasteurization

The blue cheese production process begins with the selection of the perfect milk. Cow's milk is usually used, but sheep and goat milk can also be used, each of which creates different flavor profiles and textures. The selection is crucial; each milk type brings its native microbial ecosystem, protein composition, fat composition, and minerals and vitamins which will influence the cheese's final taste.

Then it is important that the milk be pasteurized. This step serves two purposes: first, to ensure food safety by eliminating pathogenic microorganisms and, second, to create a blank microbial canvas. Pasteurization conditions (typically at 72 °C for 15 s (HTST)) must be carefully calibrated to ensure pathogen death while preserving beneficial enzymes that contribute to flavor and texture development during cheese maturation [7].

3.2 Inoculation with Starter Cultures

After pasteurization, the milk is cooled to an optimal temperature for the inoculation of starter cultures (*see* **Note 1**). Inoculation with starter cultures introduces beneficial bacteria into the pasteurized milk to begin the fermentation process. Starter cultures usually consist of specific strains of lactic acid bacteria selected for their acidification profiles and ability to produce a desired flavor. These cultures serve a dual purpose: they lower the pH of the milk to create conditions unfavorable for growth of harmful bacteria, and they begin the initial stages of flavor development. The inoculation process must be carried out under aseptic conditions to prevent the introduction of contaminants [8].

3.3 Starter Cultures for Blue Cheese

Typically, blue cheese production involves a combination of mesophilic and thermophilic lactic acid bacteria. The common strains used include:

Lactococcus lactis subsp. *lactis:* This strain grows well at moderate temperatures and is crucial for producing lactic acid, which aids in lowering pH and developing the initial flavors of the cheese.

Lactococcus lactis subsp. *cremoris:* Similar to *Lactococcus lactis* subsp. *lactis*, it also produces lactic acid and contributes to flavor but has different growth kinetics and flavor outcomes.

Lactococcus lactis subsp. *biovar diacetylactis:* Known for producing diacetyl, a compound that gives a buttery flavor to the cheese.

Leuconostoc mesenteroides: This strain is involved in the production of carbon dioxide gas, which can contribute to the formation of small holes or eyes in the cheese. It is also another producer of diacetyl.

Streptococcus thermophilus (used in some varieties): A thermophilic strain that can endure higher temperatures and contributes to acid production.

For the development of the characteristic blue veins, *P. roqueforti* is added (*see* **Note 2**). This mold is specifically responsible for the development of the characteristic blue or green veining, flavor profile, and texture that blue cheeses are known for.

P. roqueforti spores (or freeze-dried cultures) (*see* **Note 3**) are typically added to the milk after the starter culture bacteria have been inoculated and before the rennet is added to the milk and sets into curds, although the exact point of addition can vary.

In essence, the role of *P. roqueforti* is distinct and different from the starter cultures. Lactic acid bacteria are important in starter cultures in terms of acidity regulation and initial flavor components, while *P. roqueforti* is essential for the later stages of ripening, where it determines the unique characteristics of blue cheese [2, 9, 10].

3.4 Culture Preparation and Addition

For direct vat inoculation (DVI), freeze-dried cultures are used. It must be ensured that they are evenly distributed throughout the milk for consistent fermentation. This is often done by first sprinkling the DVI culture over the surface of the milk, and after allowing it to rehydrate for a few minutes, it is thoroughly stirred into the milk.

When using DVI, the culture is standardized by the supplier and clear dosing instructions are provided. This could be in the range of 20 to 50 DCU per 1000 liters of milk, but it always depends on the specific product being used.

For bulk starter cultures, the live cultures are typically grown in a sterile environment in a small amount of milk until active (reaching a particular pH level indicating sufficient growth (*see* **Note 4**)). This bulk starter is then added to the main vat of milk, ensuring even distribution.

For bulk starter cultures, about 1% to 2% of the total milk volume will be used as an inoculum. It means that 1 to 2 liters of culture should be used for every 100 liters of milk. This amount can be adjusted based on the activity level of the starter, the temperature of cultivation, and the desired rate of acid production. It is best to start with this guideline and gradually optimize it based on the growth of acidity and previous experiences (*see* **Note 5**).

Then, in addition to the bacteria, mold spores are added to the milk to create the characteristic blue mold throughout the cheese. The spores are typically not added to the starter culture. Instead, they are added separately to the main batch of milk or to the cheese curds (sprayed into the cheese during formation) at a later stage of cheese production. Lactic acid bacteria create an environment in the cheese that allows mold to grow. The mold is only needed for the development of the blue veins and distinct flavor associated with blue cheese. The amount of *P. roqueforti* is usually much lower compared to the starter culture bacteria and depends on the specific type of blue cheese to be made and the recipe. This amount might be in the range of a few milligrams per liter of milk (e.g., 0.02–0.06 grams per 100 liters). The milk should be stirred gently and thoroughly so that the mold spores are evenly distributed throughout the milk. This can be done directly or by first mixing the spores in a small amount of water to create a slurry, which is then added to the milk [2, 8].

3.5 Rest and Acidification

After proper distribution of cultures and mold spores in the milk, the mixture should be left undisturbed for a certain period of time to allow the bacteria to start fermenting lactose and producing lactic acid. At this phase, the pH of the milk decreases and the conditions for the formation of curds are provided after the addition of rennet (*see* **Note 6**). The acidification process and milk temperature must be carefully controlled to maintain the desired fermentation rate. Inoculation progression and effectiveness depend on several factors, including milk quality, culture activity, and environmental conditions. Therefore, control, cleanliness, and precision are paramount in ensuring successful acid development and curd formation. The practices and specific strains used can vary depending on regional practices and the desired characteristics of the final cheese product.

3.6 Coagulation

The next step is the addition of rennet, which contains the enzyme chymosin. This enzyme catalyzes the coagulation of milk and turns it from a liquid state to solid curds and liquid whey. The exact moment of adding the rennet, its concentration, and the duration of the coagulation phase are very important because they determine the texture and moisture content of the final product. Coagulation must be monitored using instrumentation to ensure that the reaction proceeds to its optimal extent. The cheesemaker should check for a clean break, that is, when the curd is firm enough to be split cleanly by inserting a finger or curd knife [11, 12].

3.7 Rennet Addition Timing

The optimal time to add rennet is typically when the milk has reached a pH of approximately 6.5 to 6.4, which occurs after the rest and acidification phase mentioned earlier (*see* **Note 7**). This usually happens within 30 min to an hour after the starter culture has been added, but the actual time can vary based on the milk's initial pH, temperature, and the activity level of the starter culture (*see* **Note 8**) [2, 13].

3.8 Rennet Concentration

For blue cheese, a typical range might be 18 to 25 ml of rennet per 100 liters of milk, but this amount can be adjusted according to the specific activity of the rennet and the characteristics of the milk. The supplier's recommendations for rennet concentration should also be taken into account and adjusted if necessary through trial and production monitoring (2, 11).

3.9 Coagulation Time

The time required for coagulation depends on the temperature, the characteristics of the milk, and the amount of rennet used. Typically, it takes around 30 min to an hour for a firm curd to form. The curd should be firm enough to hold its shape when cut, but not so firm that it becomes crumbly. The goal is to achieve a texture that allows blue veins to develop properly during aging (2, 11).

3.10 Cutting and Stirring Curds

Once coagulated, the curd is cut. This step increases the surface area of the curds, accelerating whey expulsion. The size the curd is cut into affects the moisture content of the cheese; smaller curds generally yield a firmer cheese, while larger ones result in a softer texture. For blue cheese production, the goal is often to produce a curd that retains more moisture, which helps in the development of the blue mold throughout the cheese. Before handling the curds, hands, tools, and work surface should be sterilized. Cut the curd mass into large, even cubes (usually about 1/2 inch to 1 inch). Start by making vertical cuts on the curd mass. If using a curd knife, make parallel cuts across the width of the vat, spaced evenly. Rotate the vat 90 degrees and repeat to make a checkerboard pattern. Then, make horizontal cuts by inserting the curd knife at an angle to ensure even-sized cubes. Alternatively, use a curd harp to cut horizontally. Allow the cut curd to rest for a few minutes to firm up, which will make the curd less likely to break during stirring [14].

The stirring process that follows is just as essential; this prevents the curds from matting together and also ensures a consistent release of whey. Begin slowly stirring the curd. This is done slowly and carefully to avoid breaking the cubes too much which would release too much whey and harden the curds. While stirring, the firmness of the curds should be checked. During the stirring process, the target temperature according to the recommended recipe (typically around 30 °C/86 °F) should be maintained. The stirring time can be from 20 min to an hour depending on the specific recipe and desired texture. During the stirring period, the whey is released and makes the curds firmer (*see* **Note 9**).

3.11 After Stirring

Some cheese-making processes require washing the curd. For this, a portion of the whey should be removed and replaced with water of the same temperature. This step can reduce the lactose content and influence the final acidity of the cheese. In traditional blue cheese production, curd washing is not usually done. This step is usually not part of the process in the production of blue cheeses such as

Roquefort, Gorgonzola, and Stilton, in order to preserve the natural sugars in the milk, as this helps to develop a strong blue veining and a distinct flavor profile during the aging process. However, depending on the cheese being produced, if a particular blue cheese recipe calls for washing the curds (which could be the case with less traditional, milder-flavored blue cheese recipes), here is how it can generally be done:

Portion of the whey should be removed. The amount removed could be anywhere from one-third to one-half of the whey. Water should be added at the same temperature as whey (around 30 °C) to avoid "shocking" the curds. Stir gently to mix the curd with fresh water (*see* **Note 10**). This process can help create a milder cheese because it dilutes the lactose in the whey, creating a less acidic environment (*see* **Note 11**).

3.12 Draining and Removal of Whey

When the stirring is complete, the curds should stop being agitated and allowed to settle to the bottom of the vat; this will take about 5–10 min. During this time, the curds will cluster together and continue to release whey. Then the whey should be carefully removed from the top without disturbing the curd. If working on a small scale, this could be done using a ladle or small cup. On a large scale, vats are typically equipped with valves that can be opened to drain the liquid. Enough whey should be left to cover the curds; this prevents them from drying out too quickly.

During draining, periodically check the pH of the curds using a pH meter to make sure they are reaching the acidity levels specific to your cheese recipe. For blue cheese, the target pH is usually around 4.5–4.6 at the end of the aging process but will be higher during the draining process. Typically, the curd should be transferred to molds when the pH is between 6.0 and 6.4. At this point, the curd has some acidity but is relatively sweet because some lactose is still present in it. As the curd continues to acidify after being transferred to the mold, excess lactose ferments to lactic acid, gradually lowering the pH.

Once sufficient whey has been drained and the curds are at the appropriate pH level, the curds should be transferred to cheese molds covered with cheesecloth (if necessary) to facilitate further draining and shaping. The molds should be filled evenly and care should be taken that the curds are not disturbed or broken during transfer. The curds should be placed in the molds with enough gentle pressure to ensure that they knit together during the pressing phase without too much whey escaping prematurely. Once the curds are in the molds, allow the whey expulsion and curds knitting at a suitable temperature, often room temperature or slightly warmer. This could take several hours; during this time, the acidity continues to increase as the lactic acid bacteria ferment the lactose in the curds. After this draining period, the molds will often be turned to ensure even drainage and a uniform cheese shape. As the curds continue to acidify and expel whey, they will gradually form into the final shape and texture of the cheese.

3.13 Salting and Molding

Salting and molding are important steps because the salt helps draw out the whey, develop flavor, inhibit unwanted bacteria, and create a rind.

After the curds from the drain have reached the target pH level (typically around 6.0 to 6.4 for blue cheese), they must be transferred to the molds. Gently ladle or spoon the curd into the molds (*see* **Note 12**). The molds should be filled evenly and care should be taken that the curds are not disturbed or broken during transfer. They are not yet salted and continue to acidify within the molds. The purpose of this step is to allow at the temperature specified, often around room temperature or slightly warmer, the curds to progressively expel whey as the pH drops. Flip the molds over several times to make sure the curd knits together properly and the whey drains evenly. Once the pH of the curds has dropped sufficiently, typically reaching a pH of around 4.9 to 5.4, salt (either by dry salting or brine) is then added. This is usually done by evenly applying cheese salt over the curd surface (*see* **Note 13**). The amount of salt used can be around 2–3% by weight of the curds. Accurate measurement of the curds and salt is important. To ensure even distribution, the salt should be mixed gently with the curds. This should be done delicately so as not to damage the structure of the curd. The curd will continue to expel whey due to the presence of salt.

After salting, pressing can occur if it is part of your blue cheese recipe. Typically in the process of producing blue cheese, the curd is often not pressed or is lightly pressed to retain moisture needed for mold growth. Additionally, the presence of salt and pressing lightly should help the cheese maintain a texture that is firm enough to handle and ripe without being too dense for *P. roqueforti* to grow. If the curds need to be pressed to achieve the desired texture, place the molds into a cheese press and apply a specified weight. The duration of pressing time can vary from a couple of hours to overnight (*see* **Note 14**) [15]. Follow your recipe. As the curds continue to expel whey, they gradually take on the final shape and texture of the cheese (*see* **Note 15**). To ensure that the desired characteristics are achieved, the pH of the cheese must be controlled. Once the cheese has reached the correct firmness and pH (which would continue to drop and should be close to 4.9–5.2 at this stage), remove it from the mold. At this point, the cheese should have a defined shape and be firm enough to handle [16].

3.14 Needling (Aeration)

The needling stage in blue cheese production is a carefully orchestrated dance that transforms the cheese from a solid mass into a canvas for the development of its iconic blue veins. After the initial molding, the cheese undergoes a controlled needling process where thin metal rods are delicately inserted, creating a network of channels that serve a dual purpose. Firstly, these channels allow oxygen to permeate the interior of the cheese. This introduction of air creates a hospitable environment for the growth of *Penicillium* mold cultures, specifically *Penicillium roqueforti*. The mold requires oxygen to thrive and, as it grows, imparts the characteristic blue veins that define blue cheese. The density and placement of these channels play a crucial role in determining the intricate patterns and distribution of the mold within the cheese. Secondly, the needling process influences the cheese's texture. By creating channels, cheesemakers influence the internal structure, promoting the development of air pockets. This aeration contributes to the cheese's crumbly and creamy texture, juxtaposing the firmness of the unaltered interior. Cheesemakers exhibit finesse during the needling process, ensuring that the channels are evenly distributed throughout the cheese. The depth of penetration and the spacing between channels are calculated to balance mold growth, preventing an overly concentrated or sparse distribution of blue veins. The timing of needling is equally crucial. Typically conducted after the initial molding and before the cheese enters the aging cellar, it sets the stage for the subsequent maturation phase. Cheesemakers must strike a delicate balance between aeration and maintaining the structural integrity of the cheese. As the blue cheese ages, the impact of the needling process becomes more apparent. The interplay of oxygen, mold growth, and the evolving chemistry within the cheese leads to a complex tapestry of flavors and textures. The needling stage is, therefore, a transformative moment, where the artisan's skill and the science of cheese-making converge to create a product that is not only visually striking but also a sensorial delight for aficionados of blue cheese [17, 18].

3.15 Aging and Maturation

The aging and maturation phase in blue cheese production is akin to the slow evolution of a masterpiece, where time becomes an essential ingredient. As the cheeses are carefully placed in dedicated aging rooms or caves, the complex orchestration of environmental factors initiates a symphony of biochemical reactions that contribute to the distinct character of each blue cheese variety. Precise control of temperature and humidity during aging is paramount. Blue cheeses thrive in environments where the temperature hovers between 48 and 55 °F (9 and 13 °C) and humidity is maintained at a level conducive to mold growth and enzymatic activity. This controlled atmosphere serves as the canvas upon which the cheesemaker crafts the flavor profile of the cheese. Within the cheese wheels, enzymes break down proteins (proteolysis) and fats (lipolysis), transforming the raw ingredients into a mosaic of flavors and textures. Proteolysis, in particular, results in the breakdown of casein proteins into peptides and amino acids, contributing to the savory and umami notes characteristic of well-aged blue cheeses.

Simultaneously, lipolysis releases free fatty acids, enhancing the creamy texture and adding depth to the overall taste. As the cheese ages, the carefully introduced *Penicillium* mold cultures work their magic. The mold spores grow and spread, creating the iconic blue veins that define blue cheese. The channels created during needling provide pathways for the mold to flourish, not only contributing to the cheese's appearance but also infusing it with the distinctive tangy flavor associated with blue cheeses. The duration of aging varies depending on the desired characteristics of the final product. Some blue cheeses undergo a relatively short aging period, resulting in a milder flavor profile and a more crumbly texture. In contrast, others may be aged for several months, developing a more robust flavor, creamier texture, and pronounced complexity that appeal to aficionados of aged cheeses. In artisanal cheese production, the expertise of the cheesemaker is evident during affinage, the art of refining and caring for the cheeses during the aging process. Regular turning and handling ensure that the cheese matures uniformly, preventing deformities and encouraging the consistent development of flavors. The artisanal touch extends to the careful selection of each cheese wheel, with experienced cheesemakers discerning subtle variations in aroma, texture, and appearance. Just as wines express the terroir of their vineyards, blue cheeses, too, reflect the characteristics of their origin. The specific conditions of the aging environment, the quality of milk, and the unique combination of mold strains contribute to the terroir of each cheese. This individuality is celebrated by cheese enthusiasts who appreciate the nuances that make each wheel of blue cheese a singular expression of craftsmanship. In essence, the aging and maturation of blue cheese is a transformative journey where time, environment, and the cheesemaker's skill converge to create a gastronomic masterpiece. The result is a cheese that captivates the palate with its complex interplay of flavors, a testament to the art and science inherent in the production of high-quality blue cheeses [2, 8, 19].

3.16 Turning and Handling

Turning and handling represent a pivotal phase in the meticulous craft of blue cheese production. As the cheese wheels undergo the aging process, the conscientious act of turning and handling plays a crucial role in guaranteeing uniform development and preventing potential deformities. During the initial stages of aging, when the cheese wheels are still malleable, they are turned regularly. This process ensures that each part of the cheese is exposed uniformly to the surrounding environment. The goal is to distribute moisture, encourage consistent microbial activity, and facilitate the breakdown of proteins and fats. Expert cheesemakers carefully orchestrate this turning, acknowledging that even the slightest imbalance can result in uneven maturation and affect the final product's flavor and texture. The handling aspect involves a delicate touch as cheesemakers interact with the developing cheeses. This not only includes the physical turning of the wheels but also gentle assessments to detect any irregularities. Careful hands navigate the aging room, feeling for changes in texture, moisture levels, and the emergence of the characteristic blue veins. The skilled artisan recognizes the subtle cues that guide decisions on the ideal moment to turn, ensuring that each cheese matures into a harmonious blend of flavors. Turning and handling are not only about promoting even maturation but also preventing deformities. Uneven pressure, air pockets, or other anomalies can arise during the aging process. The attentive cheesemaker identifies and addresses these issues promptly. The application of consistent pressure and handling techniques helps maintain the cheese's structural integrity, ensuring that it develops into a visually appealing and flavorful product. Turning and handling elevate the craft of blue cheese production to an art form. It requires an intuitive understanding of the cheese's evolving nature and a deep appreciation for the fusion of science and creativity. The artisanal touch comes to the forefront as cheesemakers, often guided by years of experience, use their senses to navigate the subtle changes in each wheel, making decisions that influence the cheese's destiny. The dedication to turning and handling is not merely a routine; it is a commitment to excellence. Uniformity in these processes ensures that each wheel of blue cheese meets the highest quality standards. By preventing irregularities and deformities, cheesemakers guarantee that consumers experience the intended texture, flavor complexity, and visual appeal with every bite. In the realm of blue cheese production, turning and handling stand as a testament to the craftsmanship and dedication involved. As each wheel matures, the careful orchestration of these processes by skilled hands transforms a simple combination of ingredients into a gastronomic masterpiece, embodying the rich tradition and nuanced artistry of blue cheese creation [14, 20–22].

3.17 Controlling Temperature and Humidity for Ripening

Ripening is a transformative phase where the cheese undergoes profound changes, developing its unique flavors and textures. Central to this metamorphosis is the meticulous control of temperature and humidity, creating an environment that nurtures the complex biochemical processes essential for crafting exceptional blue cheese. Temperature control is an artistry in the ripening stage, influencing enzymatic reactions and microbial activities crucial to flavor development. Typically, blue cheeses thrive in cooler conditions, with temperatures ranging from 45 °F to 55 °F (7 °C to 13 °C). This controlled chill allows the cheese to age gradually, encouraging the breakdown of proteins and fats while fostering the growth of desirable microorganisms. During the initial days of ripening, a slightly higher temperature may be maintained to initiate enzymatic activities. As the process unfolds, adjustments are made to create an environment conducive to the progression of specific flavor profiles. Cheesemakers skillfully navigate temperature nuances, ensuring the harmonious orchestration of biochemical reactions that define the cheese's character. Humidity levels are as critical as temperature, influencing moisture content, microbial activity, and texture. Blue cheeses thrive in environments with elevated humidity, typically ranging from 80% to 95%. This moisture-rich ambiance supports the growth of surface molds and encourages the development of the cheese's characteristic creamy texture and complex flavors. Maintaining optimal humidity levels prevents excessive drying, ensuring that the cheese retains its moisture and avoids undesirable textural characteristics. The interplay between temperature and humidity requires continuous monitoring, allowing cheesemakers to adjust environmental conditions as the cheese progresses through different stages of ripening. Temperature and humidity not only impact the cheese directly but also play a pivotal role in shaping the microbial communities within it. The controlled environment encourages the proliferation of specific bacteria and molds, contributing distinct aromas, flavors, and textures. This microbial tapestry evolves over time, guided by the deliberate management of ripening conditions. Continuous monitoring of temperature and humidity is imperative during the ripening process. Advanced technologies, such as climate-controlled aging rooms and sensors, aid cheesemakers in maintaining precise conditions. Regular checks and adjustments are made to accommodate the cheese's evolving needs, ensuring that each wheel undergoes an optimal journey toward flavor maturity. Controlling temperature and humidity for ripening is a delicate dance that marries science with artisanal skill. Cheesemakers, often relying on their intuition and experience, navigate the subtle shifts required for flavor development. This harmonious balance ensures that each wheel of blue cheese emerges with a distinct character, reflecting the expertise and dedication invested in its creation. The mastery of temperature and humidity during the ripening phase elevates blue cheese production to an art form. The controlled environment becomes a canvas, allowing the cheese to unfold its unique flavors and textures. This meticulous stage sets the foundation for the sensory delight that awaits those who savor the intricate complexities of well-ripened blue cheese [22–24].

3.18 Blue Mold Inoculation and Managing Growth

The infusion of blue mold into the heart of the cheese is an artful process, where precision and expertise converge to create the signature blue veins that define this celebrated variety. The meticulous management of mold growth is an essential aspect of crafting blue cheese, requiring a delicate balance between nature's spontaneity and human intervention. At this stage, the carefully selected *Penicillium* mold cultures, notably *Penicillium roqueforti*, are introduced into the cheese. These cultures, derived from carefully cultivated strains, bring about the distinctive blue veining. The cheese acts as a canvas for the mold to weave its intricate patterns, and the cheesemaker's skill lies in guiding this process with finesse. As the mold cultures are introduced, aeration becomes a crucial element in ensuring an even distribution of the mold throughout the cheese. The needling process, involving the insertion of thin rods, creates channels that allow air to permeate. This controlled aeration provides the necessary oxygen for the mold to thrive, facilitating the growth of the characteristic blue veins. The spacing and depth of these channels influence not only the visual appeal but also the flavor development within the cheese. The subsequent aging period is where the magic truly unfolds. Cheesemakers must navigate the delicate balance of environmental conditions, including temperature and humidity, to foster optimal mold growth. The cheeses are placed in carefully controlled ripening rooms, where the evolving atmosphere encourages the mold to flourish. The artisanal touch lies in the ability to manage these conditions, coaxing the mold to spread consistently and uniformly. Throughout the aging process, constant monitoring is imperative. Cheesemakers assess the growth of the mold, adjusting conditions as needed to ensure a harmonious development. This might involve tweaking temperature and humidity levels or fine-tuning aeration to address any irregularities. The experienced eye of a cheesemaker becomes attuned to the nuances of mold growth, allowing for timely interventions to maintain the cheese's quality. The growth of *Penicillium* mold is not merely aesthetic; it plays a pivotal role in shaping the flavor and texture of the cheese. The mold secretes enzymes that break down proteins and fats, contributing to the complex taste profile of blue cheese. The interplay between mold growth and biochemical transformations during aging results in the creation of savory, tangy, and sometimes spicy notes that characterize blue cheese. Achieving an even distribution of mold requires a skilled hand and a deep understanding of the cheese's development. The cheesemaker's artistry lies in orchestrating this symphony of mold growth, ensuring that each wheel of blue cheese emerges with a harmonious balance of flavors and a visually stunning array of blue veins. This attention to detail distinguishes artisanal blue cheeses and elevates them to a level of culinary art. The blue mold inoculation and managing growth stage is a pivotal chapter in the saga of blue cheese production. It is where the cheesemaker's craftsmanship meets the unpredictable beauty of microbial life, resulting in a masterpiece that captivates the senses and delights palates worldwide. The intentional infusion and meticulous management of blue mold contribute not just to the appearance but to the very essence of what makes blue cheese an enduring symbol of culinary excellence [8, 22, 25, 26].

3.19 Flavor Development

Flavor development is the culmination of an intricate and dynamic interplay of biochemical processes during the aging and maturation of blue cheese. This phase is akin to a symphony of reactions orchestrated by the combined efforts of enzymes, microbes, and environmental factors, ultimately shaping the nuanced and complex taste profile that defines each variety of blue cheese. Proteolysis and lipolysis, the breakdown of proteins and fats, are central to flavor development. Proteolysis involves the enzymatic cleavage of proteins into peptides and amino acids. This process, facilitated by proteolytic enzymes from both the milk and microbial sources, generates savory and umami notes, contributing to the cheese's depth of flavor. Simultaneously, lipolysis, the breakdown of fats, releases free fatty acids and esters. These compounds contribute to the development of buttery, nutty, and creamy flavors within the cheese. The specific enzymes involved, such as lipases produced by microbes, play a crucial role in shaping the character of the cheese. The diverse microbial community within the cheese, including bacteria and molds, further contributes to flavor development. During aging, these microorganisms metabolize nutrients, producing a myriad of volatile compounds that impart distinctive aromas and flavors to the cheese. The presence of *Penicillium* mold cultures, introduced intentionally or occurring naturally, adds earthy and tangy notes, contributing to the characteristic taste of blue cheese. The growth of *Penicillium* mold within the channels created during the needling process is not only aesthetic but also pivotal to flavor development. The mold's metabolic activity results in the production of compounds such as ketones and lactones, which contribute to the cheese's complexity. The blue veins act as conduits for flavor diffusion, creating pockets of intense flavor that contrast with the creamier surrounding areas. Achieving the perfect balance of flavors in blue cheese requires a delicate equilibrium of factors. Cheesemakers, through their expertise, carefully monitor and adjust parameters such as temperature, humidity, and microbial activity. The duration of aging is a critical factor, as an extended maturation allows for the gradual development and integration of flavors, producing a well-rounded and harmonious taste profile. The environmental conditions of the cheese aging facility, often referred to as the terroir of the cheese, also play a role in flavor development. The specific geography, climate, and microbial ecology of the region contribute unique characteristics to the cheese, making each batch a product of its terroir. This concept is particularly relevant in artisanal and traditional cheese-making practices. In essence, the flavor development stage is a journey that transforms the raw ingredients into a sensory masterpiece. It requires a deep understanding of the biochemical processes at play, a keen appreciation for the influence of microbial communities, and the artistry to guide the cheese through a nuanced maturation process. The result is a blue cheese that captivates the palate with its intricate flavor profile, embodying the craftsmanship and expertise of those who nurture it to perfection [8, 11, 26–30].

3.20 Sensory Evaluation

Sensory evaluation stands as the apex of the artisanal journey in blue cheese production, where the expertise of cheesemakers is finely honed to discern every nuance of the cheese's character. This phase goes beyond the technicalities of the production process, delving deep into the realms of taste, aroma, texture, and visual appeal. The first aspect of sensory evaluation involves a detailed visual examination of the cheese. Cheesemakers scrutinize the exterior for characteristics such as color, texture, and the distribution of blue veins. The presence of an even, well-distributed mold network is a testament to successful aeration and mold growth during the needling process. Any deviations from the expected appearance are carefully noted, signaling potential variations in flavor and texture. The aroma of blue cheese is a symphony of complex compounds that evolve during the aging process. Cheesemakers delicately inhale the bouquet, identifying subtle notes that hint at the cheese's origin, age, and unique flavor profile. The interplay of earthy, nutty, and sometimes pungent aromas contributes to the overall olfactory experience. Tasting is an art unto itself in the world of blue cheese production. Cheesemakers sample different sections of the cheese, from the creamy center to the outer blue-veined regions. The taste experience evolves from the initial creamy, mild flavors to the more intense, complex notes near the blue veins. The balance of sweetness, saltiness, and umami is carefully evaluated, ensuring a harmonious blend that captivates the palate. The texture of blue cheese is a crucial element that distinguishes exceptional products. Cheesemakers assess the firmness, creaminess, and mouthfeel, considering factors such as moisture content and the distribution of blue veins. The ideal texture strikes a delicate balance between crumbly and creamy, with the blue veins providing a satisfying contrast. Beyond the basic taste components, the evaluation extends to the overall flavor complexity. Cheesemakers discern the intricacies brought about by the breakdown of proteins and fats during aging. Notes of nuttiness, earthiness, and a subtle tanginess from the blue mold contribute to a multidimensional flavor profile that distinguishes each batch. Consistency across batches is a hallmark of a skilled cheesemaker. Sensory evaluation includes comparing the current batch with previous ones to ensure a recognizable and consistent flavor profile. Any deviations prompt a thorough review of the production process, materials, and environmental conditions to maintain the brand's integrity. Sensory evaluation is not merely a checklist of attributes but an art form that requires years of experience and a deep connection with the craft. Expert cheesemakers rely on intuition and a refined palate, making split-second decisions that shape the final product. This artisanal expertise is the essence of creating blue cheeses that transcend the ordinary and become culinary treasures. In the realm of sensory evaluation, cheesemakers elevate their craft to an art, ensuring that each wheel of blue cheese is a testament to their skill, passion, and dedication. The culmination of this sensory journey results in cheeses that not only meet stringent quality standards but also carry the distinctive mark of artisanal excellence [8, 11, 31].

3.21 Packaging and Storage

The final stage of blue cheese production, packaging and storage, is a critical juncture where the culmination of meticulous craftsmanship is preserved to deliver a premium product to consumers. This stage not only protects the cheese from external influences but also plays a pivotal role in maintaining its texture, flavor, and overall quality. Selecting appropriate packaging materials is an art in itself. Cheesemakers choose materials that provide a protective barrier against contaminants while allowing the cheese to breathe. Traditional options include specially designed cheese paper or wraps that balance moisture retention and controlled airflow. The packaging must shield the cheese from external odors and moisture, preserving its unique qualities. For some artisanal blue cheeses, controlled atmosphere packaging (CAP) is employed. This advanced technique involves modifying the mix of gases within the packaging to extend shelf life and preserve the cheese's quality. Adjusting the levels of oxygen and carbon dioxide helps slow down the aging process, ensuring that the cheese reaches consumers with its intended flavor profile intact. Maintaining precise temperature and humidity conditions during storage is paramount. Dedicated cheese cellars or aging rooms are often equipped with climate control systems to ensure the optimal environment for the cheese. Temperature influences the rate of biochemical reactions, while humidity prevents the cheese from drying out. These controlled conditions are crucial for preserving the cheese's texture and preventing undesired alterations. In certain varieties of blue cheese, the post-production ripening process continues even after packaging. This additional maturation period allows flavors to further evolve and mature, resulting in a product with enhanced complexity. Careful monitoring during this phase ensures that the cheese maintains its peak condition until it reaches consumers. Before the final seal of approval, cheesemakers conduct rigorous quality assurance checks. These may include visual inspections, sensory evaluations, and even microbial analyses to ensure that the cheese meets the highest standards. Any deviations from the expected quality are identified and addressed, underscoring the commitment to delivering a consistently exceptional product. Once packaged and quality-checked, blue cheese embarks on its journey to consumers. During shipping and distribution, maintaining appropriate conditions is crucial to prevent temperature fluctuations and physical damage. Proper handling ensures that the cheese arrives at its destination in the best possible condition, ready to be savored by enthusiasts and culinary aficionados alike. The packaging of blue cheese goes beyond functional considerations; it is an integral part of its identity. Artisanal producers often invest thought into the visual presentation, choosing packaging that reflects the essence of their brand. Labels, descriptions, and even the aesthetic design of the packaging contribute to the overall experience, inviting consumers to indulge in the story behind the cheese. The packaging and storage of blue cheese represent the final act in a carefully orchestrated symphony of craftsmanship. From the controlled environments of aging cellars to the hands of consumers, each element of this stage is crafted with precision and care to ensure that the essence of the cheese, painstakingly developed throughout the production process, remains intact and delightful to the last bite. The packaging becomes the vessel that carries not just the cheese but the story of its creation and the passion of its makers [13, 32–35].

4 Notes

1. The ideal temperature for inoculation of starter cultures in cheese-making, including blue cheese, usually ranges from 20 °C to 32 °C (68 °F to 90 °F). For blue cheese specifically, inoculation temperatures are often on the warmer side of this spectrum, around 30 °C (86 °F). This temperature range provides an ideal environment for the starter cultures to begin acidifying the milk effectively, leading to the ideal environment for proper cheese development. It is important to note that the exact temperature can vary based on the specific recipe, the blend of starter cultures, and the specific practices of the cheesemaker. Careful control of temperature and other conditions during this stage is essential because it affects the rate of acid development, flavor compounds, curd formation, and ultimately the quality and characteristics of the finished cheese.
2. Incorporation of *P. roqueforti*: The *P. roqueforti* mold can be added at the same time as the lactic acid bacteria or slightly afterward. It is usually added as a powder and sprinkled in the milk and mixed, or in some cases it may be added directly into the curds later after draining.
3. The mold used in the production of blue cheese is usually in the form of mold spores. Spores are dormant and will germinate in cheese under the right conditions. Freeze-dried cultures are often activated more quickly because the freeze-drying process can stimulate spore germination after rehydration.
4. The proper pH level is usually considered to be around 4.5 to 4.6.

5. As a cheesemaker, it is very important to record the amounts and results with each batch made, as it is a valuable reference for fine-tuning the quantities for future cheese-making, ensuring the consistency and quality of the final product. In cases of uncertainty, consulting a dairy scientist can provide more personalized and precise instructions.
6. During cheese ripening, the pH reaches about 4.5 to 4.6, which is optimal for the growth of *P. roqueforti*. However, this low pH is achieved gradually, not in the early stages of cheese-making.
7. For example, if you add rennet at a pH of 4.5 to 4.6, the milk will become overly acidic, and the resulting curds will be too firm and dry, resulting in cheese with an undesirable texture.
8. The exact timing and pH levels can vary depending on many factors such as the cheesemaker's recipe, ambient conditions, and milk characteristics. Therefore, cheesemakers often have to adjust their process based on experience and the behavior of the particular batch of milk they are working with.
9. Periodically the curds should be tested to ensure they have reached the desired firmness. When the curds are firm enough to hold their shape, but still have some softness, the stirring should be stopped. They should not be rubbery or overly dry.
10. The curd washing process is usually brief, involving only the mixing of the curds with water and a short rest period afterward.
11. Remember, for most types of blue cheese, the steps after stirring and firming the curds usually involve allowing the curds to settle and then draining the curds, and then transferring the curds to molds for initial shaping, without putting them under the whey or washing them. When making blue cheese, maintaining the traditional cheese-making method is generally considered the best way to achieve the correct flavor and texture, unless experimenting with a new recipe or looking for a milder flavor profile.
12. Make sure the cheese molds are clean and sterilized to inhibit any unwanted microbial contamination. If using cheesecloth, line each mold with a piece large enough to fit over both sides.
13. The important thing is to achieve the desired final texture and moisture content for blue cheese, meaning that the curd is slightly moist, which helps the development of the blue vein of mold. For blue cheese, the curd must be kept moister than for harder cheeses.
14. Salting after molding helps control future microbial activity, which theoretically can reduce acidification, but the cheese

eventually reaches a target pH of 4.5 to 4.6 during the aging process.

15. Please note that different producers of blue cheese may have variations in these steps based on experience, tradition, and desired characteristics of the final product. Light pressing primarily helps the curd knit together and retain the shape of the mold, rather than to expel excess whey. Remember that pressing is minimal in the production of blue cheese because we want to maintain a certain level of moisture in the cheese so that the blue mold can grow properly during aging.

References

1. Mayo B, Alonso L, Alegría Á (2013) Chapter 19: Blue cheese. In: Handbook of cheese in health: production, nutrition and medical sciences. Wageningen Academic, pp 277–288
2. Cantor MD, van den Tempel T, Hansen TK, Ardö Y (2017) Blue cheese. In: Cheese. Elsevier, pp 929–954
3. Coton E, Coton M, Hymery N, Mounier J, Jany J-L (2020) Penicillium roqueforti: an overview of its genetics, physiology, metabolism and biotechnological applications. Fungal biology reviews. 34(2):59–73
4. Albillos SM, García-Estrada C, Martín J-F (2011) Spanish blue cheeses: functional metabolites. In: Cheese: types, nutrition and consumption, pp 89–105
5. Chávez R, Vaca I, García-Estrada C (2023) Secondary metabolites produced by the blue-cheese ripening mold *Penicillium roqueforti*; biosynthesis and regulation mechanisms. Journal of Fungi. 9(4):459
6. Ropars J, López-Villavicencio M, Snirc A, Lacoste S, Giraud T (2017) Blue cheese-making has shaped the population genetic structure of the mould *Penicillium roqueforti*. PLoS One 12(3):e0171387
7. Wilson GS. The pasteurization of milk. 1942.
8. López-Díaz TM, Alegría Á, Rodríguez-Calleja JM, Combarros-Fuertes P, Fresno JM, Santos JA et al (2023) Blue cheeses: microbiology and its role in the sensory characteristics. Dairy 4(3):410–422
9. Larsen MD, Kristiansen KR, Hansen TK (1998) Characterization of the proteolytic activity of starter cultures of *Penicillium roqueforti* for production of blue veined cheeses. Int J Food Microbiol 43(3):215–221
10. Addis E, Fleet G, Cox J, Kolak D, Leung T (2001) The growth, properties and interactions of yeasts and bacteria associated with the maturation of Camembert and blue-veined cheeses. Int J Food Microbiol 69(1–2):25–36
11. Diezhandino I, Fernández D, Sacristán N, Combarros-Fuertes P, Prieto B, Fresno J (2016) Rheological, textural, colour and sensory characteristics of a Spanish blue cheese (Valdeón cheese). LWT-Food Sci Technol 65: 1118–1125
12. Li L, Chen H, Lü X, Gong J, Xiao G (2022) Effects of papain concentration, coagulation temperature, and coagulation time on the properties of model soft cheese during ripening. LWT. 161:113404
13. Joyner HS, Francis D, Luzzi B, Johnson JR (2018) The effect of storage temperature on blue cheese mechanical properties. J Texture Stud 49(3):309–319
14. Morales MBL, Ardö Y, Berthier F, Karatzas KAG, Bintsis T (2017) Blue-Veined cheeses. In: Global cheesemaking technology: cheese quality and characteristics, pp 415–435
15. Ferroukhi I, Bord C, Lavigne R, Chassard C, Mardon J (2023) Exploring alternative salting methods to reduce sodium content in blue-veined cheeses. Int Dairy J 138:105555
16. Fox PF, Guinee TP, Cogan TM, McSweeney PL, Fox PF, Guinee TP et al (2017) Salting of cheese curd. In: Fundamentals of cheese science, pp 251–277
17. Bates M, Clark S (2023) Mold-ripened cheeses. In: The sensory evaluation of dairy products. Springer, pp 545–570
18. Finoli C, Vecchio A, Galli A, Dragoni I (2001) Roquefortine C occurrence in blue cheese. J Food Prot 64(2):246–251
19. Fox PF, Guinee TP, Cogan TM, McSweeney PL, Fox PF, Guinee TP et al (2017) Microbiology of cheese ripening. In: Fundamentals of cheese science, pp 333–390
20. Yunita D, Dodd CE (2018) Microbial community dynamics of a blue-veined raw milk cheese

from the United Kingdom. J Dairy Sci 101(6): 4923–4935

21. Maixner F, Sarhan MS, Huang KD, Tett A, Schoenafinger A, Zingale S et al (2021) Hallstatt miners consumed blue cheese and beer during the Iron Age and retained a non-Westernized gut microbiome until the Baroque period. Curr Biol 31(23): 5149–5162. e6
22. Fox PF, Guinee TP, Cogan TM, McSweeney PL (2017) Fundamentals of cheese science. Springer
23. Bord C, Guerinon D, Lebecque A (2017) Heated or raw Blue cheeses: what are the drivers influencing consumer preferences? Int J Food Sci Technol 52(9):1959–1970
24. Bord C, Guerinon D, Lebecque A (2016) Impact of heating on sensory properties of French Protected Designation of Origin (PDO) blue cheeses. Relationships with physicochemical parameters. Food Sci Technol Int 22(5):377–388
25. Judacewski P, Los PR, Benvenutti L, Alberti A, Simões DR, Nogueira A (2016) Quality assessment of white mold-ripened cheeses manufactured with different lactic cultures. J Sci Food Agric 96(11):3831–3837
26. Dumas E, Feurtey A, Rodríguez de la Vega RC, Le Prieur S, Snirc A, Coton M et al (2020) Independent domestication events in the blue-cheese fungus Penicillium roqueforti. Mol Ecol 29(14):2639–2660
27. Drake MA, Delahunty CM (2017) Sensory character of cheese and its evaluation. In: Cheese. Elsevier, pp 517–545
28. Bord C, Guerinon D, Lebecque A (2019) Application of two sensory methods to investigate the impact of heating on the flavor perception of a French blue cheese. J Sens Stud 34(5): e12509
29. Kilcawley K, O'Sullivan M (2017) Cheese flavour development and sensory characteristics. In: Global cheesemaking technology: cheese quality and characteristics, pp 45–70
30. Martín J, Coton M (2017) Blue cheese: microbiota and fungal metabolites. In: Fermented foods in health and disease prevention. Elsevier, pp 275–303
31. Cleere MM, Novodvorska M, Geib E, Whittaker J, Dalton H, Salih N et al (2024) New colours for old in the blue-cheese fungus *Penicillium roqueforti*. npj Sci Food 8(1):3
32. Duval P, Chatelard-Chauvin C, Gayard C, Rifa E, Bouchard P, Hulin S et al (2018) Changes in biochemical and sensory parameters in industrial blue-veined cheeses in different packaging. Int Dairy J 77:89–99
33. Khoshgozaran S, Azizi MH, Bagheripoor-Fallah N (2012) Evaluating the effect of modified atmosphere packaging on cheese characteristics: a review. Dairy Sci Technol 92:1–24
34. Marcuzzo E, Peressini D, Sensidoni A (2013) Shelf life of short ripened soft cheese stored under various packaging conditions. J Food Process Preserv 37(6):1094–1102
35. Duval P, Chatelard-Chauvin C, Gayard C, Rifa E, Bouchard P, Hulin S et al (2016) Microbial dynamics in industrial blue veined cheeses in different packaging. Int Dairy J 56: 198–207

Chapter 10

Ricotta Cheese

Marzia Albenzio and Antonella Santillo

Abstract

Ricotta cheese has been classified as whey cheese within the class of dairy products although it was originated by the recovery of the whey from cheese-making as a by-product of the dairy industry. Thus, the production of ricotta cheese has been considered to be one of the economical ways to use whey and create an additional source of income for all cheese-makers. The production protocol presents different steps that need to be rigorously observed to obtain ricotta cheese with high quality characteristics in terms of hygienic features, yield, structure, flavor, color, and shelf life (prolonged). This chapter covers *method and protocols* to support the experimenters, highlighting critical points in the ricotta cheese production process and allowing the prediction and resolution of difficulties that may occur during processing.

Key words Whey cheese, Heat coagulation, Acid coagulation, Whey protein

1 Introduction

The market share of recotta cheese is constantly increasing both in Europe and worldwide due to the consumers' request of natural and healthy foods with high nutritional value; from the producer's point of view, the interest in this type of cheese lies in the possibility of using whey from the processing of different types of cheese as raw material. Ricotta cheese is defined as unripened acid-heat coagulated soft whey cheese and belongs to the group of the whey dairy products. The peculiar processing procedure originated from Italy where ricotta cheese is traditionally made by the procedure of "cooking twice" (from the Latin *re cocta*) the whey from sheep or goat milk cheese, or a mixture of both, and also from cow and buffalo whey. For Italian legislation, ricotta is not a cheese, as it is not obtained by coagulation of casein from milk or cream but consists of separating, by coagulation, the proteins that remain in the whey of cheese-making. Ricotta cheese produced by traditional technology is therefore obtained exclusively by the action of heat from an acid substrate, consisting of whey, to recover the valuable proteins that are still present in the whey. The protein ratio in whey

Adriano Gomes da Cruz et al. (eds.), *Dairy Foods Processing*, Methods and Protocols in Food Science,
https://doi.org/10.1007/978-1-0716-4144-6_10,

protein/casein is undoubtedly higher than 1, so this cheese falls under the international definition of ricotta as whey cheese in agreement with the *Codex Alimentarius* [1].

In Italy, more than 25 varieties of traditional ricotta cheese have been classified, differing in raw material origin, structure, texture and taste, aroma, intended use, and production and preservation technology [2]. At present, two ricotta cheeses recognized as Protected Designation of Origin (PDO) are produced in Italy: Ricotta Romana cheese, produced only with sheep whey, and Ricotta di Bufala Campana produced exclusively with buffalo whey or with whey added with buffalo milk or cream.

Ricotta cheese is typically prepared by heating the whey and acidifying the hot liquid to coagulate whey proteins. The agglomerated curd mass that floats to the surface is collected and transferred into perforated baskets to drain the exhausted whey, called "scotta whey." Whey proteins are constituted by protein monomers or polymers characterized by an isoelectric pH value of about 6 that are able to precipitate by flocculation. Separation is allowed by heating, or acidification, or salt adjunct or a combination of these interventions. Ricotta cheese is considered a healthy food and is recognized as a source of functional and bioactive peptides [3].

Whey protein composition of cow, sheep, goat, and buffalo milk is presented in Table 1. The whey protein profile is comparable among the different milk sources, although the percentage distribution accounts for differences in the amount of each whey protein fraction.

The present chapter aims to present step-by-step approach protocols and methodologies used in ricotta cheese production. Furthermore, the ingredients, equipment, and flowcharts used to obtain ricotta cheese at different scale are illustrated to reproduce the production protocol at different scale.

Table 1
Whey protein composition of cow, sheep, goat, and buffalo (%)

Whey proteins	Cow	Sheep	Goat	Buffalo
β-Lactoglobulin	53.3–66.0	73.1	46.6	40.4
α-Lactalbumin	15.0–20.0	17.9	18.3	24.5
Serum albumin (SA)	6.0–7.0	2.7	5	3.5
Immunoglobulins	11.0–13.3	–	–	–
Lactoferrin	0.7–3.3	1.6	2	–
Lactoperoxidase	0.5–1.0	–	–	–
Enzymes	0.5	–	–	–

Ricotta cheese process can be distinguished into artisanal and industrial on the basis of the production technology. The fresh ricotta cheese has a compact mass, a granular and creamy texture, with a slightly milk flavor, and a uniform color that varies from white to yellowish-white depending on the milk source. The shelf life of ricotta cheese varies from 1–2 days in artisanal plant to 7 days in small–medium industrial plants to 20–40 days in medium–large industrial plants according to the processing and packaging technology.

2 Protocol List of Materials and Reagents

There are several types of whey, depending mainly on the processing sequence resulting in casein removal from fluid milk. Most of the whey originates from milk casein coagulation by rennet or coagulating enzymes producing a type of whey characterized by pH 6.5, referred as sweet whey. Another type of whey originates from fermentation processes or by addition of organic or mineral acids, referred as acid whey.

The core phase of ricotta cheese production is based on the principle of hot coagulation of lacto-albumins in acid environment. The acid conditions are able to destabilize the whey protein, and the high temperature drains the fat material into the coagulum; the fat material is naturally presented in the whey or derives from the milk or cream adjunct.

List of Materials for Ricotta Cheese Production Protocol

- Whey from different milk cheese sources obtained from enzyme or acid coagulated milk.
- Addition/non-addition of whole milk (up to 25% of the vat volume).
- Addition/non-addition of milk cream.
- Salt.

Standard Formulation for Ricotta Cheese

Whey 100%.

Whole milk up to 25%.

Milk cream 5–15%.

Salt 0.5–1.5%.

Coagulant

- Monohydrate citric acid, lactic acid, and acetic acid (1.5/2.5% diluted solution).
- Natural milk or whey graft obtained from previous cheese production (characterized by high acidity >25 °SH/50).

Other Ingredient

Magnesium sulfate or other saline mixture.

Equipment and Plants for Ricotta Cheese Production

- Large jacketed vat.
- Perforated mold.
- Cold rooms.

3 Step-by-Step Procedures: How to Perform the Experiments

1. Timeline flowchart of ricotta cheese production.

Time 1	Whey and/or cream mixed in the vat at pH > 6.2The optimum mixture acidity is 2.0–2.4 °SH/50; if higher than 2.4 °SH, the acidity correction is performed with alkaline solution (e.g., 44 g NaOH to neutralize 100 g of lactic acid)
Time 15 min	Heating of the mass with direct steam injection at low pressure in double jacketed vat, in the lower part, up to a temperature of 82–87 °C—Without stirring (*see* **Note 1**)
Time 1 hour	Adjunct of milk heated at 65 °C and/or salt before reaching the final vat temperature
Time 1 h and 5 min	Adjunct of diluted coagulant to obtain a final pH of the mixture of 5.5–5.9 (*see* **Notes 2–6**)
Time 1 h and 10 min	Gradual reduction of direct steam injection and pause for coagulum firming
Time 1 h and 20 min	Coagulum collection with a perforated mold or using perforated basket to obtain a prompt elimination of scotta whey. The mold is disposed on a table for draining
Time 1 h and 30 min	Cooling of ricotta cheese in cold room (4–10 °C) (*see* **Notes 7–9**) for 6–12 h and packaging

Chemical composition of commercial ricotta cheese made using whey from different species with and without adjunct of milk and cream is shown in Table 2.

Table 2
Chemical composition of commercial ricotta cheese made using whey from different species with and without adjunct of milk and cream

Whey source	Fat	Protein	Ash
Cow	3.95–15.72	6.25–11.65	0.67–2.07
Sheep	13.3–23.12	7.13–8.51	0.39–0.67
Goat	2.68–20.80	6.65–6.69	0.96–1.04
Buffalo	6.67–24.5	5.63–11.67	1.98–2.20

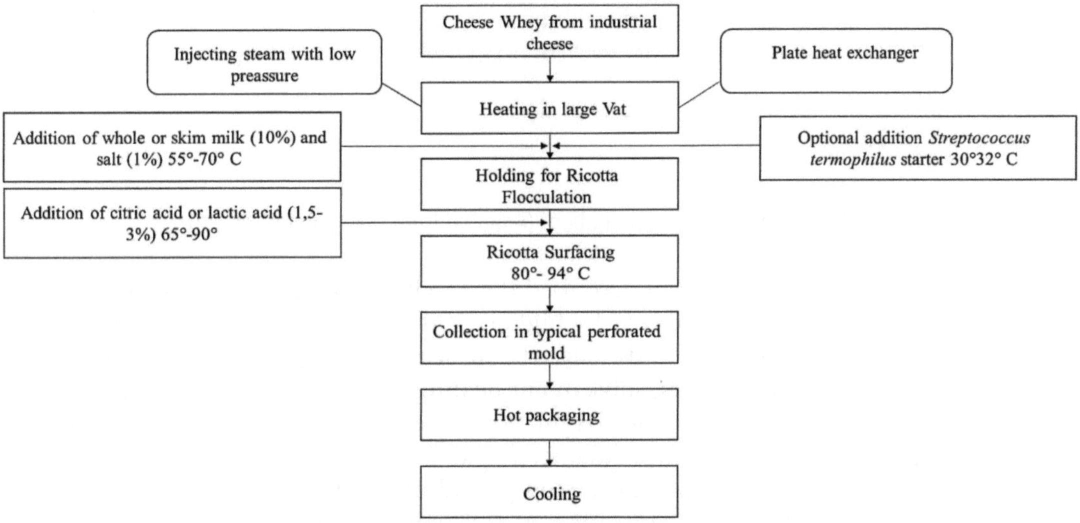

Fig. 1 Protocol of artisanal ricotta cheese production. (Adapted from Mangione et al. [4])

Protocol of artisanal and industrial ricotta cheese production is presented in Figs. 1 and 2, respectively.

Protocol of industrial ricotta cheese production with the use of ultrafiltration is demonstrated in Fig. 3. This production protocol differs from the artisanal and industrial process for the application of ultrafiltration technology to concentrate whey mixture prior to the acidification and coagulation phase. The main difference with the classic process consists in applying ultrafiltration to the whey and milk mixture to obtain the separation of the permeate and formation of retentate characterized by a level of solid matter of about 30%. Therefore, the separation of the liquid part is performed before the key processes of heating and coagulation.

Figure 4 presented the continuous production equipment for ricotta cheese production that guarantees high quality level and yield. In the equipment, the serum vat, ingredients are mixed and then heated continuously with steam coil. A belt conveyor, made of fine mesh perforated metal, conveyor belt picks up the clot from the coagulation tank and to drain, transporting it to the filling machine.

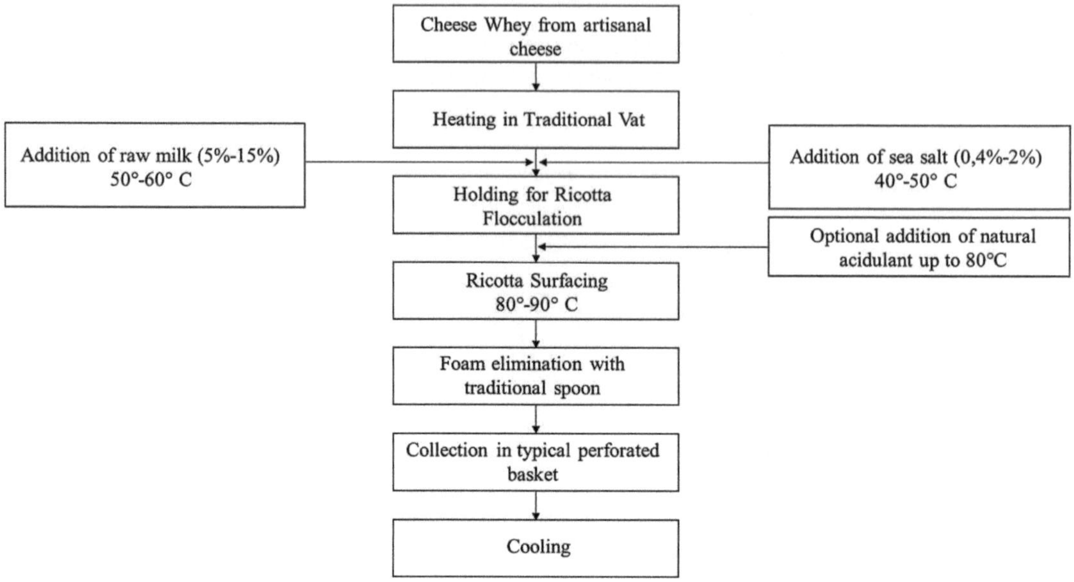

Fig. 2 Protocol of industrial ricotta cheese production. (Adapted from Mangione et al. [4])

Figure 5 described the Modler continuous system. The process is characterized by higher yield with a recovery of whey protein of about 94%. Briefly, the process provides for the initial mix of ingredients, subsequent heating with steam injection, and addition of acid for the coagulation. The acid mixture is forced into a spiral pipe (4 cm in diameter and 30 meters in length) with a flow of 4.5 L/min and a stop period of about 7–8 min. During the stop time, the separation of solid-phase flocculated protein and scotta whey takes place. At the end of the separation, the ricotta cheese is recovered by a fine mesh belt moved by a motorized roll. Warm air is directed on the product to force the ricotta drying.

4 Quality Analysis Related to Milk, Whey, and Ricotta Cheese

4.1 Milk and Whey Analyses

1. pH by using calibrated pH meter with liquid probe; the measurement is taken when the liquid is at room temperature (22 °C).
2. Milk fat, protein, and lactose content of milk by using Fourier transform infrared (FTIR) analysis. The mid-infrared region of the spectrum allows determination of basal parameters as fat 5.7–3.5 μm, protein 6.5 μm, and lactose 9.6 μm.
3. Titrable milk acidity is determined by titrating a known volume of milk with standard alkali to the point of an indicator, usually phenolphthalein. The titrable acidity test measures the amount of alkali which is required to change the pH of milk from its initial value of about 6–6 to 6.8 to the pH of the color change

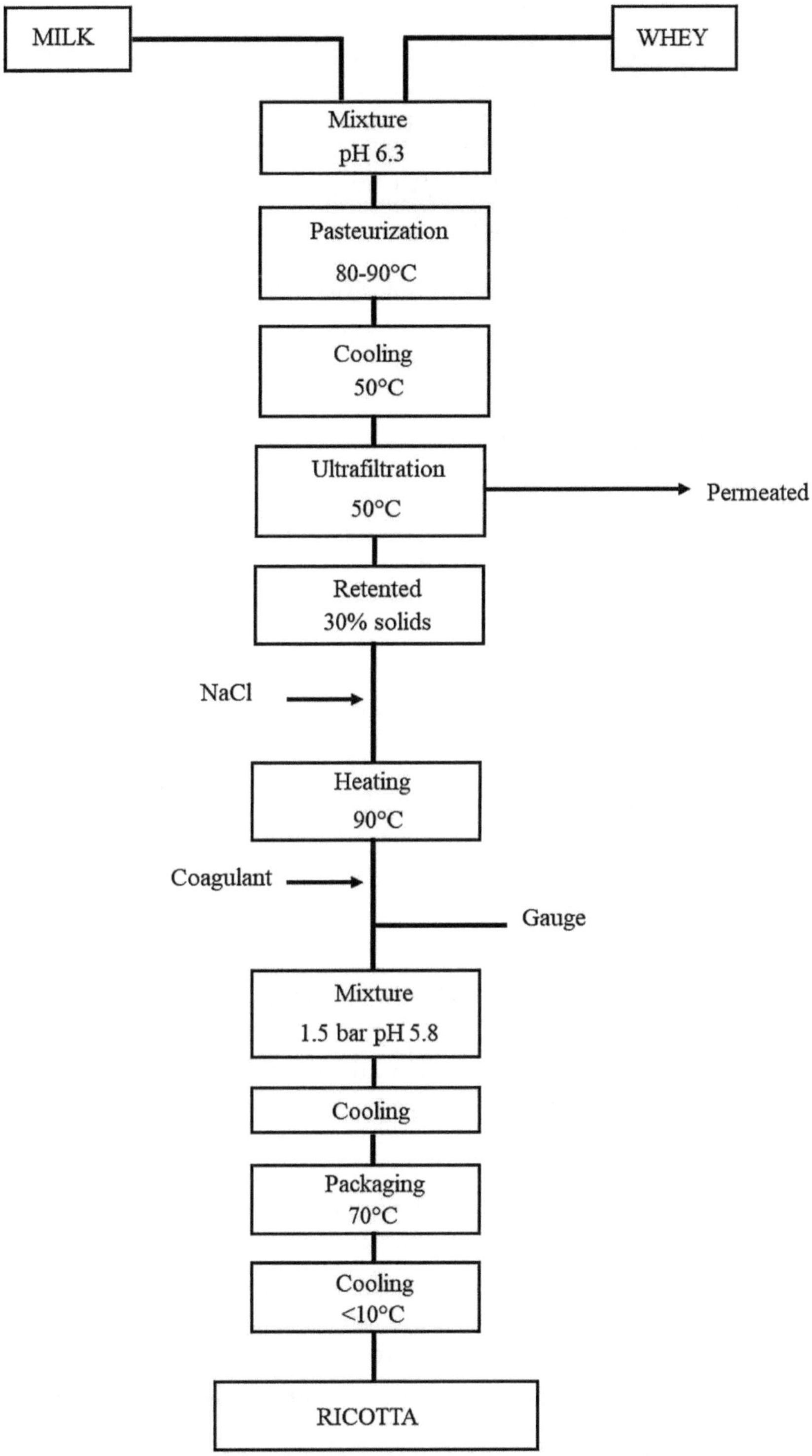

Fig. 3 Protocol of industrial ricotta cheese production with the use of ultrafiltration. (Adapted from Salvadori del Prato [5])

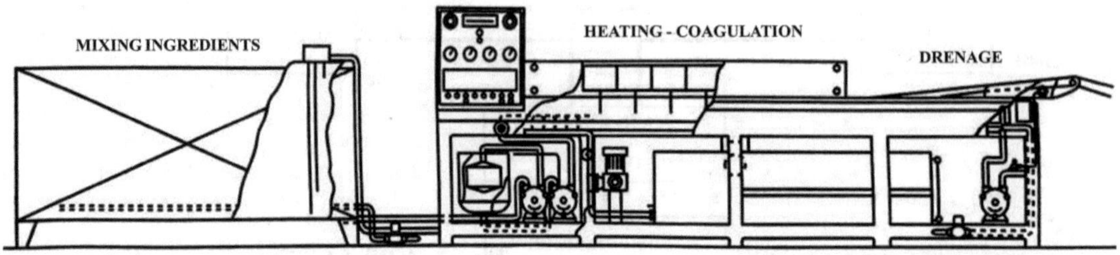

Fig. 4 Continuous production equipment for ricotta cheese production. (Adapted from Salvadori del Prato [5])

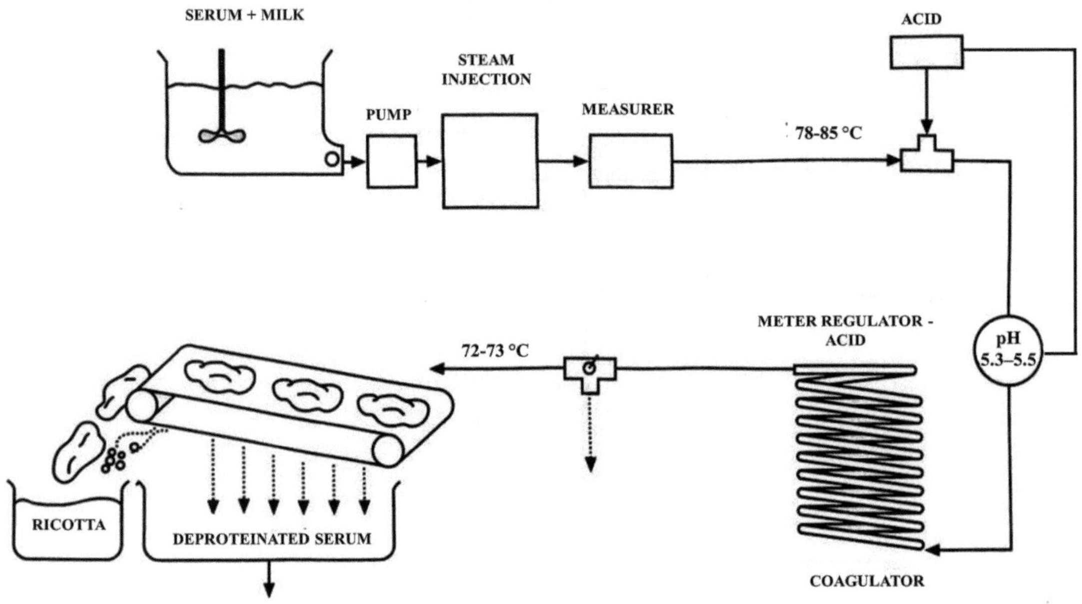

Fig. 5 Modler continuous system. (Adapted from Salvadori del Prato [5])

of phenolphthalein added to milk to indicate the end point (pH 8.3).

Calculation:

$$\%\text{Lactic acid} = \frac{\text{N}^\circ\,\text{mL of 0.1 N NaOH solutions required for neutralization} \times 0.009}{\text{Weight of sample (Weight of sample} = \text{Volume of milk} \times \text{specific gravity)}} \times 100$$

4. Microbial detection for total bacterial count and total and fecal coliforms by using a classic culture medium with standard procedures (*see* **Note 10**).
5. Adjunct of prebiotics and probiotics (*see* **Note 11**).

4.2 Ricotta Cheese Analyses

1. Titrable acidity was determined by mixing the ricotta cheese with ethanol (95%), filtering, and titrating an aliquot of the filtrate with NaOH (0.1 mol/L) using phenolphthalein as the indicator, according to the appropriate standard methods, and expressed as percentage of lactic acid.

2. Ash is determined gravimetrically, at least in duplicate. Samples (10 g) are weighed in a crucible and incinerated at 550 °C for 3 h in furnace. Afterward, crucibles are cooled and weighed for ash calculation. Results are expressed in g per 100 g of ricotta cheese.
3. Fat is determined using a milk butyrometer after digesting the ricotta cheese with H_2SO_4 (D = 1.5 g/mL) at 250 °C. The digested solution is mixed with isoamyl alcohol and the digested material is centrifuged in the butyrometer at 1200 rpm.
2. Moisture is determined gravimetrically by weighing 3 g of ricotta cheese in a metal container with dry sand and a stirring glass rod. Cheese and sand are homogenized by means of the glass rod and then dried at 102 °C until constant weight. Results are expressed as percentage (w/w).
3. Water activity (a_w) is determined at 25 °C by using a water activity analyzer.
4. Total nitrogen (N), pH 4.6-soluble N, and water-soluble N are determined by the micro-Kjeldahl method.
5. Fat is determined by the Soxhlet method using diethyl ether.
6. Lactose is determined by an enzymatic method.
7. pH is measured by using a solid-phase electrode. Measurements should be carried out in triplicate on the core of the product kept at room temperature.
8. Salt is determined in duplicate with the Volhard method [6] () on the ricotta cheese samples. Samples are heated and digested with nitric acid and potassium permanganate in the presence of a known number of moles of silver nitrate; this allows the liberation of the chloride that reacts with the silver to form AgCl. Salt content is determined by back-titration of the remaining unreacted silver using potassium thiocyanate with a ferric ammonium indicator.
9. Texture is determined by cutting a slice from the central diameter of the cheese; then rectangular parallelepipeds, 1 × 1 cm thick and 2 cm long, are obtained from the slice. Cheese samples are equilibrated at 20 °C before testing. Texture profile analysis is evaluated using a texturometer with a compression device that avoids transversal elongation of the samples. Sample undergoes two cycles of 80% compression; force × time data are used to calculate the hardness, cohesiveness, and springiness parameters. The mean values are obtained from the analysis of six rectangular parallelepipeds.
10. Cheese yield is calculated as follows:

$$100 \times \text{Ricotta weight (kg)}/[\text{Milk volume (L)} \times \text{density (kg/L)}]$$

11. Microbial quality is evaluated by total bacterial count, total and faecal coliforms count using standard culture media and procedures.
12. For color measurement by colorimeters, most used are Minolta Chromameter and Hunter Lab colorimeter. The minimum number of technical replicates is three and it is advisable to measure a homogeneous slice of cheese (without structure defects, as holes or fractures). The parameter L* defines lightness or darkness, a* redness or greenness, and b* yellowness or b* blueness.

5 Notes

1. The temperature is the parameter (maximum temperature for whey heating and holding time) that mainly influences the aggregation phase of the whey protein and the final yield. The range of temperature in ricotta cheese production varies from 75 °C to 90 °C depending on the whey origin and the different distribution of the whey proteins into the whey. The α-lactalbumin and proteose-peptone denaturation occurs at higher temperature almost close to 90 °C. As temperature is increased, so are the α-lactalbumin denaturation and ricotta yield, resulting in a product with a grainier and sandier structure and less moisture and smoothness. Furthermore, overheating of the mixture confers a cooked note to ricotta.
2. The dosage of coagulants is a critical factor and varies along with both the total solids content of the mixture and the coagulation temperature. It is suggested that the adjunct of the coagulant take place when the mixture has reached the minimum coagulation temperature in order to obtain a coagulum without structure defects. The use of natural milk or whey graft requires continuous adjustment in the dosage used due to inconsistency of the daily acidity level.
3. Table 3 offers a synthesis of the optimal parameters of pH, acidity, and coagulation temperature for the production of ricotta cheese.

Table 3
Optimal parameters of pH, acidity, and coagulation temperature for the production of ricotta cheese

	pH	Acidity °SH/50	Coagulation temperature °C
Whole milk	5.7–6.0	7.0–7.5	82–85
Milk + serum	5.5–5.8	8.0–8.5	84–87
Serum	5.4–5.7	8.5–9.5	>87

4. Excessive acidity reduces the coagulum yield, whereas insufficient acidity leads to inconsistent and fine coagulum.
5. Coagulation temperature has to be set with regard to the level of acidity developed after addition of coagulants in order to maximize the recovery of the whey protein.
6. Higher value of acidity of the whey leads to a fractionated precipitation of albumin as very fine clots, which are not recoverable prior to the optimal coagulation point.
7. The excessive acidification of the mixture leads to retention of albumin fraction in the liquid phase reducing ricotta yield.
8. Ricotta cheese yield varies along with the origin of whey used: it is about 3–4% in bovine and 7–8% in sheep.
9. In order to obtain fine and homogenous ricotta structure, it is advisable to have a fat/protein ratio in the mixture of about 1 before the processing phase.
10. Ricotta cheese is a dairy product that displays suitable conditions for a large variety of microorganisms: the maintenance of high hygienic quality of the product is the main goal during shelf life.

    10.1. The presence of spore-forming bacteria mostly belonging to *Bacillus*, *Paenibacillus*, and *Clostridium* genera has been described [7]. The processing of raw materials to eliminate spores and residual microflora is essential for improving the hygienic quality and the safety of the ricotta and to extend the shelf life.

    10.2 Microbial contamination of ricotta cheese also leads to flavor defects like sour, fermented, and fruity flavor. Moreover, excessive gas formation may cause blowing of the lid of the container. Defects ascribed to ricotta cheese discoloration are also due to spoilage microorganisms: *Bacillus mycoides* has been associated to pink discoloration, and toxigenic strain belonging to the *Bacillus cereus* group is also involved in discoloration-defective ricotta [7].

    To prevent or reduce microbial contamination consisting of heat-resistant and post-contamination microorganisms that alter the original characteristics and reduce shelf life, it is advisable to proceed with a:

    (i) Second heat treatment of ricotta after the scotta whey drainage (e.g., use of heat exchanger at temperatures between 85 and 95 °C).

    (ii) Direct hot packing, immediately after the ricotta cheese production.

11. It is possible to add prebiotics and probiotics in the mixture for ricotta cheese production, e.g., adjunct of phytosterols [8],

inulin and *Lactobacillus paracasei* [9], *Bifidobacterium* BB-12 [10], *L. acidophilus* La-05 [11], added as free or microencapsulated [12].

For the microencapsulated probiotic, the beads are prepared as follows: 18 mL of cell suspension was added to 2 g of Na alginate and mixed gently for 1–2 min. Then, the gel is manually extruded through a sterile 50 mL syringe, and the capsules (ca. 2 × 3 mm) are dipped in a CaCl2 solution for ca. 5 min and then air-dried for 30 min on blotting paper. It is advisable to reach initial cell counts of probiotic bacteria of about 8–9 log cfu/mL [13].

References

1. Alimentarius C (2011) Standard for whey cheeses. Codex Stan 284–1971. In: Codex Alimentarius. Milk and milk products. WHO and FAO, Rome
2. Mucchetti G, Carminati D, Pirisi A (2002) Fresh ricotta from cow and sheep milk: notes on production technologies and product. Il latte 27:154–166
3. Pires AF, Marnotes NG, Rubio OD, Garcia AC, Pereira CD (2021) Dairy by-products: a review on the valorization of whey and second cheese whey. Food Secur 10:1067–1091
4. Mangione G, Caccamo M, Natalello A, Licitra G (2023) Graduate student literature review: history technologies of production, and characteristics of ricotta cheese. J Dairy Sci 106: 3807–3826
5. Salvadori del Prato (2001) Dairy technology, Manual ed. Edagricole Calderini s.r.l, Milan
6. AOAC International (ed) (2016) Method number 935.43. Official methods of analysis, 20th edn. AOAC International, Gaithersburg
7. Sattin E, Andreani NA, Carraro L, Fasolato L, Balzan S, Novelli E, Squartini A, Telatin A, Simionati B, Cardazzo B (2016) Microbial dynamics during shelf-life of industrial ricotta cheese and identification of a bacillus strain as a cause of a pink discolouration. Food Microbiol 57:8–15. https://doi.org/10.1016/j.fm.2015.12.009
8. Nzekoue KF, Alesi A, Vittori S, Sagratini G, Caprioli G (2021) Development of a functional whey cheese (ricotta) enriched in phytosterols: evaluation of the suitability of whey cheese matrix and processing for phytosterols supplementation. LWT 139:110479. https://doi.org/10.1016/j.lwt.2020.110479
9. Niro S, Succi M, Fratianni A, Cinquanta L, Tremonte P, Sorrentino E, Panfili G (2013) Production of functional ricotta cheese. Agro Food Ind Hi-Tech 24:56–59
10. Fritzen-Freire CB, Prudencio ES, Pinto SS, Munoz IB, Muller CMO, Vieira CRW, Amboni RDMC (2013) Effect of the application of bifidobacterium BB-12 microencapsulated by spray drying with prebiotics on the properties of ricotta cream. Food Res Int 52: 50–55. https://doi.org/10.1016/j.foodres.2013.02.049
11. Sameer B, Ganguly S, Khetra Y, Sabikhi L (2020) Development and characterization of probiotic buffalo milk ricotta cheese. Lebensm Wiss Technol 121:108944. https://doi.org/10.1016/.lwt.2019.108944
12. Lopes L, Pimentel TC, Carvalho R, Madruga MS, Galvao MS, Bezerra T, Barao CE, Magnani M, Stamford T (2021) Spreadable goat ricotta cheese added with *Lactobacillus acidophilus* La-05: can microencapsulation improve the probiotic survival and the quality parameters? Food Chem 346:128769. https://doi.org/10.1016/j.foodchem.2020.128769
13. Santillo A, Albenzio M, Bevilacqua A, Corbo MR, Sevi A (2012) Encapsulation of probiotic bacteria in lamb rennet paste: effects on the quality of pecorino cheese. J Dairy Sci 95: 3489–3500. https://doi.org/10.3168/jds.2011-4814

Chapter 11

Mini-Cheese Models

Guillermo H. Peralta and Carina V. Bergamini

Abstract

Cheese making is an intricate process in which the milk is converted into a fermented semisolid product. Using optimized cheese models on a laboratory scale is fundamental for evaluating new ingredients for manufacturing cheese, especially for reproducibility issues, cost, and time-consuming. In this chapter, after discussing the pros and cons of several cheese models, one of the latest protocols for cheese manufacture on a laboratory scale will be outlined. This cheese model was useful for assessing the performance of potential adjunct cultures, but it could also be used to study other parameters. The main advantage of this protocol is the large number of 100 g cheeses that can be produced simultaneously. Thus, it allows researchers to assess several strains or other factors in a real-cheese matrix at the same time.

Key words Cheese, Cheese matrix, Miniature cheese, Cheese model, Cheese making, Miniature laboratory cheese, Miniature semihard cheese

1 Introduction

The origin of cheese goes back to the Middle East during the Agricultural Revolution (about 8000 years ago) [1]. In a nutshell, cheese making can be defined as the process that converts milk into a fermented semisolid product. This process involves several stages, which depend on the cheese variety. It is interesting to highlight that more than 1000 cheese varieties are manufactured worldwide [2]. Besides raw milk, microorganisms (starter and adjunct cultures), coagulants (calf rennet, fermentation-produced chymosin, and vegetable and microbial coagulants), and sodium chloride (NaCl) are other fundamental ingredients in cheese manufacture. The type of milk (i.e., cow, goat, sheep, or buffalo milk), its microbiological quality (i.e., levels of adventitious microorganisms, including lactic acid bacteria, coliforms, and yeasts), and its pH value and chemical composition (i.e., level of fat, protein, and calcium) have a significant effect on the manufacturing process and cheese quality. In the same direction, the flavor and texture of cheese are also influenced by these factors, especially the starter and

Adriano Gomes da Cruz et al. (eds.), *Dairy Foods Processing*, Methods and Protocols in Food Science,
https://doi.org/10.1007/978-1-0716-4144-6_11,

adjunct culture, which are the most developed ingredients in the last decades. The development of both cultures requires numerous studies to ensure their good performance in cheese manufacture and ripening. Considering that there are several parameters to control in the cheese making process, culture development can be expensive and time-consuming when a cheese making protocol on a laboratory scale is not optimized.

Several approaches—including cheese slurries, cheese-like products, aqueous systems similar to cheese, and miniature model cheese systems—to study the influence of cheese-ripening agents have been discussed in a review by Shakeel-Ur-Rehman et al. [3]. Over the last decades, numerous cheese models have been proposed and used to facilitate the evaluation and development of cultures, coagulants, and so forth. The systems specially designed for the development of both starter and adjunct culture, which are the most interesting for the dairy industry, can be boiled down to two types of models: liquid and semisolid. In this way, the incubation of resting cell suspensions in simple media containing some nutrients [4–6], in fermented milks [7–9], and in cheese extracts/slurries [10–12] have been used as liquid models of cheese to assess the potential of several strains as a starter or adjunct cultures. For example, Steganovic et al. [5] assessed the biodiversity of 10 strains of *Lacticaseibacillus casei* in a medium containing free amino acids and a minimal level of carbohydrates and supplemented with NaCl. Lo et al. [9] compared the diacetyl and acetoin-producing capacities of 13 *Lacticaseibacillus rhamnosus* strains in full-fat ultrahigh temperature-treated milk.

Similarly, Beret et al. [7] reported the ability of *Lacticaseibacillus paracasei* 90 to produce diacetyl and acetoin in fermented milk. Budinich et al. [10] used a cheese-ripening model system based on a medium prepared from ripening Cheddar cheese extract to evaluate the growth of *Lacticaseibacillus paracasei* ATCC 334. Also, Peralta et al. [12, 13] used a sterile extract prepared from a Cremoso cheese after 20 d of ripening to assess the performance of flavor-producing bacteria. The common preparation of these types of sterile cheese extracts is schematized in Fig. 1a. Different components, including salts, amino acids, carbohydrates, and so forth, can be added to these liquid models. These sterile models are generally incubated with the potential cultures at a temperature between 30 and 37 °C for a few days. Thus, one of the advantages of these liquid models is related to the possibility of sterilizing them and the fast outcomes they provide. In addition, it is interesting to highlight that the control of composition is one of the strengths of liquid models. For example, incubating resting cells of a potential adjunct strain in a specific buffer with one amino acid could clarify the ability of this strain to metabolize this specific amino acid. However, the negative side of these models is that they do not represent the real aqueous phase of cheese, and usually, further

Fig. 1 (**a**) General schematic diagram of preparing of a sterile cheese extract from a ripened cheese. (**b**) Miniature Cheddar cheese (20 g) (photo courtesy of Dr. Elisa Ale (CONICET, Santa Fe, Argentina)). (**c**) Miniature Cremoso cheese (230 g) (photos from the authors' private collection). (**d**) Miniature curd (25 g) (from Vélez et al. [25]). (**e**) Microplate and stirring device used for the manufacturing of 96 individual micro cheeses (170 mg) (from Bachmann et al. [27]). (**f**) Cheese-mimicking matrix distributed in 24-well plates (from Garnier et al. [28]). (**g**) Hispanic-style fresh cheeses (300 mg) (from Van Tassell et al. [29]). (**h**) Miniature semihard cheese (100 g) (photo from the authors' private collection)

studies should be carried out to confirm the preliminary outcomes. A similar situation occurs when fermented milks are used as cheese models due to the high levels of lactose, pH, and water activity, which differ from the cheese matrix. In the same direction, liquid model incubations differ from the ripening conditions. Even though the shortcomings of these models, they can contribute enormously to understanding the mechanisms or metabolic pathways of a potential culture. Of course, experiments with cheese matrices should be carried out to confirm the outcomes.

Regarding the semisolid models, several miniature cheeses and curds have been reported for different applications during the last decades. One of the earliest protocols of manufacture at the laboratory was reported by Shakeel-Ur-Rehman et al. [14], who developed a mini-Cheddar cheese protocol using wide-mouthed plastic centrifuge bottles (200 mL) as vats, useful for screening the effect of starters, adjuncts, or enzymes on cheese ripening. This model was used in several international experiments [15–17]. For instance, O'Reilly et al. [15] used this protocol to determine the contributions of different individual strains of *Lactococcus* to cheese ripening in miniature Cheddar-type cheeses. Ciocia et al. [16] used the model to screen 10 lactobacilli strains for their performances as adjuncts. Ale et al. [17] used this model to study the influence of adding an exopolysaccharide extract on the texture, primary proteolysis, and microbiologic characteristics of Cheddar cheese. A picture of this model is shown in Fig. 1b.

Furthermore, this model was adapted to several varieties of cheese. For example, Hynes et al. [18] modified the cheese model of Shakeel-Ur-Rehman et al. [14] to make miniature washed curd cheeses under controlled microbiological conditions. Also, Trancoso-Reyes et al. [19] adapted this Cheddar cheese model to manufacture a miniature version of Chihuahua cheese. Another cheese model was reported by Milesi et al. [20]. The authors reported a protocol for a mini-Cremoso cheese that was later used in several experiments [21–24]. For example, Peralta et al. [22] used this model to assess the impact of incorporating disrupted cultures in Cremoso cheese. A picture of this model is shown in Fig. 1c. Another approach was reported by Vélez et al. [25] who proposed and validated a new mini curd model of young hard cheese (Fig. 1d). This mini curd was reconstituted from frozen curd grains and whey from standard Reggianito cheese making. In general, the advantage of miniature cheeses is that several units can be manufactured simultaneously and represent better than liquid models the real cheese matrix composition and ripening paraments. On the other hand, the negative point of the miniature cheese is that, in general, there is not enough sample for sensory studies, and due to the high surface/volume ratio, they should be packaged under a vacuum to avoid the fast-drying process [26].

For specific purposes, cheese making can be carried out on an extremely small scale. For example, Bachmann et al. [27] developed a protocol that allows the parallel manufacturing of approximately 600 cheeses in individual cheese vats, each with individual process specifications (Fig. 1e). Similarly, Garnier et al. [28] developed an antifungal activity screening method using a cheese-mimicking matrix distributed in 24-well plates (Fig. 1f). Van Tassell et al. [29], using 2 mL microcentrifuge tubes as mini vats to manufacture mini cheeses, developed a protocol for testing the incorporation of novel anti-listeria components (Fig. 1g). The advantages and disadvantages of these models are the large number of cheese units manufactured and the small quantities of samples, respectively.

One of the latest protocols of manufacture at a laboratory scale was used by Peralta et al. [30, 31], who made a miniature Pategras cheese (semihard cheese) using 1000 mL of milk for each cheese (Fig. 1h). Using two thermostatic baths with the capacity of heating 12 glass beakers (vats), the protocol allows the production of 24 mini-cheese (100 g) simultaneously. It is interesting to highlight the large number of 100 g cheese allows researchers numerous chemical and microbiological analyses to be carried out, and it could even be possible to carry out sensory studies. In this chapter, this protocol will be outlined.

2 Materials

2.1 Elements

- Raw milk—a volume of 1 L for each cheese (*see* **Note 1**).
- Bacterial culture (*see* **Note 2**).
- 20% $CaCl_2$ (w/v)—a volume of 1 mL for each cheese (*see* **Note 3**).
- Coagulant—chymosin (*see* **Note 4**).
- 20% NaCl (w/v)—a volume of 1.5 L for four cheeses (*see* **Note 5**).

2.2 Utensils and Other Materials

- Cheese vats—1 L glass beakers.
- Cheese molds—100 mL perforated containers (*see* **Note 6** and Fig. 2a).
- Perforated lids (*see* Fig. 2a).
- Curd cutter and stainless-steel vertical knife (*see* Fig. 2a).
- Aluminum pot.
- Paint coating.
- Vacuum seal bags.

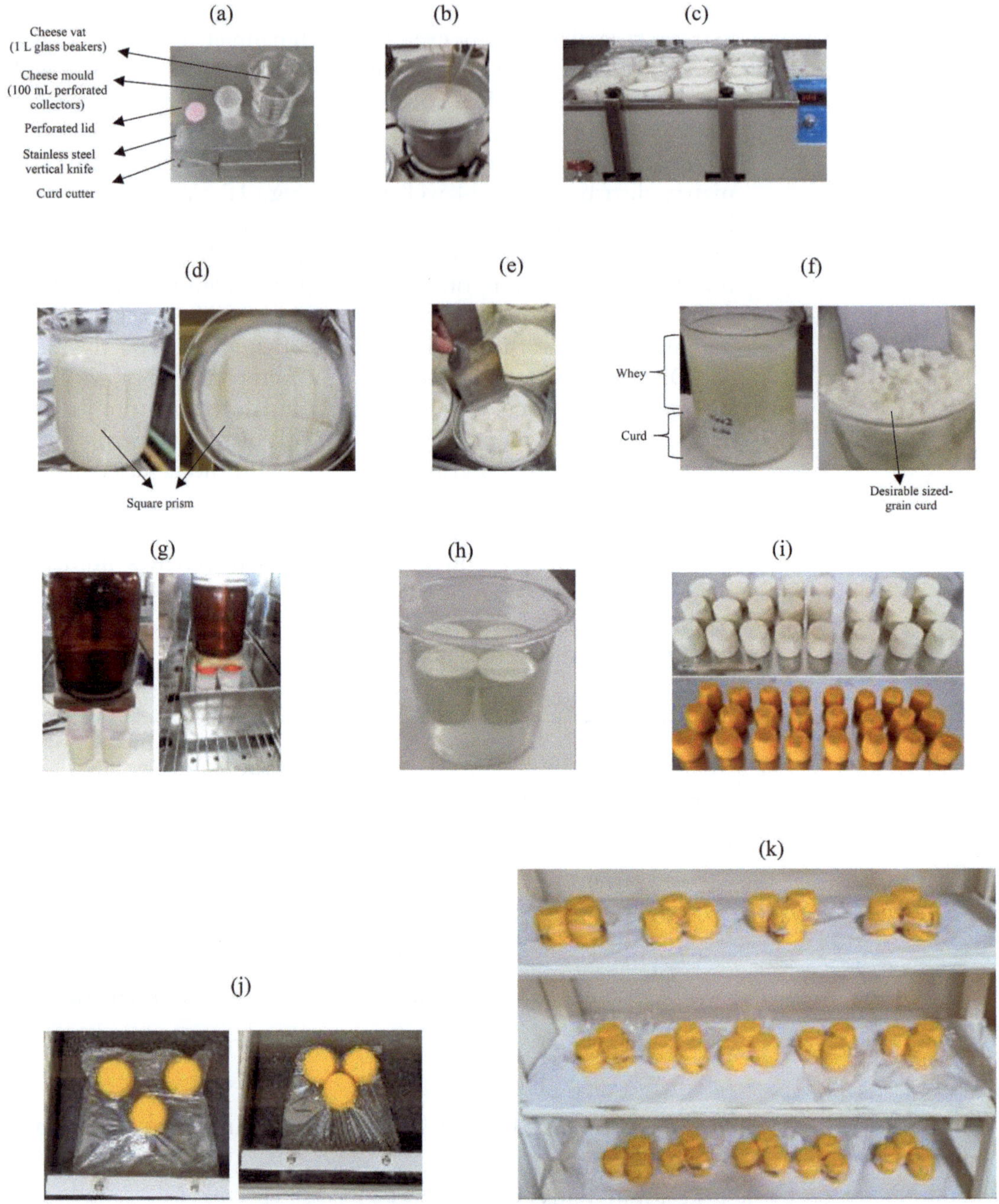

Fig. 2 (**a**) Cheese vat, cheese mold, perforated lids, curd cutter, and stainless steel. (**b**) Pasteurization in an aluminum pot. (**c**) Thermostatic bath to hold the temperature of the milk at 38 °C during the coagulation. (**d**) Rectangular prisms of curd produced after the first cutting. (**e**) Large cubes of curd produced after the second cutting. (**f**) The final desirable size of curd grain before molding. (**g**) Pressing and acidifying in a chamber at 45 °C. (**h**) Salting by submerging the cheese in a cold brine. (**i**) Painting of the cheeses with a commercial paint coating. (**j**) Vacuum packing of the cheeses in plastic bags. (**k**) The ripening process a 12 °C. (All the photos are from the authors' private collection)

- Thermometer.
- Glass rods.
- Pipettes.

2.3 Equipment

- Laboratory balance.
- Thermostatic bath (*see* **Note 1** and Fig. 2c).
- Incubator/acidification chamber (*see* **Note** 7).
- pH meter.
- Vacuum sealer.
- Maturation chamber (temperature: 10 °C–12 °C; relative humidity: 80%).

3 Methods

3.1 Pasteurization

1. Heat the total volume of raw milk for cheese making at 63 °C and hold it for 30 min using an aluminum pot (*see* Fig. 2b).
2. Once the heating time is finished, put the pot with the pasteurized milk in a bath with cold water to reach a temperature of the milk of 38 °C.
3. Weigh 1000 g of pasteurized milk in each cheese vat (1 L glass beakers).
4. Add 1 mL of the 20% $CaCl_2$ (w/v) solution to supplement milk with 0.02% (w/v) of calcium chloride.

3.2 Addition of Cultures

1. Weigh and suspend in 30 mL of pasteurized milk at 37 °C the appropriate amount of freeze-drying culture to achieve a cell suspension of 1×10^9 cells per milliliters.
2. Holding the temperature of the milk at 38 °C (Fig. 1c), add 1 mL of the previous cell suspension in each cheese vat to achieve 1×10^6 cells per milliliters (*see* **Note 2**).
3. Gently stir the milk with a glass rod to mix the inoculum and wait 10 min to adapt the bacteria into the milk before adding the coagulant.

3.3 Coagulation

1. Dilute the coagulant with deionized water and immediately add it to the milk (*see* **Note 4**).
2. Gently stir the mixture using a glass rod and subsequently stop the agitation.
3. Hold the milk quiescently for 45 min to achieve a proper gel (*see* **Note 8**).

3.4 Cutting

1. Gently insert the stainless-steel vertical knife into the coagulum and draw them along the gel mass. Repeat this process to produce rectangular prisms as seen in Fig. 2d, and then hold the vats quiescently for 3 min.
2. Gently insert the vertical knife into the rectangular prisms and then draw them along several directions to produce large cubes, as can be seen in Fig. 2e. Then hold quiescently the vats for 3 min.
3. Reduce the size of the curd grain using the curd cutter to achieve the size of corn grain, as can be seen in Fig. 2f (*see* **Note 9**).

3.5 Cooking

1. Insert a glass rod into the vat and stir the mixture of grains and whey to avoid the agglutination of the grains.
2. Keeping gently stirring the mixture, replace the water of the thermostatic bath with heating water to heat the curd particles in the whey from 38 °C to 45 °C at a rate of 1 °C/min (*see* **Note 10**).
3. Keep at 45 °C for 10 min with continuous stirring and then hold quiescently the mixture for 5 min to allow the agglutination of the grains and facilitate the elimination of the whey.

3.6 Draining and Molding

1. Gently incline the vat to eliminate most of the whey and then hold quiescently the vat for half a minute. Repeat this process to achieve a curd mass entering the cheese mold (*see* **Note 11**).
2. Cover the curd with the perforated lids.

3.7 Pressing and Acidifying

1. Place a weight on the top of each of the molded cheeses, as shown in Fig. 2g.
2. Flip the cheese into the mold 5 min later, and press again.
3. Place the molded cheeses together to their pressing system inside an acidification chamber at 45 ° C.
4. During the acidification, flip the cheese into the mold and assess the pH values every 30 min until it reach a pH of 5.3.

3.8 Salting

1. Quickly take out the cheese from the molds and weigh it to determine the salting time in brine.
2. Submerge the cheese in a cold sterile brine (36 min per 100 g of cheese) (Fig. 2h).
3. Remove the cheese from the brine and put it in the mold again.

3.9 Storing and Painting

1. Store the cheese inside the molds in a maturation chamber between 10 and 12 °C for 5–7 days. Once a day, flip the cheeses into the mold, and store them again.

2. Remove the cheese from the molds and coat them with a commercial paint coating (Fig. 2i).
3. Store the cheese in the maturation chamber until the next day and repeat the painting process once or twice if necessary.

3.10 Packaging and Ripening

1. Vacuum pack the cheeses in plastic bags (Fig. 2j).
2. Return the cheeses to the maturation chambers to ripe for a least 120 days (Fig. 2k).

4 Notes

1. This protocol was optimized for cow milk; however, other types of milks could be used (i.e., goat, sheep, and buffalo milk). Regarding the volume of milk that should be pasteurized, it will depend on the thermostatic bath that is available in the lab and the experimental design, but always be sure to pasteurize more than the minimum volume to fill the cheese vats due to loss of water by evaporation during the heating. In addition, an extra volume of pasteurized milk is necessary to resuspend the freeze-drying cultures. For example, to produce 24 cheeses simultaneously (using two thermostatic baths), a volume of 26 L should be pasteurized. Furthermore, ensure that the water level in the thermostatic batch reaches the same level as the milk in the vats to ensure adequate temperature maintenance during the cheese making.
2. This protocol was optimized to assess the performance of potential adjuncts using *Streptococcus thermophilus* as a starter culture. However, other starters can be used. Regarding the specific doses of the starters, it is advisable to ask the supplier company for their products to avoid over-acidification. In general, a dose of 1×10^6 cells per milliliters of milk is sufficient to reach a pH value of 5.3 in 2 h in the acidification chamber. Furthermore, it is important to highlight that the freeze-drying cultures should be resuspended just 5 min before adding them to cheese vats as was described in Subheading 3.2. Regarding the specific doses of the adjuncts, a dose between 1×10^4 and 1×10^6 cells per milliliters of milk is generally used, but it will depend on the strain. For example, Peralta et al. [13] reported that the dose of a spray-dried adjunct culture of *L. paracasei* 90 used in a miniature cheese significantly affected the level of acetoin and diacetyl. The authors reported that the key volatile compounds were produced at a high level, only in the cheeses at the higher dose of this strain.
3. A high concentration of calcium in the milk significantly affects the coagulation process's speed especially in the second stage of

rennet coagulation [32]. The needed volume of the solution of 20% $CaCl_2$ (w/v) will depend on the quantity of cheeses to be made. For example, dissolving 6 g of calcium chloride in 30 mL of deionized water would be more than enough to make 24 cheeses (1 mL for each cheese).

4. A commercial chymosin (a milk-clotting enzyme produced via fermentation with the use of recombinant DNA techniques) was used in this manufacturing protocol. However, other coagulants [33] could be used. Considering that the commercial chymosins have different concentrations, ask the suppliers for advice on the best dose of their products. Moreover, it is crucial to emphasize that chymosin should be diluted just before adding it to cheese vats, as described in Subheading 3.3.

5. The brine volume (20% NaCl) will depend on the number of cheeses you made. A volume of 1.5 L will be enough for salting four cheeses simultaneously (*see* Fig. 2g). Considering that the brine solution should be cold (5 °C), preparing it 1 or 2 days before the cheese making day is advisable. Furthermore, it is recommended to reach a pH of 5.4 with addition of an adequate volume of diluted lactic acid. Finally, it is recommended to add a little amount of calcium chloride (0.05–0.10% w/v) to the brine when it is just prepared to avoid the demineralization of the cheese rind. The low temperature is critical to rapidly stop the starter culture's activity, avoiding over-acidification.

6. A miniature cheese mold can be easily designed using an ordinary urine-collection container (Fig. 2a). Several perforations must be made in the containers, which must be well distributed to ensure the same whey draining in all cheeses.

7. The microbiological incubators can maintain the temperature between 45 and 50 °C with a temperature variation of ±1 °C function.

8. The cut time depends on several factors, including milk pH and composition, type and doses of coagulant, temperature, and so forth. Asking the supplier company for their coagulant's properties is highly advisable.

9. This protocol can be used to make long-ripening cheese. For this case, reduce the size of the curd grain to achieve rice or wheat size.

10. To manufacture long-ripening cheese replace the water of the thermostatic bath with heating water to heat the curd particles in the whey from 38 °C to 50 °C at a rate of 1 °C/min.

11. The remotion of the cheese whey can also be performed by centrifugation using wide-mouthed centrifuge bottles (200 mL), according to Shakeel-Ur-Rehman et al. [14].

References

1. Fox PF, Uniacke-Lowe T, McSweeney PLH, O'Mahony JA (2015) Dairy chemistry and biochemistry. Springer, Cham
2. Fox PF, Guinee TP, Cogan TM, McSweeney PLH (2017) Fundamentals of cheese science. Springer, Boston
3. Shakeel-Ur-Rehman, Fox PF, McSweeney PLH et al (2001) Alternatives to pilot plant experiments in cheese-ripening studies. Int J Dairy Technol 54:121–126
4. Kieronczyk A, Skeie S, Olsen K, Langsrud T (2001) Metabolism of amino acids by resting cells of non-starter lactobacilli in relation to flavour development in cheese. Int Dairy J 11: 217–224
5. Stefanovic E, Thierry A, Maillard M-B et al (2017) Strains of the *Lactobacillus casei* group show diverse abilities for the production of flavor compounds in 2 model systems. J Dairy Sci 100:6918–6929
6. van de Bunt B, Bron PA, Sijtsma L et al (2014) Use of non-growing *Lactococcus lactis* cell suspensions for production of volatile metabolites with direct relevance for flavour formation during dairy fermentations. Microb Cell Factories 13:176
7. Beret MV, Peralta GH, Vera-Candioti L et al (2021) Culture media based on effluent derived from soy protein concentrate production for *Lacticaseibacillus paracasei* 90 biomass production: statistical optimisation, mineral characterization, and metabolic activities. Anton Leeuw Int J 114:2047–2063
8. Bancalari E, Savo Sardaro ML, Levante A et al (2017) An integrated strategy to discover *Lactobacillus casei* group strains for their potential use as aromatic starters. Food Res Int 100: 682–690
9. Lo R, Ho VTT, Bansal N, Turner MS (2018) The genetic basis underlying variation in production of the flavour compound diacetyl by *lactobacillus rhamnosus* strains in milk. Int J Food Microbiol 265:30–39
10. Budinich MF, Perez-Díaz I, Cai H et al (2011) Growth of *Lactobacillus paracasei* ATCC 334 in a cheese model system: a biochemical approach. J Dairy Sci 94:5263–5277
11. Milesi MM, Bergamini CV, Hynes E (2011) Production of peptides and free amino acids in a sterile extract describes peptidolysis in hard-cooked cheeses. Food Res Int 44:765–773
12. Peralta GH, Wolf IV, Perotti MC et al (2016) Formation of volatile compounds, peptidolysis and carbohydrate fermentation by mesophilic lactobacilli and streptoccocci cultures in a cheese extract. Dairy Sci Technol 96:603–621
13. Peralta GH, Bergamini CV, Audero G et al (2017) Spray-dried adjunct cultures of autochthonous non-starter lactic acid bacteria. Int J Food Microbiol 255:17–24
14. Shakeel-Ur-Rehman, McSweeney PLH, Fox PF (1998) Protocol for the manufacture of miniature cheeses. Lait 78:607–620
15. O'Reilly CE, O'Connor PM, Murphy PM et al (2002) Effects of high-pressure treatment on viability and autolysis of starter bacteria and proteolysis in Cheddar cheese. Int Dairy J 12: 915–922
16. Ciocia F, McSweeney PLH, Piraino P, Parente E (2013) Use of dairy and non-dairy *Lactobacillus plantarum*, *Lactobacillus paraplantarum* and *Lactobacillus pentosus* strains as adjuncts in cheddar cheese. Dairy Sci Technol 93:623–640
17. Ale EC, Ciocia F, Reinheimer J, Binetti A, McSweeney P (2016) Influence of an exopolysaccharide (EPS) extract obtained from *Lactobacillus fermentum* Lf2 on the texture parameters, starter viability and primary proteolysis of Cheddar cheese. Abtract presented at the XXIII Congreso Latinoamericano de Microbiología & XIV Congreso Argentino de Microbiologia, Asociación Argentina de Microbiología & Asociación *Latinoamericana de Microbiología*, Rosario, 26–30 September 2016
18. Hynes E, Ogier J-C, Delacroix-Buchet A (2000) Protocol for the manufacture of miniature washed-curd cheeses under controlled microbiological conditions. Int Dairy J 10: 733–737
19. Trancoso-Reyes N, Gutiérrez-Méndez N, Sepulveda DR, Hernández-Ochoa LR (2014) Assessing the yield, microstructure, and texture properties of miniature Chihuahua-type cheese manufactured with a phospholipase A1 and exopolysaccharide-producing bacteria. J Dairy Sci 97:598–608
20. Milesi MM, Candioti M, Hynes E (2007) Mini soft cheese as a simple model for biochemical studies on cheese-making and ripening. LWT Food Sci Technol 40:1427–1433
21. Milesi MM, McSweeney PLH, Hynes ER (2008) Viability and contribution to proteolysis of an adjunct culture of *Lactobacillus plantarum* in two model cheese systems: Cheddar cheese-type and soft-cheese type. J Appl Microbiol 105:884–892
22. Peralta GH, Bergamini CV, Hynes ER (2019) Disruption treatments on two strains of

Streptococcus thermophilus: levels of lysis/permeabilisation of the cultures, and influence of treated cultures on the ripening profiles of Cremoso cheese. Int Dairy J 92:11–20

23. Giménez P, Peralta GH, Guglielmotti D, Audero G, Páez R, Hynes ER, Bergamini CV (2021) Preventing undesired eye formation in soft cheese. Int Dairy J 116:1–7
24. Peralta GH (2014) Bioformación de flavour en quesos vía metabolismo de aminoácidos por fermentos lácticos primarios y adjuntos. Universidad Nacional del Litoral, Santa Fe
25. Vélez MA, Perotti MC, Rebechi SR, Hynes ER (2015) Short communication: a new minicurd model system for hard cooked cheeses. J Dairy Sci 98:3679–3683
26. Carafa I, Stocco G, Nardin T et al (2019) Production of naturally γ-aminobutyric acid-enriched cheese using the dairy strains *Streptococcus thermophilus* 84C and *Lactobacillus brevis* DSM 32386. Front Microbiol 10:93
27. Bachmann H, Kruijswijk Z, Molenaar D et al (2009) A high-throughput cheese manufacturing model for effective cheese starter culture screening. J Dairy Sci 92: 5868–5882
28. Garnier L, Salas ML, Pinon N et al (2018) Technical note: high-throughput method for antifungal activity screening in a cheese-mimicking model. J Dairy Sci 101:4971–4976
29. Van Tassell ML, Ibarra-Sánchez LA, Takhar SR et al (2015) Use of a miniature laboratory fresh cheese model for investigating antimicrobial activities. J Dairy Sci 98:8515–8524
30. Peralta GH, Beret V, Bürgi M et al (2023) Impact of media culture, freeze-drying and storage conditions on preservation of *Lacticaseibacillus paracasei* 90: viability and metabolic potential as a secondary culture in semi-hard cheese. Int Dairy J 147:105763
31. Peralta GH, Aguirree A, Bürgic MDM et al (2023) Effect of high-pressure homogenization on metabolic potential of *Lacticaseibacillus paracasei* 90: in vitro and in situ studies in fermented milk and semi-hard cheese. Int J Dairy Technol 76(3):583–596
32. McSweeney P (2007) What factors affect rennet coagulation time? In: McSweeney P (ed) Cheese problem solved. Woodhead Publishing, Cambridge, pp 60–62
33. Andrén A (2021) Milk-clotting enzymes. In: Kelly AL, Larsen LB (eds) Agents of change: enzymes in Milk and dairy products. Springer, Cham, pp 349–362

Chapter 12

Concentrated, Isolated, and Hydrolysate Whey Protein Obtained by Membrane Separation Processes

Giordana Demaman Arend, Mariane Carolina Proner, and Katia Rezzadori

Abstract

For many years, whey represented a challenge for the dairy industry due to its high concentration of organic material, which, if not treated and disposed of correctly, can cause major environmental damage. However, with the advances in membrane processes, it was possible to add value to this waste, obtaining compounds of high nutritional value and great commercial interest, popularly known as whey protein (WP). This chapter describes the protocol of whey protein production, outlining important factors related to sample handling, membrane selection, and the operation process for obtaining different kinds of whey protein using membrane technology.

Key words Microfiltration, Ultrafiltration, Nanofiltration, Whey, Dairy industry waste

1 Introduction

Whey is a by-product of the dairy industry generated during milk coagulation to produce cheese and other dairy products. Although whey is considered a residue from the food industry, its composition contains proteins, lactose, and minerals, making it a highly nutritional compound [1–4]. The whey is mainly constituted of lactose (0.18–60 kg.m^{-3}), protein (1.4–33.5 kg.m^{-3}), and fats (0.08–10.6 kg.m^{-3}) [5]. The American Dairy Products Institute (ADPI) (2016) [6] defined whey protein (WP) as the product obtained by the removal of sufficient nonprotein constituents, with a final protein concentration higher than 25%.

WP has attracted commercial interest due to its nutritional and functional properties, such as its ability to form foam and gelification and emulsification properties due to its protein-rich composition [7, 8]. Thus, the advancement of separation technologies has allowed the effective treatment of whey to obtain concentrated and purified protein fractions for applications in food and pharmaceutical products [9].

Adriano Gomes da Cruz et al. (eds.), *Dairy Foods Processing*, Methods and Protocols in Food Science,
https://doi.org/10.1007/978-1-0716-4144-6_12,

In this context, the membrane separation process (MSP) is one of the main methods used in the cheese whey treatment to obtain commercial products. We can define the membrane as a selective barrier that can separate the compounds of a solution. During the process, it is possible to concentrate the solution on one side of the membrane, referred to as the concentrate, and the solution that permeates the other side, known as the permeate. The main advantages of MSP include easy application without the need to use additives and/or thermal processes. Furthermore, it allows separation and concentration operations to be carried out in a single step. Thus, MSP guarantees a high-quality final product with the desired characteristics, low degradation, and off-flavor, reducing the loss of the molecule's bioactivity [10–13].

In the whey treatment, the MSPs commonly used are those that use the pressure differential as a driving force: microfiltration (MF), ultrafiltration (UF), nanofiltration (NF), and reverse osmosis (RO). Among these, microfiltration and ultrafiltration processes are the most common in the dairy industry [9].

MF retains particles between 0.1 and 10 μm, which allows it to remove bacteria, fats, and caseins present in whey, mainly used as a pretreatment for the UF process [9, 14–16]. UF retains particles between 0.01 and 0.1 μm and is the most used MSP due to its ability to separate, fractionate, and concentrate the proteins in whey. UF is the main method of obtaining commercially available whey protein. In 2007, the global membrane area installed by UF in the dairy industry reached 350,000 m^2 [2, 9, 14, 15, 17, 18].

MSPs are affected by the fouling phenomenon and are the main limiting factor in the process. In ultrafiltration (UF), a gel layer develops on the membrane surface, influencing its subsequent behavior. Meanwhile, in microfiltration (MF), pore plugging and gel layer formation occur. In solutions containing proteins, such as whey, steric effects guide the choice of MF membranes with larger pore sizes, posing the potential risk of significant pore plugging and leading to the eventual formation of a gel layer [19]. Therefore, it is important to work with the best conditions of temperature, pressure, flow, and type of membrane, combined with cleaning cycles, to mitigate the effects of fouling and maintain process efficiency.

It is worth noting that three types of WP are available on the market. Whey protein concentrate (WPC) contains 34–80% protein and lactose, whey protein isolate (WPI) contains more than 90% protein and a low concentration of lactose, and whey protein hydrolysate (WPH) has 80% protein hydrolyzed by enzymes to facilitate absorption [2, 9, 17].

In this way, it is clear that advances in MSP were important in making it possible to treat whey, producing products with important benefits in the food and pharmaceutical industries, and thus becoming the main operation in obtaining whey protein widely applied in functional foods, nutritional supplements, and medicines

[2, 9, 18]. Membrane technologies for protein whey recovery are well established and widely used on an industrial scale. Therefore, these protocols outline important factors related to sample handling, membrane selection, operation process, and membrane performance.

1.1 Concentrated Whey Protein (WPC)

Whey is mainly obtained from cheese production and is rich in protein and fat (*see* **Note 1**). Due to the size of these molecules, they are mainly recovered by ultrafiltration [20]. According to Kelly (2019) [19], the principal challenge during whey processing is to sustain a sufficient retentate flow rate once the volume is reduced during advanced stages of the UF process.

In addition, it is well known that continuous alterations in feed composition over time may still affect the retentate flow rate and composition if these effects are not monitored [20]. Davey et al. (2004) [21] created a model that predicts filtration processes' efficacy. In summary, the model is based on the equilibrium height of the solid component obtained through centrifuge tests with solutions of WPC and the pressure drop of the membrane. The proposed model can enhance ultrafiltration systems' cycle times, pressures, and various other operational parameters.

Continuous ultrafiltration plants employed in manufacturing whey protein concentrates indicated that a conventional 12-loop commercial ultrafiltration plant is suitable for meeting the desired concentration [19]. In addition, due to the high concentrations in the retained product, diafiltration (DF) is employed as a protein purification method, ensuring high purification levels while maintaining efficient process performance (*see* **Note 2**). Notably, the incremental addition of small DF volumes multiple times proves more effective than introducing a large volume in a single instance [22]. This process is suggested by Yee et al. (2007) [23] and can be seen in Fig. 1. Below are described the process for producing whey protein concentrated, with more than 80% of protein, using membrane technology.

1.2 Isolated Whey Protein (WPI)

Isolated whey protein (WPI) is concentrated with low lactose and fat content obtained by processing whey in multistage filtration and with different types of membranes. As a result, more carbohydrates and fats are removed, resulting in a product with fewer calories than WPC and with at least 90% protein in its composition. Due to its high degree of purity, WPI is generally used in nutritional supplements and protein-enriched foods [9, 17, 24].

The main benefits of WPI are its high protein concentration (at least 90%) and the low fat and carbohydrate content (including lactose) due to the process carried out to obtain isolated proteins, which makes it an ideal product for consumption by people with lactose intolerance or who wish to restrict their intake of macronutrients. It is easily absorbed, assists in muscle recovery, and is rich

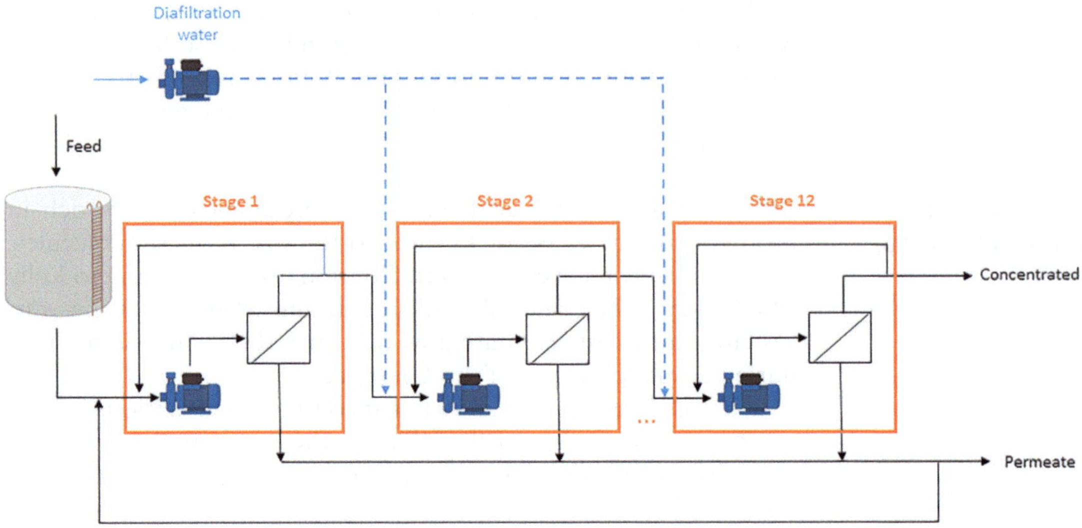

Fig. 1 Continuous whey ultrafiltration process with diafiltration to obtain concentrated whey protein

in essential amino acids fundamental for muscle repair and growth [9, 17, 24].

To obtain WPI, a combination of filtration steps involves ultrafiltration and microfiltration, which is considered a key step in the process. UF generally concentrates whey and MF to retain caseins, aggregated proteins, and fat globules [23]. The combination of UF and MF is typically applied in two ways: (i) MF followed by multiple stages of UF with diafiltration [10, 25, 26] and (ii) UF followed by MF and then multiple stages of UF with diafiltration [23, 24]. In this case, diafiltration guarantees a higher purity of proteins present in the UF concentrate [24]. Each procedure is described in Subheadings 1.2.1 and 1.2.2.

1.2.1 MF as the First Step (Method 1)

According to processes suggested by Reig et al. (2021) [10], Daufin et al. (2001) [25], and Neville et al. (2001) [26], the whey first goes through the filtration stage through a microfiltration membrane with a pore size of 0.1 μm and ceramic material (alumina, zirconia dioxide, or titanium dioxide). The main objective of this first stage is to clarify the whey, removing caseins and residual fat globules, ensuring a product with low-fat content, and allowing the passage of all soluble proteins and other compounds (such as lactose and minerals). The permeate (clarified whey with low-fat content) is used as feed in the second filtration stage with UF (*see* **Note 3**).

In UF, a 10 kDa membrane is used to concentrate soluble proteins and allow the passage of lactose and minerals. To obtain a more purified whey, UF and diafiltration cycles are carried out to increase the process efficiency and the degree of purification, guaranteeing at the end of the cycle the obtaining of WPI with at least

90% protein and low lactose content. The suggested process scheme can be seen in Fig. 3.

The joint use of MF and UF to obtain WPI does not involve using chemicals, the need to lower pH or heat treatment, or avoiding chemical modifications of the protein. In other words, when using soft processing conditions, a product with low protein denaturation is obtained at the end and preserves its natural characteristics (*see* **Note 4**).

1.2.2 Use of UF Alternated with MF (Method 2)

The process suggested by Blais et al. (2022) [23] and Carvalho et al. (2009) [24] involves the whey filtration in a UF membrane, pore size of 10 kDa, as the first stage of the process (*see* **Note 1**). According to the authors, this first step aims to focus on soluble whey proteins. The retentate from the first stage is utilized as the feed for the second processing stage, which involves the MF of concentrated whey. A membrane with a pore size of 0.1 μm and ceramic materials (such as alumina, zirconia dioxide, or titanium dioxide) is employed for microfiltration. MF is applied to eliminate residual fat globules, ensuring low-fat content (*see* **Notes 3** and **4**). The MF permeate is utilized as the feed for ultrafiltration (UF) and diafiltration cycles. UF cycles, conducted simultaneously with diafiltration, aim to purify the concentrated whey, particularly by removing lactose, resulting in a final product with a high protein concentration (greater than 90% protein) and low fat and lactose content. In this process, diafiltration ensures higher protein purity in the UF concentrate. The suggested process for obtaining whey protein isolate (WPI) using UF alternated with MF is illustrated in Fig. 4.

1.3 Whey Protein Hydrolysate

Whey protein hydrolysate (WPH) is the product obtained from the hydrolysis of whey protein concentrated or isolated. Normally, hydrolysis is carried out by enzymes that break the peptide bonds of proteins and transform them into short peptide chains, easily absorbed by the body (*see* **Note 5**). In summary, it is a predigestion process that allows proteins to be absorbed quickly in the intestine and in a much higher concentration than WPC or WPI, considered the highest quality form of whey protein [27, 28].

Among the main WPH uses is its inclusion in food formulations that wish to have the nutritional value of whey protein but avoid allergic reactions that intact proteins can trigger. Also used in sports nutrition, due to its characteristics, WPH provides faster muscle recovery after exercise (hydrolyzed protein is digested and absorbed more quickly than long-chain proteins) [27].

For WPH production, Peng et al. (2009) [29] suggest using WPI as whey and Alcalase as enzymes. After obtaining WPI in the membrane filtration steps, hydrolysis is carried out with the Alcalase addition, using an enzyme: WPI ratio (E/S) 2:100, pH 8.5, and temperature of 65 °C. The process takes place under agitation to ensure homogenization of the entire system for 5 h. Once the hydrolysis process is complete, the pH is adjusted to 7.0, and the drying process is carried out to obtain the WPH powder (*see* **Note 6**). Figure 5 shows the steps in obtaining the WPH proposed by Peng et al. (2009) [29] from WPI.

2 Materials

2.1 WPC

1. Sweet whey with total solid content next to 6%.
2. Ultrafiltration polyethersulfone membrane 10 kDa.
3. Storage tank.
4. Feed tank.
5. Pump.
6. Housing for spiral membrane module.
7. Manometers.
8. Tap water.
9. Sodium bicarbonate ($NaHCO_3$).
10. Vacuum pump.
11. Rotary filter.
12. Plate heat exchanger.
13. Centrifuge.
14. Spray dryer.

2.2 WPI

2.2.1 WPI (Method 1)

1. Sweet whey with total solid content next to 6%.
2. Microfiltration ceramic membrane 0.1 μm.
3. Ultrafiltration polyethersulfone membrane 10 kDa.
4. Storage tank.
5. Feed tank.
6. Pump.
7. Housing for module spiral polymeric membrane.
8. Housing for module tubular ceramic membrane.
9. Manometers.
10. Tap water.
11. Vacuum pump.
12. Rotary filter.
13. Centrifuge.
14. Spray dryer.

2.2.2 WPI (Method 2)

1. Sweet whey with total solid content next to 6%.
2. Microfiltration ceramic membrane 0.1 μm.
3. Ultrafiltration polyethersulfone membrane 10 kDa.
4. Storage tank.
5. Feed tank.
6. Pump.
7. Housing for module spiral polymeric membrane.
8. Housing for module tubular ceramic membrane.
9. Manometers.
10. Tap water.
11. Vacuum pump.
12. Rotary filter.
13. Centrifuge.
14. Spray dryer.

2.3 Whey Protein Hydrolysate

1. Sweet whey with total solid content next to 6%.
2. Microfiltration ceramic membrane 0.1 μm.
3. Ultrafiltration polyethersulfone membrane 10 kDa.
4. Storage tank.
5. Feed tank.
6. Pump.
7. Housing for module spiral polymeric membrane.
8. Housing for module tubular ceramic membrane.
9. Manometers.
10. Tap water.
11. Vacuum pump.
12. Rotary filter.
13. Centrifuge.
14. Spray dryer.
15. Alcalase.
16. 1 N sodium hydroxide (NaOH).
17. Mechanical stirrer.

3 Methods

3.1 Concentrated Whey Protein (WPC)

1. Test the material to determine the solid content on the feed solution.
2. Fill the storage tank with the sweet whey.

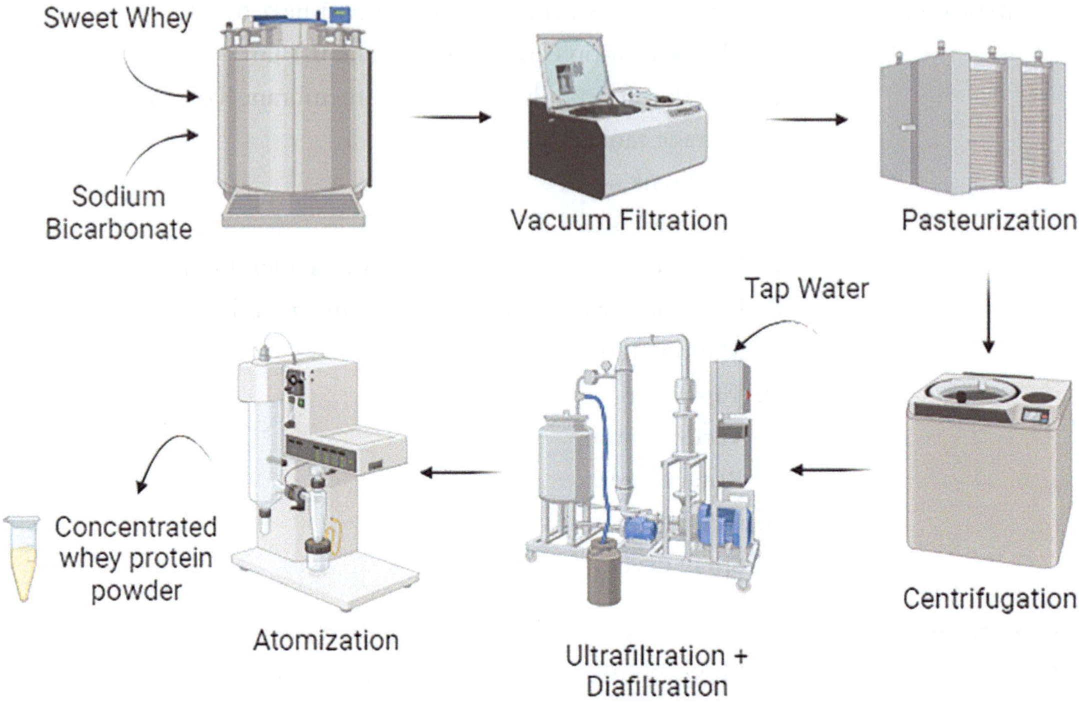

Fig. 2 Flow diagram of an experimental unit for the fabrication of concentrated whey protein

3. If necessary, reduce the acidity of the whey with sodium bicarbonate ($NaHCO_3$).
4. Remove the fine solids in the vacuum filtration through a rotary filter.
5. Heat the whey in a plate heat exchanger for pasteurization (*see* **Note** 7).
6. Reduce the fat in a centrifuge (*see* **Note 8**).
7. Cool the pasteurized whey in the plate heat exchanger.
8. Fill the processing tank with pretreated whey.
9. Set up the membrane system so that the tap water is used as feed together with sweet whey (Fig. 2).
10. Start the membrane system with total removal of the permeates and recirculation of the final concentrate to the processing tank.
11. Set the transmembrane pressure to two bars.
12. Regulate the permeate flow to assess membrane performance and predict the conclusion of the process.
13. Assess the protein content in both the final concentrated solution and permeate.

14. The Volume Reduction Factor (VRF) is determined by the ratio of the initial volume (L) of the sweet whey used in the feed to the final volume (L) of the concentrate after diafiltration–ultrafiltration. The permeate flux (J) ($L\,h^{-1}\,m^{-2}$) during filtration can be calculated as J = VptA, where Vp (L) is the permeate collected during the period t (h), and A (m^2) is the permeation surface area of the membrane. The filtration process performance was assessed based on the protein content present in the concentrate (*see* **Notes 9**, **10**, **11**, and **12**).
15. The concentrated is forwarded directly to the atomizer, where the powder is produced (*see* **Note 13**).
16. After the filtration and atomization, the equipment is cleaned with an alkaline solution (0.1%) and peracetic acid solution (0.1%) according to the manufacturer's instructions.

3.2 Isolated Whey Protein (WPI)

3.2.1 WPI (Method 1)

1. Test the material to determine the solid content on the feed solution.
2. Fill the storage tank with the sweet whey.
3. Remove the fine solids in the vacuum filtration through a rotary filter.
4. Reduce the fat in a centrifuge.
5. Fill the processing tank with pretreated whey.
6. Set up the membrane system so that the microfiltration permeate is used as a feed solution for the ultrafiltration membrane (Fig. 3).
7. Start the membrane system with total removal of the permeate and recirculation of the final concentrate to the processing tank.
8. Set the microfiltration transmembrane pressure to one bar.
9. Set the ultrafiltration transmembrane pressure to two bars (*see* **Note 14**).
10. Regulate the permeate flow to assess membrane performance and predict the conclusion of the process.
11. Assess the protein content in both the final concentrated solution and permeate.
12. The Volume Reduction Factor (VRF) is determined by the ratio of the initial volume (L) of the sweet whey used in the feed to the final volume (L) of the concentrate after diafiltration–ultrafiltration. The permeate flux (J) ($L\,h^{-1}\,m^{-2}$) during filtration can be calculated as J = VptA, where Vp (L) is the permeate collected during the filtration time t (h), and A (m^2) is the permeation surface area of the membrane. The filtration process quality was assessed based on the protein content present in the fractions of interest.

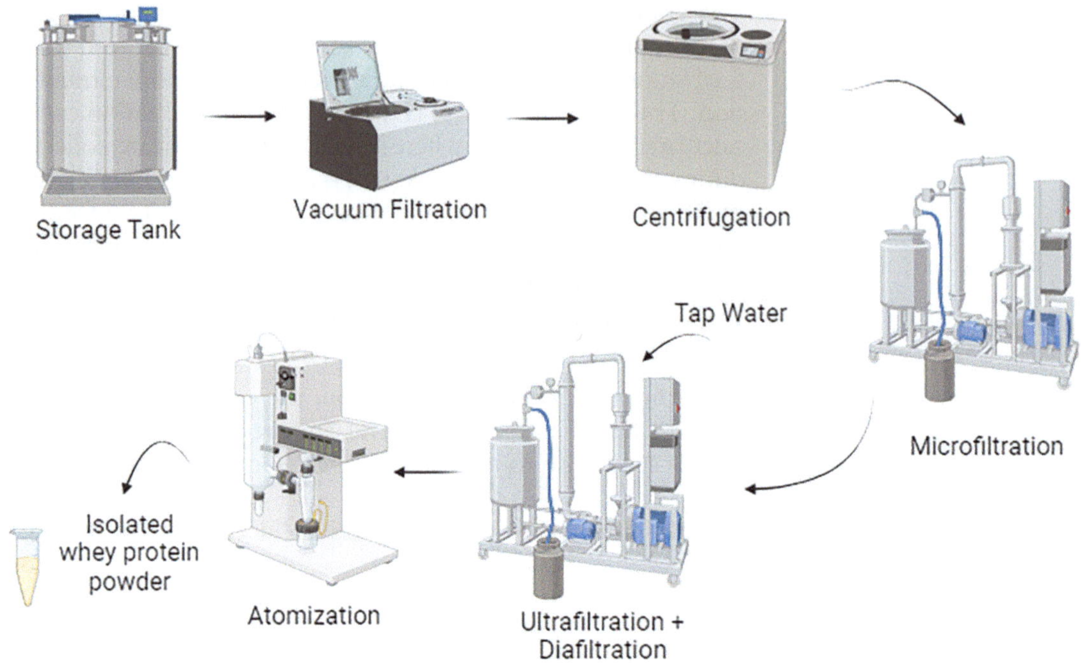

Fig. 3 Outline of the whey protein isolated obtaining process with microfiltration as the first step

13. The ultrafiltration concentrate is forwarded directly to the atomizer, where the powder is produced (*see* **Note 15**).
14. After the filtration and atomization, the equipment is cleaned with an alkaline solution (0.1%) and peracetic acid solution (0.1%) according to the manufacturer's instructions.

3.2.2 WPI (Method 2)

1. Test the material to determine the solid content on the feed solution.
2. Fill the storage tank with the sweet whey.
3. Remove the fine solids in the vacuum filtration through a rotary filter.
4. Reduce the fat in a centrifuge.
5. Fill the processing tank with pretreated whey.
6. Configure the membrane system so that the first step occurs in ultrafiltration, the second step occurs in microfiltration, and the third step occurs in ultrafiltration with diafiltration (Fig. 4).
7. Set up the membrane system so that the ultrafiltration permeate is used as the feed solution for the microfiltration membrane, and the microfiltration permeate is used as the feed solution for the third step with the ultrafiltration membrane.

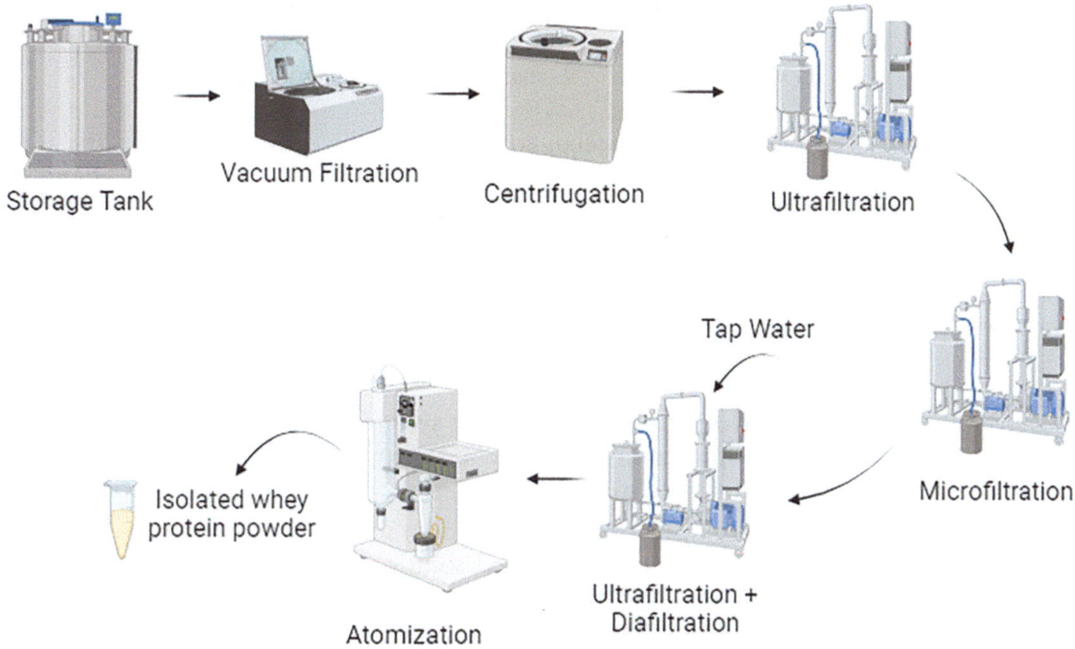

Fig. 4 Scheme for industrial production of whey protein isolated using ultrafiltration alternated with microfiltration

8. Start the membrane system with total removal of the permeates and recirculation of the final concentrate to the processing tank.
9. Set the ultrafiltration transmembrane pressure to two bars.
10. Set the microfiltration transmembrane pressure to one bar.
11. Regulate the permeate flow to assess membrane performance and predict the conclusion of the process.
12. Assess the protein content in both the final concentrated solution and permeate.
13. The Volume Reduction Factor (VRF) is determined by the ratio of the initial volume (L) of the sweet whey used in the feed to the final volume (L) of the concentrate after diafiltration–ultrafiltration. The permeate flux (J) ($L\,h^{-1}\,m^{-2}$) during filtration can be calculated as J = VptA, where Vp (L) is the permeate collected during the filtration time t (h), and A (m^2) is the permeation surface area of the membrane. The filtration process quality was assessed based on the protein content present in the fractions of interest.
14. The ultrafiltration concentrate is forwarded directly to the atomizer, where the powder is produced.

15. After filtration and atomization, the equipment is cleaned with an alkaline solution (0.1%) and peracetic acid solution (0.1%) according to the manufacturer's instructions.

3.3 Whey Protein Hydrolysate

1. Test the material to determine the solid content on the feed solution.
2. Fill the storage tank with the sweet whey.
3. Remove the fine solids in the vacuum filtration through a rotary filter.
4. Reduce the fat in a centrifuge.
5. Fill the processing tank with pretreated whey.
6. Configure the membrane system so that the first step occurs in ultrafiltration, the second step occurs in microfiltration, and the third step occurs in ultrafiltration with diafiltration.
7. Set up the membrane system so that the ultrafiltration permeate is used as the feed solution for the microfiltration membrane, and the microfiltration permeate is used as the feed solution for the third step with the ultrafiltration membrane.
8. Start the membrane system with total removal of the permeates and recirculation of the final concentrate to the processing tank.
9. Set the ultrafiltration transmembrane pressure to two bars.
10. Set the microfiltration transmembrane pressure to one bar.
11. Regulate the permeate flux to assess membrane performance and predict the conclusion of the process.
12. Assess the protein content in both the final concentrated solution and permeate.
13. The Volume Reduction Factor (VRF) is determined by the ratio of the initial volume (L) of the sweet whey used in the feed to the final volume (L) of the concentrate after diafiltration–ultrafiltration. The permeate flux (J) ($L\,h^{-1}\,m^{-2}$) during filtration can be calculated as J = VptA, where Vp (L) is the permeate collected during the filtration time t (h), and A (m^2) is the permeation surface area of the membrane. The filtration process quality was assessed based on the protein content present in the fractions of interest.
14. The enzyme is added to the ultrafiltration concentrate to hydrolyze whey protein isolate (Fig. 5).
15. Add Alcalase (enzyme) to the whey protein isolate solution at a ratio of 2:100 (E/S).
16. Adjust pH 8.5 with 1 N NaOH.
17. Set the hydrolysis temperature to 65 °C.
18. Keep the solution under agitation during hydrolysis.

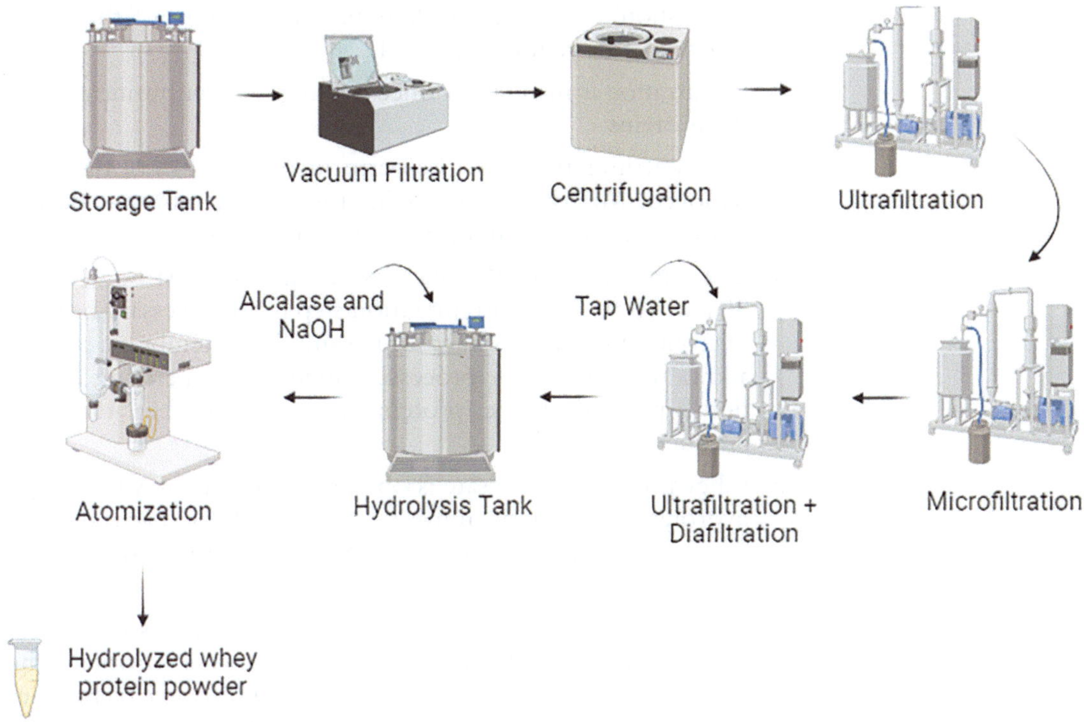

Fig. 5 Steps to obtain whey protein hydrolysate from whey protein isolated through enzymatic hydrolysis

19. The degree of hydrolysis (DH) of WPI is calculated based on the pH-stat method, which is the ratio of hydrolysis equivalent (h) to total hydrolysis equivalent (h_{tot}). The hydrolysis equivalent (h) is calculated by h = B * Nb * 1/a * 1/MP, where B is the NaOH consumption (mL), Nb is the NaOH concentration (1 M NaOH), 1/a is the calibration factor for pH-stat (1/a = 1.01 for Alcalase) and MP is protein mass (g). The total hydrolysis equivalent (h_{tot}) for whey is 8.8 mmol g^{-1} protein.
20. The ultrafiltration concentrate and hydrolysate are forwarded directly to the atomizer, where the powder is produced.
21. After the filtration and atomization, the equipment is cleaned with an alkaline solution (0.1%) and peracetic acid solution (0.1%) according to the manufacturer's instructions.

4 Notes

1. Some parameters must be considered during isolated whey protein production. Like concentrated whey, reception characteristics, such as reception temperature, must be evaluated and maintained below 4 °C. Additionally, the storage time for sweet

whey should not exceed 10 hours. The acidity of the sweet whey must be reduced to 12 °D.

2. Diafiltration is mandatory once it allows greater purification of the proteins.
3. The centrifugation must be performed until 90% of the fat is removed. Microfiltration retains the fat not removed during centrifugation and can break down emulsions.
4. The temperature of the microfiltration feed must be between 30 and 40 °C. The pasteurization step is not necessary because microfiltration retains microorganisms. The membrane process allows the whey to be processed under mild operating conditions, preserving the protein nature present in the final product.
5. Whey protein hydrolysate is obtained using an enzyme, so the hydrolysis conditions must be optimized for each type of enzyme used. Each enzyme has its optimum pH and temperature for enzyme activity.
6. After filtration, the whey protein is concentrated, isolated, and hydrolyzed to go through the drying process to obtain the powder. Typically, a spray dryer is used, but other techniques can also be used (for example, freeze-drying).
7. The pasteurization is carried out at 72 °C for 15 s.
8. The centrifugation must be performed until 90% of the fat is removed.
9. The filtration process performance depends on the membrane pore size, applied transmembrane pressure, and crossflow velocity.
10. The larger the membrane pore size, the greater the permeate flux, but fouling formation will be faster. The larger the pore size, the smaller the selectivity of the membrane. Low molecular mass proteins can pass into the permeate. Membranes with different pore sizes allow the separation of different protein fractions, which can meet specific needs.
11. The higher the pressure, the greater the permeate flux and the faster the fouling formation.
12. Higher crossflow velocity increases permeate flux. The crossflow velocity increases the shear between the fluid and the membrane surface, hindering the formation of the polarized layer, which acts as a second filtration barrier, leading to a drop in permeate flux.
13. The maximum moisture content of the powder must be 4%. Typically, a spray dryer is used, but other drying techniques can also be used.

14. For the ultrafiltration process, diafiltration is mandatory once it allows greater purification of the proteins. Diafiltration is the key process in purifying whey and increasing protein concentration. The greater the number of diafiltration and ultrafiltration cycles, the greater the protein concentration in the final product.
15. The maximum moisture content of the powder must be 4%. A spray dryer is typically used, but other drying techniques can also be used.

References

1. Buchanan D, Martindale W, Romeih E, Hebishy E (2023) Recent advances in whey processing and valorisation: technological and environmental perspectives. Int J Dairy Technol 76:291–312. https://doi.org/10.1111/1471-0307.12935
2. Ganju S, Gogate PR (2017) A review on approaches for efficient recovery of whey proteins from dairy industry effluents. J Food Eng 215:84–96. https://doi.org/10.1016/j.jfoodeng.2017.07.021
3. Silva AND, Perez R, Minim VPR et al (2015) Integrated production of whey protein concentrate and lactose derivatives: what is the best combination? Food Res Int 73:62–74. https://doi.org/10.1016/j.foodres.2015.03.009
4. Tavares T, Malcata FX (2016) Whey and whey powders: protein concentrates and fractions. In: Encyclopedia of food and health. Elsevier, Oxford, pp 506–513
5. Prazeres AR, Carvalho F, Rivas J (2012) Cheese whey management: a review. J Environ Manag 110:48–68. https://doi.org/10.1016/j.jenvman.2012.05.018
6. ADPI (2016) Whey protein concentrate (WPC) standard. American Dairy Products Institute. https://www.adpi.org/Portals/0/Standards/WPCStandard_book.pdf. Accessed 29 Dec 2023
7. Chen GQ, Qu Y, Gras SL, Kentish SE (2023) Separation technologies for whey protein fractionation. Food Eng Rev 15:438–465. https://doi.org/10.1007/s12393-022-09330-2
8. Wen-qiong W, Yun-chao W, Xiao-feng Z et al (2019) Whey protein membrane processing methods and membrane fouling mechanism analysis. Food Chem 289:468–481. https://doi.org/10.1016/j.foodchem.2019.03.086
9. Argenta AB, Scheer ADP (2020) Membrane separation processes applied to whey: a review. Food Rev Intl 36:499–528. https://doi.org/10.1080/87559129.2019.1649694
10. Reig M, Vecino X, Cortina JL (2021) Use of membrane technologies in dairy industry: an overview. Food Secur 10:2768. https://doi.org/10.3390/foods10112768
11. Pires AF, Marnotes NG, Rubio OD et al (2021) Dairy by-products: a review on the valorization of whey and second cheese whey. Food Secur 10:1067. https://doi.org/10.3390/foods10051067
12. Białas W, Stangierski J, Konieczny P (2015) Protein and water recovery from poultry processing wastewater integrating microfiltration, ultrafiltration and vacuum membrane distillation. Int J Environ Sci Technol 12:1875–1888. https://doi.org/10.1007/s13762-014-0557-4
13. Castro-Muñoz R, Orozco-Álvarez C, Yáñez-Fernández J (2015) Recovery of bioactive compounds from food processing wastewaters by ultra and nanofiltration: a review. Advances in Bioresearch 63(6):152–158
14. Mulder M (1996) Basic principles of membrane technology. Springer, Berlin
15. Cheryan M (1998) Ultrafiltration and microfiltration handbook. CRC Press, Boca Raton
16. Ye H, Lei J, Zhang Y et al (2011) Investigation of microfiltration for pretreatment of whey concentration. Desalin Water Treat 34:173–178. https://doi.org/10.5004/dwt.2011.2914
17. Abd El-Salam MH, El-Shibiny S, Salem A (2009) Factors affecting the functional properties of whey protein products: a review. Food Rev Intl 25:251–270. https://doi.org/10.1080/87559120902956224
18. Brans G, Schroën CGPH, van der Sman RGM, Boom RM (2004) Membrane fractionation of milk: state of the art and challenges. J Membr Sci 243:263–272. https://doi.org/10.1016/j.memsci.2004.06.029

19. Kelly P (2019) Manufacture of whey protein products. In: Whey proteins. Elsevier, London, pp 97–122
20. Yee KWK, Wiley DE, Bao J (2007) Whey protein concentrate production by continuous ultrafiltration: operability under constant operating conditions. J Membr Sci 290:125–137. https://doi.org/10.1016/j.memsci.2006.12.026
21. Davey M, Landman K, Perera JM et al (2004) Measurement and prediction of the ultrafiltration of whey protein. AICHE J 50:1431–1437. https://doi.org/10.1002/aic.10136
22. Baldasso C, Barros TC, Tessaro IC (2011) Concentration and purification of whey proteins by ultrafiltration. Desalination 278:381–386. https://doi.org/10.1016/j.desal.2011.05.055
23. Blais HN, Schroën K, Tobin JT (2022) A review of multistage membrane filtration approaches for enhanced efficiency during concentration and fractionation of milk and whey. Int J Dairy Technol 75:749–760. https://doi.org/10.1111/1471-0307.12884
24. de Carvalho AF, Maubois JL (2010) Applications of membrane technologies in the dairy industry. In: Engineering aspects of milk and dairy products. Taylor and Francis Group LLC, Boca Raton, pp 33–56
25. Daufin G, Escudier J-P, Carrère H et al (2001) Recent and emerging applications of membrane processes in the food and dairy industry. Food Bioprod Process 79:89–102. https://doi.org/10.1205/096030801750286131
26. Neville JR, Armstrong KJ, Price J (2001) Ultra whey 99: a whey protein isolate case study. Int J Dairy Technol 54:127–129. https://doi.org/10.1046/j.1364-727x.2001.00022.x
27. Boland M (2010) 'Designer' milks: functional foods from milk. In: Improving the safety and quality of Milk. Elsevier, Boca Raton, pp 74–93
28. Mudgil D, Barak S (2019) Dairy-based functional beverages. In: Milk-based beverages. Elsevier, Boca Raton, pp 67–93
29. Peng X, Xiong YL, Kong B (2009) Antioxidant activity of peptide fractions from whey protein hydrolysates as measured by electron spin resonance. Food Chem 113:196–201. https://doi.org/10.1016/j.foodchem.2008.07.068

Chapter 13

Powdered Milk

Patrícia Fernanda Schons, Rosicler Colet, Stephanie Silva Pinto, and Silvani Verruck

Abstract

Powdered milk is a dairy product widely consumed worldwide due to its nutritional quality and ease of use. It can be stored at room temperature for extended periods without substantial loss of quality. It also finds broad application in the food industry as an ingredient in producing other dairy products, baked goods, and infant formulas. Heat application is an essential part of the powdered milk processing technology, which includes thermal treatment (pretreatment), concentration, and dehydration of fluid milk. These heat treatments significantly influence the final product's physical and functional properties and its stability during storage. This chapter serves as a practical guide to the industrial-scale powdered milk manufacturing process, presenting the ingredients, equipment, and steps involved in its production. Additionally, it addresses how process variables influence the characteristics of the obtained product both during production and throughout its shelf life.

Key words Whole milk powder, Skim milk powder, Heat processes, Concentration, Spray drying, Roller drying, Instantization, Physical properties, Milk powder storage

1 Introduction

Powdered milk stands out as the primary commodity in the dairy products market, with the global production of 4.59 million tons of whole milk powder and 4.64 million tons of skimmed milk powder projected for 2023 [1]. This scenario reflects the versatility and convenience of the application of this product.

Powdered milk is the dehydrated product obtained through dehydration of fluid milk (whole, skimmed, or partially skimmed), which contains about 87% water in its initial composition. At the end of the process, only the dry extract and approximately 2.5–4.0% water remain [2, 3]. Thus, dehydration allows the conversion of milk, which is highly perishable, into a product that can be stored for extended periods (approximately 1 year), even at room temper-

Adriano Gomes da Cruz et al. (eds.), *Dairy Foods Processing*, Methods and Protocols in Food Science,
https://doi.org/10.1007/978-1-0716-4144-6_13,

ature, without substantial loss of quality [4]. The process, in addition to inhibiting microbial growth and enzymatic activity by reducing water activity (average values of 0.288 ± 0.035), also facilitates the product's transportation, storage, and handling [5, 6].

The product is primarily available in whole, partially skimmed, and skimmed versions, with the latter being the most common type [5, 6]. In the food industry, powdered milk is applied as an ingredient in the manufacturing of a wide variety of products, such as dairy beverages, fermented milks, ice creams, infant formulas, baked goods, confectionery products, snacks, meat products, milk chocolate, among others [6, 7]. For consumers, powdered milk is a ready-made food product, offering an alternative to the consumption of fluid milk, contributing to ensuring food security [7]. Furthermore, reconstituted powdered milk is used in scientific research as a medium for growing starter cultures at dairies [8] and as an encapsulating agent for starter and probiotic bacteria [9–11].

Applying thermal processes is essential for the industrial-scale production of powdered milk, occurring in the stages of thermal treatment (or preheating), concentration, and dehydration [7]. However, powdered milk's physical and rehydration properties, such as solubility, wettability, and dispersibility, vary considerably according to the intensity of the employed thermal treatment (from pasteurization methods to ultrahigh temperature (UHT) conditions). The more intense the applied thermal treatment, the higher the percentage of serum protein denaturation, reducing the solubility of powdered milk in water [2, 5]. Additionally, the equipment used in the concentration process (for example, type and number of evaporator effects) and dehydration (drum dryer or spray dryer), as well as the parameters applied in these processes, also exert influence [2, 4, 5]. Therefore, understanding the characteristics of the obtained product is crucial for directing its industrial application.

In addition to processing technology, appropriate packaging and storage conditions are fundamental to preventing negative changes. The increase in temperature and relative humidity stimulates various chemical reactions and processes, such as the Maillard reaction, lactose crystallization, and lipid oxidation, which can degrade the quality of powdered milk [6, 7].

Finally, optimizing the milk powder processing plant through the prioritization of energetically efficient equipment and effectively controlling process variables is essential to achieve a more economical and sustainable process [12]. Therefore, this chapter addresses the ingredients and equipment for powdered milk production, its manufacturing process, and the main causes of alterations during storage.

2 Materials

2.1 Ingredients for Powdered Milk Production

The legal requirements for different types of powdered milk can be observed in Table 1 [13]. Powdered milk can be produced using the following ingredients, additives, and coadjuvants:

- Milk (*see* **Note 1**).
- Emulsifier: soy lecithin (*see* **Note 2**).
- Anticaking agents (*see* **Note 3**).
- Minerals: calcium (calcium carbonate), iron (ferric pyrophosphate), zinc (zinc sulfate).
- Vitamins A, C, D, and E (*see* **Note 4**).
- Coadjuvants: inert gases (nitrogen and carbon dioxide) (*see* **Note 5**).

2.2 Equipments

The main equipment required for powdered milk production is listed below:

- Receiving tank.
- Storage silos.
- Pumps for milk flow.
- Milk filter.
- Skim milk centrifuge.
- Clarifying centrifuge.
- Fat standardizer.
- Homogenizer.

Table 1
Legal requirements for powdered milk [13]

Requirement	Whole	Partially skimmed	Skimmed
Fat (% m/m)	≥26.0	1.5 a 25.9	<1.5
Moisture (% m/m)	Maximum 5.0	Maximum 5.0	Maximum 5.0
Protein content (% m/m)	Minimum 34.0	Minimum 34.0	Minimum 34.0
Titratable acidity (mL NaOH 0.1 N/10 g nonfat solids)	Maximum 18.0	Maximum 18.0	Maximum 18.0
Insolubility index (mL)	Maximum 1.0	Maximum 1.0	Maximum 1.0
Burnt particles (max)	Disk B	Disk B	Disk B
For instant powdered milk			
Wettability (s)	Maximum 60	Maximum 60	Maximum 60
Dispersibility (% m/m)	Minimum 85	Minimum 90	Minimum 90

- Pasteurizer.
- Cooler.
- Ingredient mixer.
- Tubular heaters.
- Concentrator (evaporators, membranes, cryoconcentrators).
- Drying chamber (spray dryer, whether coupled or not to fluidized milk; cylinders; rollers; drum dryer; just-hatmaker; freeze dryer).
- Fluidized bed vibrators.
- Dosing machines.
- Packaging machines.

3 Methods

The processing of powdered milk depends on the type of product to be produced, which can be whole milk powder, partially skimmed, skimmed, instant, or non-instant. The stages include raw material reception, pasteurization, ingredient mixing, concentration, dehydration, powder separation, and packaging (Fig. 1).

1. The collection of milk and its transportation to the company is carried out by isothermal trucks under ideal conditions of hygiene and quality. Upon reception, the raw material undergoes various sensory, microbiological, and physicochemical analyses (*see* **Note 1**).
2. The milk is filtered to remove impurities or foreign material, cooled to temperatures below 4 °C, and stored in isothermal tanks (to hinder thermal exchange with the environment).
3. The milk undergoes clarification through a centrifugal clarifier (2000–3000 rpm), an optional step used for the physical removal of impurities such as somatic cells, dead bacteria, blood clots, leukocytes, pus, protein-rich bacterial pellets, and other impurities that cannot be removed by filtration (*see* **Notes 6** and **8**).
4. Then the fat and/or protein content standardization occurs, with these parameters aligning with the intended milk classification and current legal standards. Fat standardization usually involves total skimming and subsequent incorporation until the desired content is achieved. The protein content can be adjusted by adding milk retentate, milk permeate, or lactose (*see* **Notes** 7 and **8**).
5. Next, the milk undergoes thermal treatment, with rapid pasteurization being the most commonly used method (72–75 °C for 15–20 s) to ensure microbiological safety and inactivation of undesirable enzymes (*see* **Notes 9** and **10**).

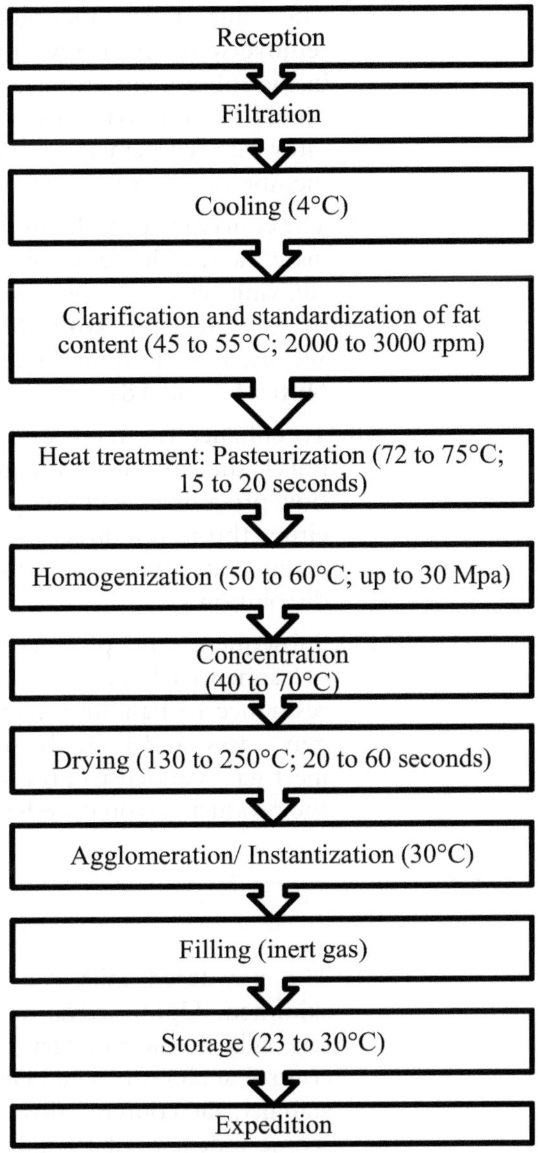

Fig. 1 General flowchart of powdered milk processing

6. Homogenization is performed to reduce the size of milk fat globules (0.4–2 μm), improving the stability of the fat emulsion and preventing flocculation (particle aggregation). This process occurs in homogenizers with pressures of up to 30 MPa (*see* **Note 11**).
7. Following homogenization, optional ingredients (emulsifier, anticaking agents, minerals, vitamins) are added and mixed to ensure nutritional value and product quality. This phase may occur after the milk has been concentrated.

8. The milk then undergoes concentration/evaporation to remove water from raw milk partially, reducing energy costs in the subsequent atomization stage of concentrated milk (*see* **Notes 12** and **13**). This is achieved through a vacuum evaporator with temperatures between 40 and 70 °C (*see* **Note 14**) or membrane filtration (*see* **Note 15**).
9. The concentrated milk undergoes dehydration/drying to eliminate the remaining moisture content through energy transfer, allowing water evaporation and obtaining powdered milk (3–4% water) (*see* **Note 16**). Spray drying is the most industrially used method (with temperatures medium 180 °C) (*see* **Notes 17** and **18**).
10. To enhance the reconstitution ability of the powder (instant milk), the product undergoes agglomeration and instantization processes (wettability and dispersibility) (*see* **Note 19**), either through a direct process (*see* **Note 20**) or using a fluidized bed dryer (*see* **Note 21**), increasing its wetting and dissolution rate.
11. Subsequently, the product is packaged in metalized, laminated, or canned packages, ensuring physical protection, mechanical resistance to handling, and impermeability to moisture, oxygen, gases, and light. In some product and packaging types, inert gas is added to prevent fat oxidation (*see* **Note 22**), and the product is stored preferably at 30 °C (*see* **Note 23**).

4 Notes

1. The powdered milk can be whole, skimmed, or partially skimmed. Upon arrival at the receiving platform for chilled raw milk in the industry, it undergoes the following quality control analyses before being released for production: alizarol stability, fat content, titratable acidity, relative density, cryoscopic index, nonfat solids, total solids, research on residues of veterinary products, and investigation of substances foreign to its composition. Additionally, periodic monitoring of the received milk from producers is conducted for total protein content, lactose, standard plate count, and somatic cell count, as described in Table 2 [14–16].
2. Adding emulsifiers to milk can occur either before dehydration or during the physical agglomeration process of powdered milk particles. The function of this additive is to reduce surface tension. Emulsifiers have hydrophilic portions in their physicochemical structure, which have an affinity for water, and hydrophobic portions, which have an affinity for lipids. Therefore, when this ingredient is applied in powdered milk production,

Table 2
Legal parameters for receiving refrigerated raw milk in Brazil [14], the United States [15], and the European Union [16]

Variables	Brazil	USA	European Union
Temperature (°C)	10	–	≤5
Minimum density (g/mL)	1.028–1.034	–	1.028
Fat (%)	>3.0	–	–
Nonfat solids (g/100 g)	>8.4	–	–
Protein (%)	>2.9	>2.0	>2.8
Titratable acidity (g lactic acid/mL)	0.14–0.18	–	–
Cryoscopic index (°H)	−0.530 a − 0.550	–	–
Stability to alizarol 72% (v/v)	Stable	–	–
Total bacterial count (10^3 CFU/mL)	<300	≤ 100	≤ 100
Somatic cell count (10^3 cells/mL)	<500	<750	≤400

the fat particles are coated by this additive. When the milk comes into contact with water, there is a dispersion of the product. The market for natural emulsifiers is dominated by lecithin, with soy being the most common source (95%) due to its availability and emulsifying properties [17].

3. Anticaking agents are highly hygroscopic additives, meaning they have a high affinity for water. They are used in foods where reducing moisture absorption is desired, preventing particle clumping when exposed to ambient humidity. Their use is limited to powdered milk intended for vending machines. Examples of anticaking agents that can be applied include calcium silicates, magnesium silicates, tricalcium phosphate, silicon dioxide, calcium carbonate, and magnesium carbonate [18].
4. Sources of vitamins A (retinol acetate), C (sodium ascorbate), D (cholecalciferol), and E (DL-alpha-tocopherol acetate).
5. For the packaging technology of powdered milk, inert gases such as nitrogen and carbon dioxide can be used. The dehydrated product granules retain a small amount of oxygen in their structure, which is released into the free space of the packaging after sealing. This can lead to undesirable modifications in the food, such as oxidative rancidity. To avoid this problem, modified atmosphere packaging is used, where the oxygen content (at levels of 1–2%) in the packaging is reduced by adding inert gas [19].

6. Milk clarification involves removing smaller impurities through mechanical methods using centrifugation separation technology [20]. The milk is preheated to 45–55 °C to reduce viscosity and facilitate the removal of unwanted particles. It is then directed to the clarifying centrifuge under pressure through a stainless steel pipeline. The centrifugal force (from 2000 to 3000 rpm) in the clarifying centrifuge separates substances, keeping impurities inside the equipment, and pure milk is collected by a centrifugal impeller that directs it for cooling.
7. Standardization or partial/full skimming serves to standardize raw material by adjusting the fat content. This step is necessary because the fat content in raw milk varies according to seasons, the animal's age, the breed of cows, lactation period, diet, and management, among other factors. Brazilian legislation mandates a minimum of 26% fat for whole powdered milk, 1.5–25.9% for partially skimmed, and a maximum of 1.5% fat for skimmed powdered milk. The standardization process can be carried out using a standardizing centrifuge, which involves the controlled incorporation of part of the cream into the skimmed milk from the centrifugation process [21].
8. Milk clarification and fat standardization can occur simultaneously in a centrifugal separator, using the lower density of fat concerning skimmed milk to cause separation. The degree of fat separation varies with the rotation speed, the diameter of the centrifugation chamber, and the time the milk stays within it [22].
9. Thermal treatment reduces the pathogenic and deteriorating microbial load and inactivates enzymes, thereby increasing the durability of the final product. Thermal treatment for milk can be through low-temperature long time (LTLT; 63 °C for 30 min), high-temperature short time (HTST; 76–85 °C for 15–30 s), and ultrahigh temperature (UHT; 140–150 °C for 2–4 s) which can be indirect or direct; after the milk is followed by cooling to a temperature equal to or below 4 °C [21].
10. Depending on the thermal treatment it undergoes, skimmed milk powder can be classified into low thermal treatment, where the content of undenatured whey protein nitrogen (whey protein nitrogen index, WPNI) is greater than or equal to 6.00 mg/g (recombined products, starter cultures, dairy drinks, ice cream, yogurts, fermented skimmed milk, standardization of milk for cheese); medium thermal treatment between 1.51 and 5.99 mg/g WPNI (prepared mixes, ice cream, yogurts, confectionery, meats, recombined milk products); and high thermal treatment less than 1.50 mg/g WPNI (baking, meats, prepared mixes, and ice cream) [21]. The last

two types of milk are not recommended for direct consumption as they do not reconstitute well but are recommended for industrial use as ingredients in baking, ice cream, and meat products.

11. Homogenization is a mechanical, physical process used in whole milk, carried out in homogenizers. It involves forcing the milk at high pressures (up to 30 MPa) through a small gap at high speeds to break down fat globules into smaller diameter globules (ranging from 0.4 to 2 μm) [23]. The main objectives of this step are to delay creaming of the milk (natural separation of a large portion of fat globules from the rest of the milk components due to density differences between phases) and increase its stability by preventing the coalescence of fat globules (a colloidal destabilization phenomenon involving the fusion of fat globules) [21].
12. Heat-treated milk is concentrated under vacuum conditions at temperatures between 40 and 70 °C to avoid Maillard browning reactions, minimize the adverse impact of heat on sensitive milk components (fats and vitamins), achieve the desired viscosity to reduce the energy required for the subsequent step (with an energy cost per kilogram of evaporated water up to 20 times lower than the spray drying process), and reduce the likelihood of food deterioration due to microbial action [24].
13. Different methods can be used for milk concentration, such as membrane separation, cryoconcentration (freezing water and physical separation of pure water crystals), or evaporation using heat. When the solids concentration reaches 30–50%, it is immediately pumped to a drying tower to obtain the final moisture content [25].
14. In the multiple-effect evaporator system, which is most commonly used by the industry, temperatures range from 66 to 68 °C for the first effect (with a vapor pressure of 27.18–28.50 kPa) and 45 to 47 °C for the last effect (with a vapor pressure of 9.58–10.54 kPa), with a retention time in each evaporator of approximately 60 s. The temperature gradient of 21 °C is divided among the number of effects (usually four to seven effects) in the evaporation system, resulting in minimal physicochemical changes to the product [12, 26]. In this arrangement, several evaporators are connected in series, and the vapor generated by boiling the milk in one evaporator (or effect) is used as a heating medium for the next. There must be a progressive reduction in pressure across the effects to maintain the temperature difference between the liquid and the heating vapor [27].
15. The main membrane separation processes used to concentrate milk are microfiltration and ultrafiltration (operating with a molecular weight cutoff of 2000 to 3000 Da at pressures of

50 to 1000 × 10^3 Pa), nanofiltration, and reverse osmosis (110 to 500 Da). These processes differ in the sizes of the pores in the membranes used and operational parameters such as pressure value. Notably, they have low energy consumption, can operate in both continuous and batch systems, and do not require milk heating. However, a major disadvantage is the decline in permeate flow due to reversible polarization layer and irreversible fouling, attributed to milk protein and fat, respectively. Additionally, membrane processes have higher costs compared to evaporation [28].

16. To obtain these products on a large scale, among various moisture reduction processes, notable methods include roller or drum drying or just-hatmaker (contact with hot surfaces at 120–140 °C, leading to intense protein denaturation), spray drying, and freeze-drying (a dehydration process employing vacuum, promoting water sublimation while keeping the food at low temperatures of −30 to −80 °C, resulting in an instant final product, but not widely used due to its high energy, operational, and maintenance costs) [29, 30].
17. The atomization dehydration process is the most widely used method for dehydrating dairy products [12], transforming a fluid food into a dehydrated product. In Fig. 2, the stages of dehydration in a spray dryer can be observed. The milk enters

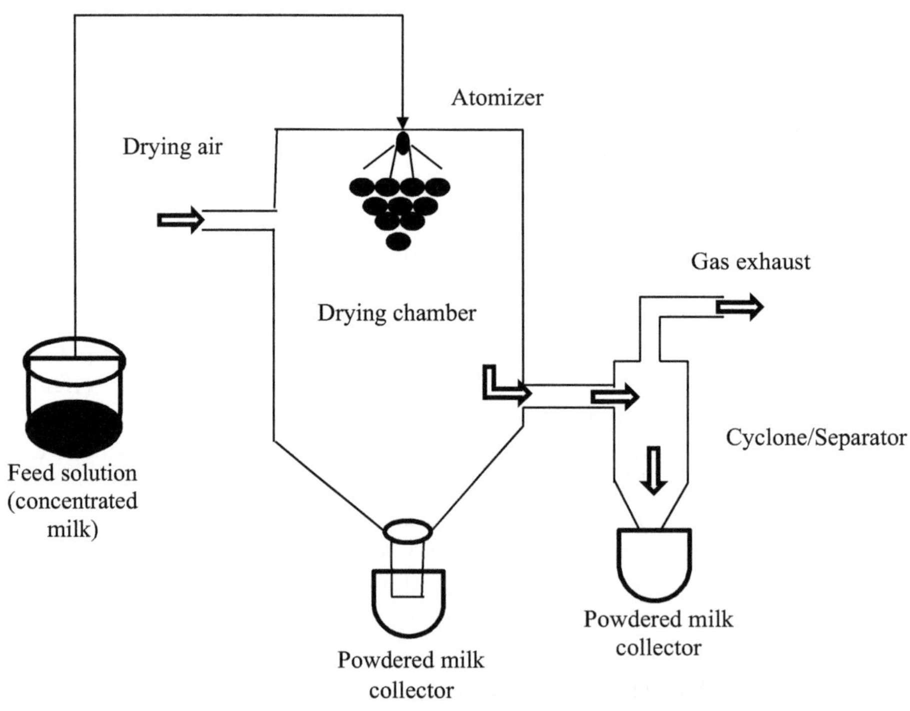

Fig. 2 Schematic representation of the dehydration process in a spray dryer

the process in small droplets, like a mist through atomizer spraying, where the food is pumped to the atomizer nozzle at high pressures (100–600 kgf/cm^2). Drying is carried out with hot air (130–250 °C, averaging 180 °C, vapor pressure of 1.6 Pa, and 0.1% relative humidity), coming into contact with the product (vapor pressure of 9.6 Pa and 100% relative humidity). The product reaches a temperature of 50–80 °C, and the process is fast (20–60 s), ensuring minimal thermal changes to the product and a diameter of 10–250 μm [31].

18. If the temperature of the air surrounding the particles is sufficiently high, overheating may occur, denaturing the protein. This influences the inadequate solubilization of powdered milk and the hardening of amorphous lactose, preventing vapor diffusion [32].
19. Instantaneization is an optional phase in milk powder production. If performed, it produces a powder that is easier and faster to reconstitute due to particle agglomeration (dispersion and wettability), without the formation of lumps upon contact with water. The solubility of milk powder depends on wettability (the powder's ability to sink in water, based on the time it takes, in seconds, for the sample to submerge in water fully) and dispersibility (the powder's ability to separate into individual particles through agitation with water) [33].
20. It can be carried out in two ways: through direct processing during the dehydration stage or after this stage using a fluidized bed dryer. In the direct process, initiated in the atomizer, collisions occur between wet and dry particles, with the fines returning from the cyclone redirected to the atomizer. The dry particles come into contact with the sprayed droplets, leading to their agglomeration. This process uses the same drying equipment for instantaneization and can represent cost savings. For whole milk powder, which has a high-fat content, adding lecithin is allowed to make it instant. Lecithin can mediate the binding of milk fat and water [12, 34].
21. In the fluidized bed dryer, milk powder is sprayed with a binding liquid, i.e., water vapor, responsible for grouping the particles into sizes ranging from 250 to 400 μm. In the subsequent stages, excess water is removed by dehydration air, and the powder is further cooled by cold air. The introduction of fluidized bed drying solved a problem for industries regarding soy lecithin allergy, eliminating the need for emulsifier addition [34].
22. For packaging powdered milk, metalized and laminated films or metal tin cans are used to meet product requirements, protecting against light, oxygen, moisture, and foreign odors, with weight according to the manufacturer and storage

temperature of 23–30 °C. Skimmed milk powder has a shelf life of approximately 3 years, while whole milk powder has a shelf life of 1.5 years due to the fat in the powder oxidizing during storage, leading to deterioration and flavor changes. Instant forms have a shelf life of 6–12 months [35]. Especially for whole and partially skimmed milk powder and in consumer packaging, modified atmosphere packaging involves removing air from the packaging and injecting inert gas, such as nitrogen or carbon dioxide [36].

23. The main defects in milk powder are oxidative and hydrolytic rancidity (for whole milk powder), Maillard reaction, and caking (difficulty in rehydration), which can occur if the product is transported and stored improperly, altering the physicochemical and functional properties of the product [37, 38]. Temperature has a significant effect on both lactose crystallization and the occurrence of the Maillard reaction. These temperature-sensitive reactions can further influence lipid oxidation, as an increase in water activity from lactose crystallization or during the early stages of the Maillard reaction will accelerate lipid oxidation due to increased molecular mobility [32, 39].

References

1. Foreign Agricultural Service/USDA (2023) Dairy: world markets and trade—July 2023. https://fas.usda.gov/data/dairy-world-markets-and-trade. Accessed 14 Dec 2023
2. Tetra Pak Processing Systems (1995) Dairy Processing Handbook. Tetra Pak Processing Systems, Lund
3. Pereira PC (2014) Milk nutritional composition and its role in human health. Nut 30:619–627. https://doi.org/10.1016/j.nut.2013.10.011
4. Sharma A, Jana AH, Chavan RS (2012) Functionality of milk powders and milk-based powders for end use applications—a review. Compr Rev Food Sci Food Saf 11:518–528. https://doi.org/10.1111/j.1541-4337.2012.00199.x
5. Pugliese A, Cabassi G, Chiavaro E et al (2017) Physical characterization of whole and skim dried milk powders. J Food Sci Technol 54(11):3433–3442. https://doi.org/10.1007/s13197-017-2795-1
6. Ryabova AE, Semipyatny VK, Galstyan AG (2023) Effects of storage conditions on milk powder properties. J Dairy Sci 106:6741–6758. https://doi.org/10.3168/jds.2022-23094
7. Li Y, Jia X, Wang Z et al (2022) Characterizing changes in Maillard reaction indicators in whole milk powder and reconstituted low-temperature pasteurized milk under different preheating conditions. J Food Sci 87(1):193–205. https://doi.org/10.1111/1750-3841.15989
8. Stulova I, Kabanova N, Krisciunaite T et al (2013) Fermentation of reconstituted milk by Streptococcus thermophilus: effect of irradiation on skim milk powder. Int Dairy J 31:139–149. https://doi.org/10.1016/j.idairyj.2013.02.004
9. Mutukumira AN, Wang F (2022) Microencapsulation of *Limosilactobacillus reuteri* DPC16 by spray drying using different encapsulation wall materials. J Food Process Preserv 46(10):1–9. https://doi.org/10.1111/jfpp.16880
10. Cence K (2020) Salame tipo italiano com chá verde e cultura *starter* e sais de cura encapsulados. Thesis. Universidade Regional Integrada do Alto do Uruguai e das Missões—URI Erechim
11. Guli O, Atalar I (2019) Different stress tolerance of spray and freeze dried Lactobacillus casei Shirota microcapsules with different encapsulating agents. Food Sci Biotechnol 28(3):807–816. https://doi.org/10.1007/s10068-018-0507-x
12. Rosa E, Prudencio ES (2023) A comprehensive approach about comparison between drying

technologies and powdered dairy products. Food Res Int:113326e. https://doi.org/10.1016/j.foodres.2023.113326

13. MAPA Instrução Normativa n° 53, de 1° de outubro de 2018, Resolução Mercosul/GMC/RES. no 07/18. https://www.gov.br/agricultura/pt-br/assuntos/insumos-agropecuarios/insumos-agricolas/fertilizantes/legislacao/in-53-2013-com-as-alteracoes-da-in-3-de-15-01-2020.pdf. Accessed 20 Nov 2023
14. Brasil. Ministério da Agricultura, Pecuária e Abastecimento (2018) Instrução Normativa n° 76, 26 Nov 2018. Regulamentos Técnicos que fixam a identidade e as características de qualidade que devem apresentar o leite cru refrigerado, o leite pasteurizado e o leite pasteurizado tipo A. Diário Oficial da União: seção 1, Brasília, DF, n. 230, p 9. https://www.in.gov.br/materia/-/asset_publisher/Kujrw0TZC2Mb/content/id/52750137/do1-2018-11-30-instrucao-normativa-n-76-de-26-de-novembro-de-2018-52749894IN%2076. Accessed 14 Dec 2023
15. U.S. Department of Health and Human Services, Public Health Service and Food and Drug Admin. Grade "A" pasteurized milk ordinance. 2017 revision. https://www.fda.gov/media/114169/download. Accessed 14 Dec 2023
16. Corrigendum to Regution (EC) no 853/2004 of the European Parliament and of the Council of 29 April 2004 Laying down specific hygiene rules for food of animal origin. Off J Eur Union, 29 April 2004. https://faolex.fao.org/docs/pdf/eur63427.pdf. Accessed 14 Dez 2023
17. Li Q, Wang Y, Wu Y et al (2019) Flexible cellulose nanofibrils as novel pickering stabilizers: The emulsifying property and packing behavior. Food Hydrocoll 88:180–189. https://doi.org/10.1016/j.foodhyd.2018.09.039
18. Romão BL, Sasaki CL (2023) Ingredientes e composição nutricional de suplementos alimentares dedicados ao esporte: uma análise de suplementos de proteína e carboidratos comercializados no Distrito Federal (DF). Scire Salutis 13(2):140–152. https://doi.org/10.6008/CBPC2236-9600.2023.002.0012
19. Cunha JMD (2021) Os métodos convencionais, as tecnologias emergentes e a reação de Maillard no processamento de leite e derivados, Trabalho de conclusão de curso. Universidade de Santa Catarina, Florianópolis
20. Brasil. Decreto n. 9.013/2017. Regulamento de Inspeção Industrial e Sanitária de Produtos de Origem Animal (RIISPOA). Brasília: Ministério da Agricultura, Pecuária e Abastecimento. Publicado no Diário Oficial da União de 29 Mar 2017
21. Cruz AG, Zacarchenco PB, Oliveira CAF, Corassin CH (2017) Processamento de leites de consumo. Coleção Lácteos – volume 2, 1ª ed. Editora Elsevier, Amsterdam, 372p
22. Serviço Brasileiro de Respostas Técnicas—SBRT (2020) Beneficiamento de leite. Instituto de Tecnologia do Paraná—TECPAR. http://www.respostatecnica.org.br. Accessed 5 Dec 2023
23. Vélez MA, Hynes ER, Meinardi CA et al (2017) Cheese milk low homogenization enhanced early lipolysis and volatiles compounds production in hard cooked cheeses. Food Res Int 96:215–225. https://doi.org/10.1016/j.foodres.2017.02.011
24. Caric M, Akkerman JC, Milanovic S et al (2009) Technology of evaporators, membrane processing and dryers. In: Dairy powders and concentrated products, 1st edn. Wiley-Blackwell, Chichester, p 53
25. Zhang Y, Munir MT, Udugama I et al (2018) Modelling of a milk powder falling film evaporator for predicting process trends and comparison of energy consumption. J Food Eng 225:26–33. https://doi.org/10.1016/j.jfoodeng.2018.01.016
26. Early R (2000) Tecnologia de los produtos lácteos. Acribia, Zaragoza, 459p.
27. Fellows PJ (2006) Tecnologia do processamento de alimentos: princípios e práticos, 2nd ed. Artmed, Porto Alegre, 602p
28. Leu M, Marciniak A, Chamberland J (2017) Effect of skim milk treated with high hydrostatic pressure on permeate flux and fouling during ultrafiltration. J Dairy Sci 100:7071–7082. https://doi.org/10.3168/jds.2017-12774
29. Oliveira OW, Petrovick PR (2010) Secagem por aspersão (*spray drying*) de extratos vegetais: bases e aplicações. Rev Bras 20(4):641–650. https://doi.org/10.1590/S0102-695X2010000400026
30. Bedir TB, Kuleasan H (2019) Determination of microbial properties of freeze dried traditional cheese. Turk J Agric Food Sci Technol 7(4):688–692. https://doi.org/10.24925/turjaf.v7i4.688-692.2126
31. Filková I, Mujumdar AS (1995) Industrial spray drying systems. In: Mujumdar AS. Handbook of industrial drying. 2nd ed. New York: Basel 1:263–304
32. Yan H, Yu Z, Liu L (2022) Lactose crystallization and Maillard reaction in simulated milk powder based on the change in water activity.

J Food Sci 87(11):4956–4966. https://doi.org/10.1111/1750-3841.16335

33. Rankin S (2023) Concentrated and dried milk products, of the sensory evaluation of dairy products. Springer, Cham, pp 345–399
34. Henning DR, Baer RJ, Hassan AN, Dave R (2006) Major advances in concentrated and dry milk products, cheese, and milk fat-based spreads. J Dairy Sci 89(4):1179–1188. https://doi.org/10.3168/jds.S0022-0302(06)72187-7
35. Higiene Alimentar (2023) Projeto aprovado e que exige validade mínima adiciona restrições a importadores de leite em pó. https://higienealimentar.com.br/projeto-aprovado-e-que-exige-validade-minima-adiciona-restricoes-a-importadores-de-leite-em-po/. Accessed 23 Nov 2023
36. Silva ACO, Cortez MAS (2021) Tecnologia de leite e derivados lácteos. Eduff, Niterói. ePUB.—(Coleção Biblioteca Básica)
37. De Oliveira LF, Stephani R, Perrone Í (2017) Problemas na fabricação de produtos concentrados e desidratados. Informe Agropecuário 38(299) https://www.researchgate.net/publication/326898105 Accessed 23 Nov 2023
38. Nasralla NN, Gomah NH, Aly MM et al (2022) Compositional characteristics of dairy products and their potential nondairy applications after shelf-life. Curr Res Food Sci 5:150–156. https://doi.org/10.1016/j.crfs.2021.12.017
39. Thomsen MK, Lauridsen L, Skibsted LH, Risbo J (2005) Temperature effect on lactose crystallization, Maillard reactions, and lipid oxidation in whole milk powder. J Agric Food Chem 53:7082–7090. https://doi.org/10.1021/jf050862p

Chapter 14

Set, Stirred, and Liquid Yogurts Obtained by Lactic Fermentation

Alice Cristina da Silva, Luan Amaral Alexandre, Ana Paula Zapelini de Melo, and Silvani Verruck

Abstract

Yogurt is obtained through the fermentation of milk with starter cultures composed of *Streptococcus thermophilus* and *Lactobacillus bulgaricus*. The yogurt production process encompasses several steps, including checking the characteristics of raw milk, standardizing fat and nonfat solid contents, heat treatment, inoculation of bacteria, incubation, cooling, and finally, packaging the product. This chapter aims to provide a detailed understanding of the three main methods of yogurt production: set, stirred, and liquid. Each method is thoroughly described, covering the necessary materials and specific procedures for implementation. This chapter will serve as a practical and reproducible guide for researchers and food industry professionals interested in yogurt production.

Key words Bacterial cultures, Dairy foods, Fermentation techniques, Yogurt production

1 Introduction

Derived from fermented milk and rooted in various cultures around the world, the popularity of yogurt grew considerably during the twentieth century, driven by advances in food science and production technologies [1]. During this period, extensive research was conducted to understand and improve the texture and flavor of yogurt and its functional properties. In addition to its rich cultural heritage, yogurt is widely recognized for its nutritional importance, being an important source of proteins, calcium, and B-complex vitamins [2, 3].

Yogurt consumption has been associated with various health benefits, including improved gut health and a strengthened immune system [2]. To validate health claims and food safety, it is recommended that yogurt contains at least 10^7 CFU/g of *Lacto-*

Adriano Gomes da Cruz et al. (eds.), *Dairy Foods Processing*, Methods and Protocols in Food Science,
https://doi.org/10.1007/978-1-0716-4144-6_14,

bacillus bulgaricus and *Streptococcus thermophilus* throughout the product's shelf life. For yogurts with mixed strains, counts refer to total lactic acid bacteria count [4].

Furthermore, probiotic cultures such as *Bifidobacterium animalis lactis* BB-12 [5], *Lacticaseibacillus casei* DN001 [6], *Lactobacillus acidophilus* LA-5 [7], and *Lactococcus lactis* 11/19-B1 [8] have also been frequently added to yogurt to enhance health benefits. To ensure beneficial effects on human health, especially gastrointestinal health, a minimum concentration of live probiotic cells must reach the intestine. The recommended minimum concentration varies between 10^8 and 10^9 CFU/dose [9]. However, lower concentrations of probiotics are sufficient for some products to claim health benefits. On the other hand, other products may require higher consumption of these concentrations, depending on the strain of microorganisms involved [10].

Pasteurized or sterilized milk in yogurt production is fermented by mixed cultures of lactic acid bacteria composed of *S. salivarius subsp. thermophilus* and *L. delbrueckii subsp. bulgaricus.* These bacteria convert the lactose in the milk into lactic acid, triggering a series of biochemical reactions that contribute to developing sensory characteristics such as aroma, taste, and texture. These characteristics can vary considerably depending on the type of milk used, its composition, the nature and activity of the fermentative bacteria, and the eventual addition of nondairy ingredients [2, 4].

The milk used must meet rigorous quality standards to ensure excellence in yogurt production. The milk should be fresh and produced under appropriate sanitary conditions, with a low presence of deteriorating microorganisms, somatic cell counts, and mycotoxins, as well as be free from pathogens and inhibitory compounds such as antibiotic residues, pesticides, or disinfectants [3]. The presence of these elements can compromise the fermentation process, resulting in the inhibition of starter cultures and negatively affecting the quality parameters of fermented products. Therefore, it is essential to subject the milk destined for yogurt production to a rigorous process of pasteurization and prior cooling [3].

In addition to the importance of milk quality in yogurt production, it is essential to understand the different classifications and methods of production of this popular food. Yogurts can be classified according to the presence or absence of additions of other ingredients, such as fruits, sugars, and flavorings [3]. Additionally, they can be produced from the milk of non-bovine species such as buffalo, goat, sheep, and camel, which show variations in protein, fat, and lactose composition, affecting their digestibility and nutritional value [11]. These characteristics favor the creation of functional yogurts, meeting the growing demand for healthier foods

[12]. Regarding chemical composition, they can be classified according to fat content, including options such as creamy, whole, partially skimmed, and skimmed [3].

Furthermore, yogurt has another classification based on the gel's production methods and physical characteristics, divided into three types: set, stirred, and liquid. This diversity of classifications reflects not only the variety of products available in the market but also the preferences and needs of consumers. Therefore, this chapter is dedicated to designing three yogurt production protocols: set, stirred, and liquid. Additionally, a detailed understanding of the steps of these protocols will be provided, covering technical and practical details as well as critical process steps.

2 Materials

2.1 Ingredients

A suggested yogurt formulation is presented in Table 1.

1. Milk (*see* **Note 1**).
2. Dairy solids (such as cream, milk powder, butter, milk fat, and other dairy-derived products).
3. Nondairy solids (such as sugars, sweeteners, fruit pieces, honey, chocolate, dried fruits, and others).
4. Starter cultures composed of *S. salivarius* subsp. *thermophilus* and *L. delbrueckii* subsp. *bulgaricus* (*see* **Note 2**).

2.2 Equipment

1. Hose and sanitary pump.
2. Expansion/refrigeration tank (such as PRV, Serap Plurinox, Brazil).
3. Centrifuge (such as CSI-230-01-772, Westfalia-GEA, Alemanha).
4. High-pressure homogenizer (such as AH55-150, Antuo Nanotechnology Suzhou Co., Ltd., China).
5. Plate heat exchanger (such as BP6, BERMO, Brazil).
6. Processing tank with agitation system.
7. Filling machine.
8. Packaging machine.
9. Incubator (such as SSRF-1600 L, SolidSteel, Brazil).
10. Fermenter with time, temperature control, and automatic agitation (such as CIE00260, Rica Nata, Brazil).
11. Refrigerator.

Table 1
Ingredients for yogurt production

Ingredients	Proportion	Observation
Milk	100 L	The volume of milk can be adjusted according to the production capacity of the industry, as well as the other ingredients
Dairy solids		
Skim powdered milk	3–4%	It avoids excess fat; improves viscosity, texture, and consistency; and reduces syneresis. Powdered milk adjusts the protein/total solids ratio (0.35) and whey protein/casein (0.22), a common method to increase nonfat solids in milk [3]
Sodium caseinate	5% [3]	It improves texture, assists in developing casein micelle networks, and increases acetaldehyde production, viscosity, and buffering capacity at low pH, besides reducing syneresis and enhancing sensory characteristics [3]
Nondairy solids		
Refined or crystal sugar	5–15%	The added sugar content varies depending on the country and target audience [3]. Before heating it, sugar should be added to the milk to eliminate potential microorganisms [13]
Fruit pulp	0.5–5.0% of the yogurt volume [13]	When using fruit pulp, it is important to consider its origin and manufacturing process to avoid yogurt contamination [13]
Starter culture		
Freeze-dried starter culture (*see* **Note 2**)	1–3% and must follow the manufacturer's instructions	First, the yeast should be diluted and homogenized in 1 L of milk to ensure uniform distribution (*see* **Note 3**). Then, add it to the fermentation tank and stir constantly for 2–4 min to distribute the yeast evenly throughout the mixture [12]
Lactic ferment (*see* **Note 4**)	1–2% (10–20 mL) for each liter of milk [13]	It is obtained by propagating freeze-dried starter culture in small portions of milk. This small amount of starter culture can be successively propagated through inoculations in increasingly larger volumes of milk until a sufficient volume is obtained to initiate milk fermentation for yogurt production then [12]

3 Methods

3.1 Milk Pretreatment

Milk pretreatment is a crucial factor in yogurt manufacturing that determines product quality [14].

1. Conduct rapid tests to observe important aspects of raw milk quality (7–10 °C), such as acidity and density (*see* **Notes 5** and **6**).
2. Using the hose and sanitary pump, transfer the milk to the expansion/refrigeration tank (4–5 °C) for initial filtration through sieves, ensuring the removal of larger particle impurities (*see* **Note** 7).
3. Through pipelines, pump the milk into the centrifuge (8500–10,000 rpm, 45–55 °C) to standardize the fat content to the desired level (*see* **Note 8**).
4. Subject the milk to homogenization (15–25 MPa, 60–70 °C) to reduce the size of fat globules (<2 μm) (*see* **Note 9**).
5. Enrich the milk with dairy solids in a processing tank with agitation to improve rheological properties (*see* **Note 10**).
6. Filter the mixture to remove any undissolved particles.
7. Heat the standardized milk in the plate heat exchanger for pasteurization (72–75 °C, 15–20 s) (*see* **Note 11**).
8. Cool the pasteurized milk in the plate heat exchanger (40–45 ° C) (*see* **Note 12**).

3.2 Inoculation of Starter Cultures

After milk pretreatment, the starter cultures (*S. salivarius subsp. thermophilus* and *L. delbrueckii subsp. bulgaricus*) should be inoculated into the pasteurized milk (*see* **Notes 13** and **14**). For set yogurt, inoculation is done directly into the packaging, while for stirred and liquid yogurts, the process is conducted in fermenters. The production protocol for the three types of yogurts (set, stirred, and liquid) is illustrated in Fig. 1.

3.3 Set Yogurt

1. After milk pretreatment, inoculate the starter cultures (*see* **Notes 13** and **14**).
2. Transfer the inoculated milk to the packaging (*see* **Note 15**).
3. Incubate the product at 42 °C for 4 to 8 h (*see* **Note 13**).
4. Cool the product to 4 °C (*see* **Notes 16**, **17**, and **18**).

Greek-style yogurt can be produced with concentrated milk or by concentrating set yogurt (*see* **Note 19**).

3.4 Stirred Yogurt

1. After milk pretreatment, add the pasteurized milk and starter cultures to the fermenter at 42 °C for 4 to 8 h (*see* **Notes 13** and **14**).

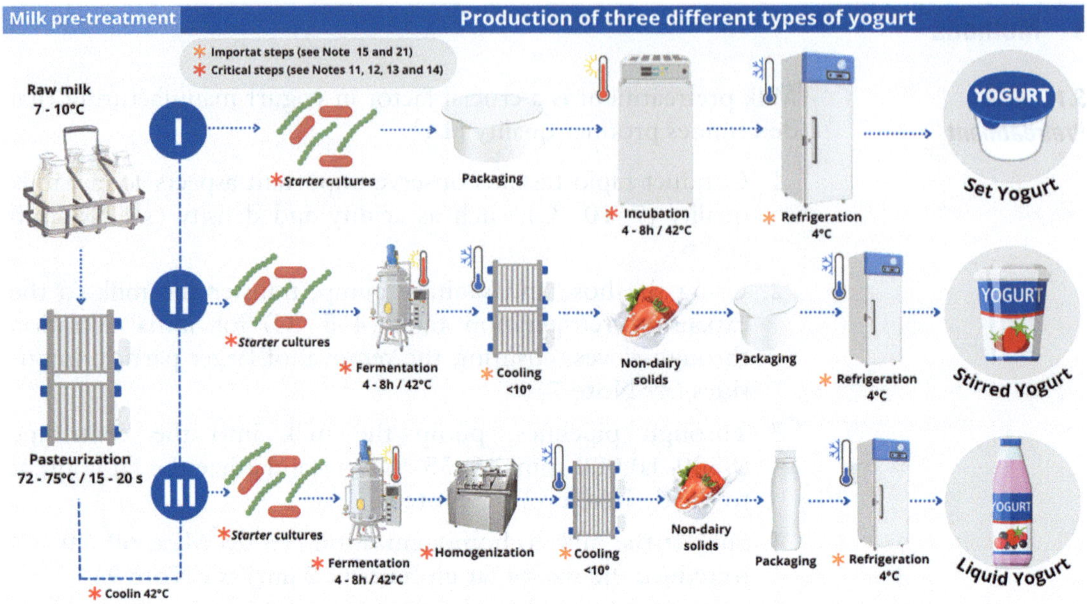

Fig. 1 Production protocols for three types of yogurts: (I) set yogurt, (II) stirred yogurt, and (III) liquid yogurt

2. After the fermentation time, cool the yogurt in a plate heat exchanger (<10 °C) (*see* **Notes 16** and **17**).
3. Add the yogurt and optional nondairy solid ingredients under agitation in the processing tank until a homogeneous mixture is obtained (*see* **Notes 20** and **21**).
4. Transfer the yogurt to the packaging stage (*see* **Note 15**).
5. Cool the product to 4 °C (*see* **Note 22**).

3.5 Liquid Yogurt

1. After milk pretreatment, add the pasteurized milk and starter cultures to the fermenter at 42 °C for 4 to 8 h (*see* **Note 13** and **14**).
2. After fermentation, subject the product to homogenization to achieve a liquid consistency (*see* **Note 17**).
3. Cool the yogurt in a plate heat exchanger (<10 °C).
4. Add the yogurt and optional nondairy solid ingredients under agitation in the processing tank until a homogeneous mixture is obtained (*see* **Notes 20** and **21**).
5. Transfer the yogurt to the packaging stage (*see* **Note 15**).
6. Cool the product to 4 °C (*see* **Note 22**).

4 Notes

1. Although cow's milk is widely used in dairy production, other types of milk, as highlighted in Table 2, are gaining prominence. This is due to increasing consumer awareness of functional foods. In some cases, non-bovine milk is superior in essential nutrients such as fatty acids, vitamins, minerals, and proteins [12].
2. Dairy producers currently use commercial starter cultures (Table 3) known as "concentrated freeze-dried dairy cultures," which have a high concentration of bacteria and can be directly inoculated into the fermenter, eliminating all preparation steps at the industrial level. The main advantages include reduced labor, lower contamination risk, and better and more uniform product quality [15].
3. Never use only a portion of the powdered culture envelope; add the entire content at once. This is essential to ensure the proper ratio of the two bacterial species in the culture in the milk as they are not evenly distributed in the powder [13].
4. Dairy starters can be prepared in the factory using heat-treated milk and added pure cultures. The starter can be stored at 7 °C and used within 20 days, propagating them at the time of use according to the volume of milk to be fermented. The process should be interrupted whenever any slight modification is observed in the yogurt produced or in the culture itself [13].

Table 2
Composition of yogurts produced with different types of milk

	Cow	Buffalo	Goat	Sheep	Camel
Total solids (%)	15.25	15.83	15.59	21.42	16.25
Proteins (%)	4.25	5.31	5.04	6.11	4.82
Fat (%)	4.55	8.25	4.77	9.35	5.56
pH	4.16	4.24	4.19	4.30	4.37

Table 3
Top suppliers of freeze-dried yogurt starter cultures

Supplier	Product	Corporate headquarters
Chr. Hansen	YF-L903, YF-L812, YC-381, YC-X11	Horsholm, Denmark
Danisco	Yo-Mix 496 LYO 100 DCU	New Century, USA
Sacco Brasil	Lyofast Y450B	Campinas, Brazil
Bioprox Cultures	YO-PROX	Levallois-Perret, France

5. The alizarol test is the most commonly used method for a quick and approximate determination of milk acidity (1 °D = 0.1 g of lactic acid/L). A 5 mL of milk is mixed with an equal amount of alizarol solution (2% alizarin dissolved in 72% alcohol) in a test tube. The interpretation of the result is as follows: (a) brick-red (reddish-purple) without coagulation = acidity between 14 and 18 °D (normal); (b) red-brown with fine coagulation = acidity of 19 to 21 °D (slightly acidic); (c) coagulated yellow = above 21 °D (very acidic); (d) violet without coagulation = <14 °D (alkalized or tampered) [15].
6. A thermolactodensimeter, along with a graduated cylinder and the milk sample, is used to determine the density of milk. The sample is added slowly to the cylinder to avoid foam formation. Then the thermolactodensimeter is placed in the cylinder, and the reading is taken, obtaining the density relative to the temperature. Using a density correction table at 15 °C, the final density of the milk can be determined. The average density of milk varies between 1.028 and 1.034 g/cm^3, which can be expressed in g/L (degrees lactodensimeter). To convert between units, subtract or add 1.0. For example, 1.032 g/cm^3 equals 32 g/L [12].
7. Until the thermal treatment, the milk should be kept in expansion/refrigeration tanks at an average temperature between 4 and 5 °C, creating an unfavorable environment for the growth of undesirable microorganisms and reducing enzymatic activity, thereby preserving its quality for a longer period [15].
8. Creamy yogurt: at least 6 g of fat/100 g; whole milk yogurt: at least 3 g of fat/100 g; partially skimmed yogurt: maximum of 2.9 g of fat/100 g; skimmed yogurt: maximum of 0.5 g of fat/100 g [3].
9. The fat globules in milk naturally do not interact with the casein gel matrix. Homogenization forms small fat globules, reducing the formation of cream layers and increasing viscosity, especially in high-fat milk [16]. When milk is not homogenized or homogenized fat is added after fermentation, the fat globules destabilize the structure, reducing yogurt viscosity [14]. Homogenization is not mandatory and involves using relatively expensive equipment employed almost exclusively in large-scale production industries [17]. High homogenization pressures increase the firmness/viscosity of the product due to the formation of a greater number of smaller fat particles. This also reduces serum separation and increases the whiteness of the product. In fermented products, homogenization reduces serum loss and prevents the formation of fat layers (cream) [3].
10. To ensure the desired consistency of set and stirred yogurts, the solids content in the milk needs to be adjusted. Standardization is done by increasing the solids content, usually using skim milk

powder (1–6%), as it provides a protein/total solids ratio similar to natural milk. Other methods include boiling the milk until it is reduced to about two-thirds of its original volume, concentration by membrane filtration, or adding other ingredients such as whey powder, whey protein (up to 5%), caseinate, or powdered buttermilk. The choice of method will depend on the availability of materials, cost, and production scale. The total solids concentration in commercial yogurts varies between 14% and 16% [17]. For liquid yogurts, the solids content of the milk is sufficient to produce a fluid product although some ingredients may be added to the formulation to generate additional viscosity [3].

11. The thermal treatment aims to eliminate microbial load in the milk and vegetative forms of pathogenic microorganisms, reduce the population of deteriorating microorganisms to acceptable levels, and denature proteins to improve product texture, preventing syneresis during storage. Additionally, thermal treatment also aids in the hydration of stabilizers after heating [17]. Thermal treatment can occur through slow or fast pasteurization. Slow pasteurization (LTHT, low-temperature holding) involves indirectly heating milk between 63 and 65 °C for 30 min, with slow mechanical agitation, using specialized equipment [18]. Fast pasteurization (HTST, high-temperature short time) heats the milk in a laminar layer between 72 and 75 °C for 15–20 s in plate pasteurizers, the industry's most commonly used heaters [18].
12. After thermal treatment, the milk should be cooled to reach 42–43 °C, the ideal temperature for *L. bulgaricus* and *S. thermophilus* growth. It is essential to properly control the temperature during this stage to ensure the quality and consistency of the final product. Temperatures that are too high can inhibit or even kill the culture microorganisms, compromising fermentation. On the other hand, temperatures that are too low prolong the fermentation time, affecting the efficiency of the process [15, 17].
13. To be officially categorized as yogurt, the product must necessarily contain cultures of *S. thermophilus* and *L. bulgaricus*, which grow symbiotically. *S. thermophilus* uses lactose as a substrate and produces lactic acid, which lowers the pH to an optimum value for the growth of *L. bulgaricus*, releasing formic acid and sulfur dioxide, which stimulate the latter's growth. In turn, *L. bulgaricus* releases amino acids and peptides from milk proteins, stimulating the growth of *S. thermophilus*. However, this symbiosis requires cultivation in a specific ratio of these microorganisms, usually 1:1, 2:1, or 3:2 [17].

14. Inoculation should range between 2% and 3% of the total milk content. To achieve the suggested minimum dose of 10^7 CFU/g, the starter culture should contain between 10^9 and 10^{12} CFU/g. The use of higher inoculation rates is not recommended due to increased production costs and excessive metabolic activity of the bacteria, which can lead to difficulties in controlling the fermentation process and result in lower yogurt quality due to the generation of a bitter taste. Additionally, higher inoculation rates increase the tendency for yogurt whey syneresis [19].
15. The most commonly used packaging material for set and stirred yogurts is plastic, which can take different forms, such as cups, bottles, bowls, or pots. For liquid yogurts, plastic bottles or carton packaging are usually used due to their liquid characteristic [3].
16. The fermentation process is completed when the pH reaches 4.6, corresponding to the isoelectric point of casein, leading to the formation of a firm curd, and the fermentation process is halted by cooling. It is recommended to start cooling when the acidity reaches about 0.8–1.0 g/100 g of lactic acid, so the fresh yogurt's acidity is between 1.2 and 1.4 g/100 g of lactic acid [19].
17. Set yogurt: During fermentation, a firm product with a suitable consistency to be consumed with a spoon is generated (Fig. 1, part I) [17]. Stirred yogurt: The semisolid texture is obtained after the product is cooled, and a thickening agent may be used (Fig. 1, part II) [17]. Liquid yogurt: Liquid yogurt differs from stirred only in the degree of lactic gel breakdown, undergoing a more intense homogenization process before cooling, thus acquiring a liquid consistency (Fig. 1, part III) [17].
18. For set yogurt, cooling can be done by spraying cold air into the incubation room or transferring the packages to climate-controlled rooms, maintaining the temperature between 2 and 4 °C [17].
19. There are three main methods for producing Greek-style yogurt on an industrial scale. (1) Use of concentrated milk: Standardized milk is concentrated in multiple evaporators. (2) Concentration by mechanical separation: The curd of natural yogurt is separated by centrifugation, allowing industrial-scale Greek-style yogurt production. (3) Concentration by membrane processes: Membrane techniques, such as ultrafiltration, drain the whey from natural yogurt after fermentation to produce Greek yogurt on a large scale. Due to the drainage process, Greek yogurt has higher total solids and lower lactose content than regular yogurt [20].

20. Commonly added are chunks or pulp of fruits, cocoa, honey, coffee, spices, sugars, or carbohydrates. Sorbic acid and salts are recommended when adding fruit pulp or fruit preparations. Additives such as gelatin, pectin, starches, gums, alginates, and microcrystalline cellulose are used for skimmed yogurts or additions [3]. For yogurts with additions, milk must represent at least 70% (w/w) of the total ingredients used [17]. Additives are prohibited in natural yogurts [3].
21. Adding sugar can interfere with the activity of lactic cultures, increasing the fermentation time and decreasing the acidity. This inhibitory effect can be attributed to milk solutes' adverse osmotic influence and water activity reduction in the medium [17].
22. The temperature throughout the distribution chain should be maintained between 2 and 5 °C, never exceeding 10 °C during intermediate phases. Generally, the products have a shelf life of 35 days [17].

References

1. Salama HH, Bhattacharya S (2022) Chapter 8: Advancement of yogurt production technology. In: Advances in dairy microbial products, pp 117–131. https://doi.org/10.1016/B978-0-323-85793-2.00021-7
2. Chandan RC, Gandhi A, Xá NP (2017) Chapter 1: Yogurt: historical background, health benefits, and global trade. In: Yogurt in health and disease prevention, pp 3–29. https://doi.org/10.1016/B978-0-12-805134-4.00001-8
3. Cruz AG, Zacarchenco PB, Oliveira CAF, Corassin CH (2017) Processamento de Produtos Lácteos: Queijos, Leites Fermentados, Bebidas Lácteas, Sorvete, Manteiga, Creme de Leite, Doce de Leite, Soro em Pó e Lácteos Funcionais, 1st edn. Elsevier, Rio de Janeiro, pp 169–193. ISBN: 978-85-352-8085-2
4. Brasil (2007) Ministério da Agricultura, Pecuária e Abastecimento. Instrução Normativa (MAPA) N° 46, de 23 de outubro de 2007. Adota o Regulamento Técnico de Identidade e Qualidade de Leites Fermentados, anexo à presente Instrução Normativa. Diário Oficial da União. Brasília, DF, 24 out. 2007. Seção 1(205):4–7
5. Abdi-Moghadam Z, Darroudi M, Mahmoudzadeh M et al (2023 Oct) Functional yogurt, enriched and probiotic: a focus on human health. Clin Nutr ESPEN 57:575–586. Epub 2023 Aug 5. https://doi.org/10.1016/j.clnesp.2023.08.005
6. Ejtahed HS, Mohtadi-Nia J, Homayouni-Rad A et al (2012) Probiotic yogurt improves antioxidant status in type 2 diabetic patients. Nutrition 28(5):539–543., ISSN: 0899-900. https://doi.org/10.1016/j.nut.2011.08.013
7. Mousavi SN, Saboori S, Asbaghi O (2020) Effect of daily probiotic yogurt consumption on inflammation: a systematic review and meta-analysis of randomized controlled clinical trials. Obes Med 18:100221., ISSN: 2451-8476. https://doi.org/10.1016/j.obmed.2020.100221
8. Nishiyama K, Kobayashi T, Sato Y et al (2018) A double-blind controlled study to evaluate the effects of yogurt enriched with *Lactococcus lactis* 11/19-B1 and *Bifidobacterium lactis* on serum low-density lipoprotein level and antigen-specific interferon-γ releasing ability. Nutrients 10:1778. https://doi.org/10.3390/nu10111778
9. WGO (2017) Probiotics and prebiotics. World Gastroenterology Organisation, Milwaukee
10. Pimentel TC, da Costa WKA, Barão CE et al (2021) Vegan probiotic products: a modern tendency or the newest challenge in functional foods. Food Res Int 140:110033. https://doi.org/10.1016/j.foodres.2020.110033
11. Zacarchenco PB (2020) Probióticos e prebióticos (livro eletrônico) – desafios e avanços. Setembro Editora, São Paulo. ISBN: 978-65-88947-01-2

12. Terzioglu ME, Bakırcı I, Oz E et al (2023) Comparison of camel, buffalo, cow, goat, and sheep yoghurts in terms of various physicochemical, biochemical, textural and rheological properties. Int Dairy J 146:105749. https://doi.org/10.1016/j.idairyj.2023.105749
13. SENAR – Serviço Nacional de Aprendizagem Rural (2009) Iogurte, bebidas lácteas e doce de leite: produção de derivados do leite. Serviço Nacional de Aprendizagem Rural, 5th edn., SENAR, Brasília, 2015. 68 p. il.; Coleção SENAR; 138. ISBN: 978-85-7664-047-9
14. Hernández A, Harte FM (Oct 2008) Manufacture of acid gels from skim milk using high-pressure homogenization. J Dairy Sci 91(10): 3761–3767. https://doi.org/10.3168/jds.2008-1321
15. Silva G, Silva AMAD, Ferreira MPB (2012) Processamento de leite. e-Tec Brasil, Colégio Agrícola Dom Agostinho Ikas (CODAI), Universidade Federal Rural de Pernambuco (UFRPE), Recife
16. Lucey JA (2004, May/Aug 2004) Cultured dairy products: an overview of their gelation and texture properties. Int J Dairy Technol 57(2/3). https://doi.org/10.1111/j.1471-0307.2004.00142.x
17. Pimentel TC, Antunes AEC, Zacarchenco PB et al (2017) Brazilian yogurt-like products. In: Yogurt in health and disease prevention, pp 331–351. https://doi.org/10.1016/B978-0-12-805134-4.00019-5
18. Brasil (2017) Decreto N° 9.013, DE 29 DE MARÇO DE 2017. Dispõe sobre o regulamento da inspeção industrial e sanitária de produtos de origem animal, que disciplina a fiscalização e a inspeção industrial e sanitária de produtos de origem animal, instituídas pela Lei n° 1.283, de 18 de dezembro de 1950, e pela Lei n° 7.889, de 23 de novembro de 1989. Artigo 255, 2017
19. Tamime AY, Robinson RK (2000) YOGHURT science and technology, 2nd edn. Woodhead Publishing Limited/CRC Press LLC, 1999, Cambridge/Boca Raton
20. Gyawali R, Ibrahim SA (2016) Effects of hydrocolloids and processing conditions on acid whey production with reference to Greek yogurt. Trends Food Sci Technol 56(Oct 2016):61–76. https://doi.org/10.1016/j.tifs.2016.07.013

Chapter 15

Traditional and Industrial Methods for Milk Kefir Production

Luan Amaral Alexandre, Alice Cristina da Silva, Ana Paula Zapelini de Melo, and Silvani Verruck

Abstract

Kefir is a fermented dairy beverage produced using the natural microbiota of kefir grains or freeze-dried starter cultures. These grains are composed of a symbiotic community of microorganisms, including *Kluyveromyces marxianus*, *Saccharomyces unisporus*, *Saccharomyces cerevisiae*, *Saccharomyces exiguus*, as well as different strains of the genera *Lactobacillus, Leuconostoc, Lactococcus*, and *Acetobacter*. When fermented, these microorganisms transform milk into a probiotic beverage rich in nutrients and bioactive compounds, promoting health benefits. There are several methods for kefir production, with the traditional and industrial methods being the most commonly used. The traditional method, widely employed by small producers, involves inoculating milk with kefir grains. On the other hand, the industrial method, used by large manufacturers, aims for reproducibility and quality control on a large scale, using freeze-dried kefir cultures added directly to the milk. Given the importance and popularity of both methods, this chapter provides a detailed description of the two main methods of milk kefir production: traditional and industrial. Each approach's key technical and practical aspects will be explored, providing a comprehensive guide for small artisanal producers and large-scale industrial operations.

Key words Dairy foods, Fermented milk, Functional food, Health effects, Kefir grains

1 Introduction

The term "kefir" derives from the Turkish word "keyif," which means "pleasure," alluding to a feeling of well-being when consumed [1, 2]. Originating from the Caucasus Mountains, kefir is a fermented dairy product that has been produced for hundreds of years. The beverage is characterized by its white-yellowish appearance, viscous texture, and sour taste, and it may contain a small amount of alcohol, ranging from 0.08% to 2%, due to the fermentation process [3].

The main difference between kefir and other fermented dairy products lies in its fermentation method, which occurs from "kefir grains." These grains are composed of a symbiotic community of microorganisms, including various species of lactose-fermenting

Adriano Gomes da Cruz et al. (eds.), *Dairy Foods Processing*, Methods and Protocols in Food Science,
https://doi.org/10.1007/978-1-0716-4144-6_15,

yeasts, such as *Kluyveromyces marxianus*, and non-lactose-fermenting yeasts such as *Saccharomyces unisporus*, *Saccharomyces cerevisiae*, and *Saccharomyces exiguus*. Additionally, there are different strains of the genera *Lactobacillus*, *Leuconostoc*, *Lactococcus*, and *Acetobacter* [4].

The symbiosis of these microorganisms during the fermentation process results in a probiotic beverage rich in nutrients and bioactive compounds that promote health benefits. Regular consumption of kefir is associated with various positive effects, including antitumor and anticancer activities, aid in lactose digestion, synthesis of B vitamins, anti-inflammatory effects, modulation of the immune system, and hypocholesterolemic properties [1]. It is important to note that the claim of beneficial effects is specific to each strain. Since kefir grains from different origins contain varied combinations of microbial species, human studies must be conducted with the same combination of microorganisms to validate the claimed effects [5].

There are various methods for kefir production, with the traditional and industrial methods being the most commonly used [4]. The traditional method involves inoculating milk with kefir grains, allowing fermentation to occur naturally over a period of 24 to 48 h. During this process, the microorganisms in the kefir grains ferment the lactose in the milk, producing lactic acid, carbon dioxide, ethanol, and various aromatic compounds that contribute to the characteristic flavor of kefir [3]. The resulting beverage from this process has a thick and creamy consistency, a mild sour taste, and a light aroma of fresh yeast. These attributes give kefir its unique sensory profile, which many consumers appreciate [4].

On the other hand, the industrial method uses freeze-dried kefir cultures for direct addition (Direct Vat Set, DVS). Typically, commercial cultures contain lactic acid bacteria and yeasts isolated from kefir grains. In this method, the fermentation process can occur in two ways: in fermentation tanks or directly in the packaging. These approaches allow for a more controlled and standardized kefir production on a large scale, ensuring product consistency [6]. Although the industrial method does not exactly replicate the microbial profile of traditional kefir, it offers a practical and efficient alternative for commercial production [1].

Besides production from the milk of various species (such as cow, goat, sheep, and buffalo), kefir can be produced from plant sources, resulting in so-called "water kefir" [7]. This is prepared with sucrose, dried, and fresh fruits. The result is a cloudy, acidic beverage with high sugar content and slightly alcoholic [1]. Despite the diversity of kefir production methods, cow's milk kefir is the most widely produced and consumed worldwide [4].

Given the growing interest in the diversity of fermented products, this chapter aims to describe the protocol for two methods of milk kefir production: traditional and industrial. Furthermore, each method's main technical and practical aspects will be explored, providing a comprehensive guide that caters to both small artisanal producers and large-scale industrial operations.

2 Materials

2.1 Ingredients

1. Milk (can be from various species such as cow, goat, sheep, and buffalo).
2. Kefir grains (*see* **Notes 1** and **2**).
3. Freeze-dried kefir cultures for direct addition (Table 1).

The main suppliers of freeze-dried DVS kefir cultures are provided in Table 1.

2.2 Equipment

1. Polypropylene filter (2 mm pore size).
2. High-pressure homogenizer (such as AH55–150, Antuo Nanotechnology Suzhou Co., Ltd., China).
3. Fermentation and maturation tank (such as MFL, INOXPA, Brazil).
4. Filling machine.
5. Packaging machine.
6. Refrigerator.

3 Methods

There are two main methods for kefir production: (I) the traditional method, which uses kefir grains, and (II) the industrial method, which uses freeze-dried kefir culture for direct addition (DVS). Figure 1 illustrates the detailed protocols for both methods.

3.1 Traditional Method

1. Heat the milk between 90 and 95 °C for 5 to 10 min.
2. Cool the milk to 20–22 °C.
3. Inoculate the milk with kefir grains (2–10%) (*see* **Note 3**).
4. Ferment at 20–22 °C for 20–24 h (*see* **Notes 4–7**).
5. Mature under refrigeration at 10–12 °C for 7–8 h.
6. Remove the kefir grains by filtration, wash them with pasteurized milk or sterile water, and store them at 4 °C until the next inoculation (*see* **Note 8**).
7. Fill the kefir into containers (*see* **Note 9**).
8. Store under refrigeration at 4–6 °C (*see* **Notes 10–15**).

Table 1
Main global suppliers of Direct Vat Set (DVS) kefir cultures

DVS kefir culture	Microorganisms	Recommended dosage	Additional information	Supplier/ corporate headquarter
FD-DVS XPL-1 eXact®	*Lactococcus lactis* subsp. *cremoris*, *Lactococcus lactis* subsp. *lactis*, *Lactococcus lactis* subsp. *lactis* biovar *diacetylactis*, *Leuconostoc*, and *Streptococcus thermophilus*	50–100 U for 500 L milk	Shake the mixture for 10–15 min to distribute the culture evenly. Recommended incubation temperature 30–35 °C	Chr Hansen/ Horsholm, Denmark
FD-DVS eXact® KEFIR 1	*Debaryomyces hansenii*, *Lactococcus lactis* subsp. *cremoris*, *Lactococcus lactis* subsp. *lactis*, *Lactococcus lactis* subsp. *lactis* biovar *diacetylactis*, *Streptococcus thermophilus*, *and Leuconostoc*	50 U for 500 L milk	Shake the mixture for 10–15 min to distribute the culture evenly. Recommended incubation temperature 30–35 °C	Chr Hansen/ Horsholm, Denmark
eXact® KEFIR 12 plus FD-DVS YF-L811 Yo-Flex®	*Bifidobacterium infantis*,*Bifidobacterium lactis*, *Lactobacillus acidophilus*, *Lactobacillus delbrueckii* subsp. *lactis*, *Lactobacillus fermentum*, *Lactobacillus paracasei*, *Lactobacillus rhamnosus*, *Lactococcus lactis subsp. cremoris*, *Lactococcus lactis* subsp. *lactis*, *Lactococcus lactis* subsp. *lactis biovar. diacetylactis*, *Leuconostoc mesenteroides*, *and Leuconostoc pseudomesenteroides*, *Lactobacillus delbrueckii* subsp. *Bulgaricus*, and *Streptococcus thermophilus*	250 U for 2500 L milk 50 U for 250 L milk	Shake the mixture for 10–15 min to distribute the culture evenly. Recommended incubation temperature 35–45 °C	Chr Hansen/ Horsholm, Denmark
CHOOZIT™ KEFIR DA LYO	*Kefir grains microflora*,*kefir yeasts*, *lactic acid bacteria*, *Lactococcus lactis* sp., *Leuconostoc* sp., *Lactobacillus* sp., and *Streptococcus thermophilus*	5 g per 1000 L milk	Recommended inoculation and incubation temperature 26 °C	Danisco/ Copenhagen, Denmark

VITAL KEFIR	*Lactococcus lactis* subsp. *lactis, Lactococcus lactis* subsp. *cremoris, Lactococcus lactis* subsp. lactis biovar. *diacetylactis, Lactobacillus acidophilus, Streptococcus thermophilus,* and Kefir grains microflora (including kefir yeasts and *Leuconostoc*)	1 sachet for 1 to 3 L of milk	Recommended inoculation and incubation temperature 20–24 °C	Danisco/ Copenhagen, Denmark
CHOOZIT MK5 LYO	*Streptococcus thermophilus, Lactobacillus delbrueckii* subsp. *bulgaricus, Lacticaseibacillus casei, Lactobacillus acidophilus, Bifidobacterium lactis, Lacticaseibacillus rhamnosus, Lactococcus lactis* subsp. *cremoris, Lactococcus lactis* subsp. *lactis, Lactococcus lactis* subsp. *lactis* biovar. *diacetylactis, Leuconostoc mesenteroides* subsp. *Cremoris,* and *Lactobacillus delbrueckii* subsp. *lactis*	0.04 g for 1 L milk	Agitate approx. 15–20 min on low speed. Recommended incubation temperature: 23–26 °C	Danisco/ Copenhagen, Denmark
Kefir DG CHOOZIT LYO	Kefir grains microflora, kefir yeasts, lactic acid bacteria: *Lactococcus lactis subsp., Leuconostoc sp., Lactobacillus sp., Streptococcus thermophilus*	5 g for 1000 L milk	Agitate approx. 20 minutes on low speed. Recommended incubation temperature: 23–26 °C	Danisco/ Copenhagen, Denmark
INFINITY Kefir Starter Industrial	*Lactococcus lactis ssp. lactis, Lactococcus lactis ssp. cremoris, Lactococcus lactis ssp. lactis biovar diacetylactis, Streptococcus thermophilus, Lactobacillus kefir, Lactobacillus parakefir, Lactobacillus delbrueckii ssp. bulgaricus, Saccharomyces cerevisiae*	25 g for 250 L milk	Stir well until completely dissolved. Recommended incubation temperature 30–32 °C/12–14 h	Bacillus Bulgaricus/ Hoboken, USA

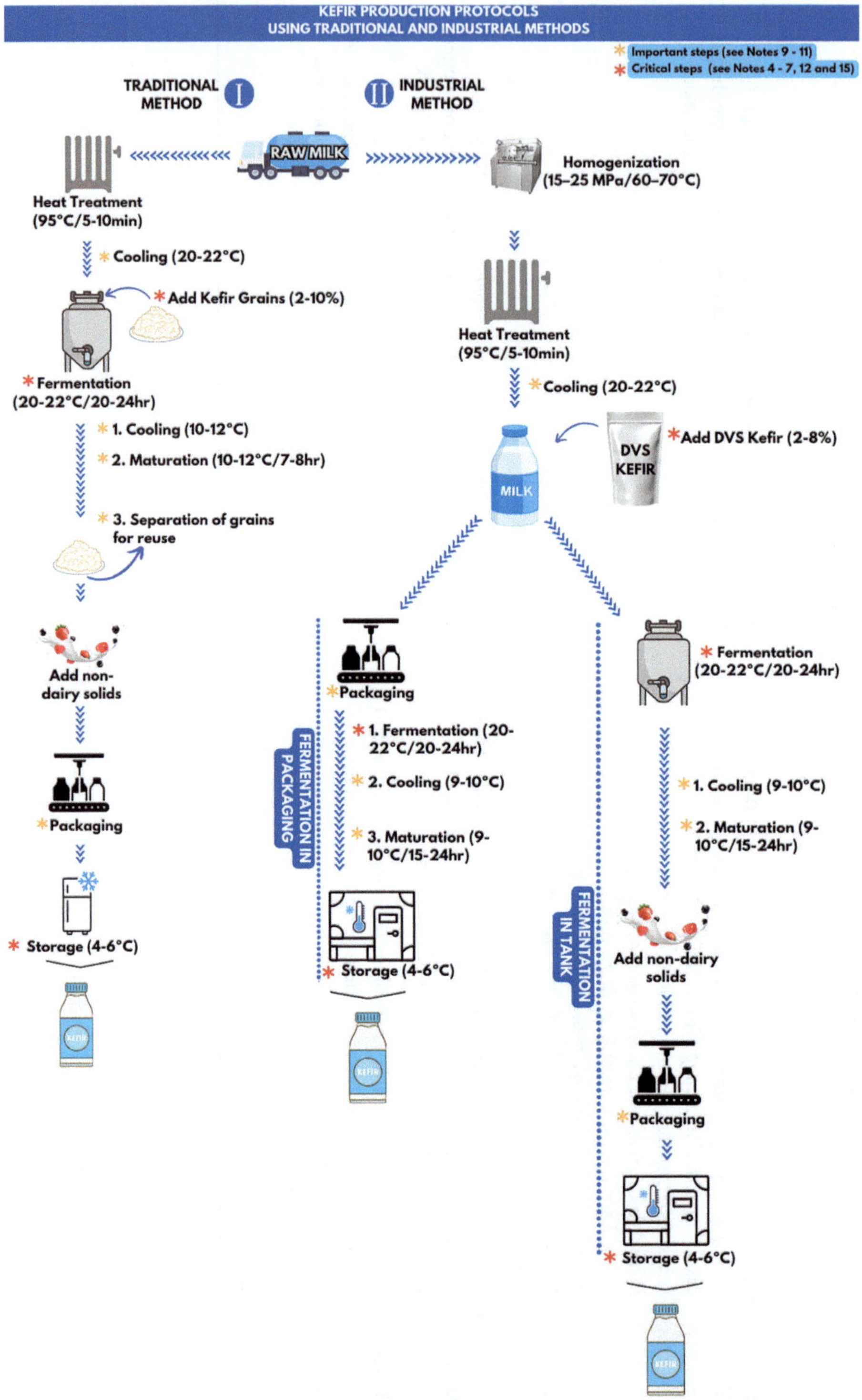

Fig. 1 Protocols for kefir production by traditional and industrial methods

3.2 Industrial Method

1. Subject the raw milk to homogenization (15–25 MPa, 60–70 °C) to reduce the size of the fat globules (*see* **Note 16**).
2. Heat the milk between 90 and 95 °C for 5–10 min.
3. Cool the milk to 20–22 °C.

After cooling the milk, the commercial freeze-dried kefir culture should be inoculated and fermented. Fermentation can occur in two ways: in the fermentation tank or directly in the sales packaging.

3.2.1 Fermentation in the Tank

1. After the cooling step, inoculate the DVS freeze-dried kefir culture (2–8% of the freeze-dried culture) in the fermentation tank (*see* **Notes 4**–7 and Table 1).
2. Cool to 9–10 °C.
3. Mature under refrigeration at 9–10 °C for 15–24 h.
4. Fill the kefir into containers (*see* **Notes 9** and **17**).
5. Store under refrigeration at 4–6 °C (*see* **Notes 10**–**15**).

3.2.2 Fermentation in the Packaging

1. After the cooling step, inoculate the DVS freeze-dried kefir culture (2–8%) (*see* **Notes 5**, 7, **17,** and Table 1).
2. Fill the kefir into containers (*see* **Note 9**).
3. Ferment at 20–22 °C for 20–24 h (*see* **Notes 3**–7).
4. Mature under refrigeration at 9–10 °C for 15–24 h.
5. Store under refrigeration at 4–6 °C (*see* **Notes 10**–**15**).

4 Notes

1. Kefir grains are gelatinous and viscous clusters, typically whitish or yellowish, with sizes ranging from 0.3 to 3 cm, resembling cauliflower due to their irregular and multilobular shape [4].
2. The chemical composition of kefir depends on various factors, including the kefir grains used, their geographical origin, temperature, fermentation conditions, and the type and quantity of milk used [8].
3. A kefir grain-to-milk ratio of approximately 1:30 to 1:50 is usually recommended for ideal fermentation and product consistency. A ratio of 1:10 produces kefir with high acidity, lower viscosity, and an intensely effervescent flavor, while a ratio of 1: 100 produces kefir with higher viscosity and lower acidity [3].
4. The fermentation process of kefir is carried out in fermentation tanks, where it is crucial to maintain uniform temperature distribution due to the tank's size. Temperature variations between the top and bottom of the tank should not exceed

1.5 °C. This strict temperature control ensures that fermentation occurs evenly throughout the tank volume, ensuring the quality and consistency of the final product [9].

5. The fermentation period of kefir can vary between 20 and 24 h, depending on the lactic acid content in the milk at the end of the incubation process, which should be approximately 0.7 mL of lactic acid per 100 mL of milk, resulting in the ideal pH of 4.5 to 4.6 [4].
6. Fermentation should be stopped when the pH reaches 4.6. At this point, immediate cooling is essential to control the acid content in the fermented milk. This cooling prevents fermentation from continuing, maintaining the kefir with the desired acidity and texture [9].
7. During kefir fermentation, lactose in the milk is converted into lactic acid, resulting in decreased pH and thickening. Approximately 30% of the milk lactose is decomposed in this process. The bacteria present in kefir can convert glucose into lactic acid, producing both D-(−)-lactic and L-(+)-lactic isomers [10].
8. After straining the kefir grains from the milk, it is recommended to rinse them with pasteurized milk or sterile water and store them at 4 °C until the next inoculation, for up to 48 h. If the kefir grains are not transferred daily to freshly prepared milk and must be stored for an extended period, they should be frozen or freeze-dried to maintain viability [6].
9. For kefir packaging, it is essential to use containers capable of withstanding the pressure generated by CO_2 production during fermentation. Sturdier materials like glass are highly recommended as they can withstand pressure without deforming. Polyethylene packaging covered with aluminum foil or polyethylene with aluminum lids is also suitable as their flexibility accommodates internal pressure, preventing leakage and preserving product integrity. This type of packaging not only maintains the sensory characteristics of kefir and ensures quality during storage and transportation [10, 11].
10. Kefir produced with commercial freeze-dried cultures can have a shelf life of up to 28 days although manufacturers usually indicate a shelf life of 15 days. In comparison, kefir produced with kefir grains has a shelf life of 3 to 12 days. However, this duration may vary among manufacturers [1, 3].
11. The Codex Alimentarius Commission establishes that the chemical composition and microbiological quality of kefir should have the following characteristics: minimum milk protein content of 2.7 g/100 g, maximum milk fat content of 10 g/100 g, and minimum titratable acidity of 0.6 mL/100 mL. Regarding the minimum quantity of microorganisms

per gram of product, a minimum count of 107 CFU/g for lactic acid bacteria and 104 CFU/g for yeast throughout the product's shelf life is required [12].

12. The flavor of kefir is often described as "fermented," with a mouthfeel that can be "tangy" and "effervescent" due to trapped CO_2 release. To increase kefir acceptability among consumers, it is common to add flavors, either directly or in the form of fruit preserves. Adding fruits and other flavorings is usually performed before filling to avoid problems such as tube fouling or component degradation during heat treatment. However, this practice poses a hygiene challenge as there is no heat treatment after adding these ingredients, increasing the risk of contamination. This is a critical control point in the production process. To ensure product safety and quality, it is essential to ensure that these ingredients are thoroughly sterilized before addition [5].
13. One of the most critical challenges in kefir production is post-acidification, mainly due to microorganism multiplication during storage. This phenomenon is influenced by various factors, including the type of starter cultures used, milk composition, temperature, pH, and processes such as homogenization, agitation, and packaging material. Rigorous control of these factors is essential to minimize post-acidification and ensure kefir quality and stability throughout its shelf life [9].
14. Problems with coagulum texture and whey separation are common in kefir. Inadequate coagulum structure can be caused by breaking bonds between proteins and polysaccharides, milk quality and composition, acid content, and microbial growth. Phage contamination can inhibit microbial growth for proper fermentation, leading to structural and sensory problems in kefir. Rigorous hygienic practices and regular replacement of starter cultures can help mitigate these issues [9].
15. The main defects of kefir are usually related to flavor, which may resemble buttermilk, and aroma, which may be excessively fermented. This intense aroma can be caused by *S. cerevisiae*, a yeast that ferments vigorously and multiplies rapidly without oxygen, often resulting in a vinegar-like or solvent-like smell. Excessive acetic acid production in kefir can be attributed to the excessive growth of *Acetobacter* spp. Furthermore, the presence of *Dekkera* spp. in kefir grains or starter culture can intensify the aroma of acetic acid [11].
16. Milk homogenization is a fundamental process in kefir production. This process involves breaking down fat globules into smaller particles through agitation. This prevents the formation of a cream layer on the milk's surface, ensuring a uniform

texture in the final product. The homogenization time varies considerably depending on the volume of milk being processed [9].

17. The pumping flow rate of fermented milk into the filler should be kept slow, below 0.5 m/s, to preserve the product structure and prevent whey separation due to mechanical force. A slow flow minimizes mechanical stress on the fermented milk, ensuring the maintenance of the desired texture and consistency and preventing degradation of the final product's sensory and structural quality [9].

References

1. Noğay NH (2019) Chapter 8: Kefir beverage and its effects on health. In: Milk-based beverages, vol 9. Woodhead Publishing, pp 273–296. https://doi.org/10.1016/B978-0-12-815504-2.00008-6
2. Cruz AG, Zacarchenco PB, Oliveira CAF et al (2017) Processamento de Produtos Lácteos: Queijos, Leites Fermentados, Bebidas Lácteas, Sorvete, Manteiga, Creme de Leite, Doce de Leite, Soro em Pó e Lácteos Funcionais, 1st edn. Elsevier, Rio de Janeiro, pp 169–193. ISBN: 978-85-352-8085-2
3. Rattray FP, O'Connell MJ (2011) Fermented Milks|Kefir. In: Encyclopedia of dairy sciences, pp 518–524. https://doi.org/10.1016/B978-0-12-374407-4.00188-6
4. Verruck S, Dantas A, Prestes AA et al (2020) Kefir e Kombucha: Alimentos Probióticos Emergentes. In: Cruz AG, Silva e Alves AT, Prudêncio E et al (eds) Probióticos e Prebióticos – Desafios e Avanços, vol 1, 1st edn. Setembro Editora, São Paulo, pp 236–260
5. Barros CP, Esmerino EA, Melo RL, da Cruz AG (2023) Psychobiotic carried by food and beverage. In: da Cruz AG, Silva MC, Pimentel TC et al (eds) Probiotic foods and beverages, Methods and protocols in food science. Humana, New York. https://doi.org/10.1007/978-1-0716-3187-4_17
6. Turkmen N (2017) Kefir as a functional dairy product. In: Dairy in human health and disease across the lifespan, [S. l.]. pp 373–383. Disponível em: https://doi.org/10.1016/B978-0-12-809868-4.00029-7
7. Guzel-Seydim ZB, Gökırmaklı Ç, Greene AK (2021) A comparison of milk kefir and water kefir: physical, chemical, microbiological and functional properties. In: Trends in food science and technology, vol 113. Elsevier Ltd., pp 42–53. https://doi.org/10.1016/j.tifs.2021.04.041
8. Farag MA, Jomaa SA, Abd El-Wahed A et al (2020) As muitas faces dos produtos lácteos fermentados com kefir: características de qualidade, química do sabor, valor nutricional, benefícios para a saúde e segurança. Forum Nutr 12:346. https://doi.org/10.3390/nu12020346
9. Ma S, Vidanarachchi JK, Ranadheera CS (2023) Probiotic fermented Milk. In: Probiotic foods and beverages: technologies and protocols. Springer, New York, pp 1–33. https://doi.org/10.1007/978-1-0716-3187-4_1
10. Farnworth ER, Mainville I (eds) (2008) Handbook of fermented functional foods, 2nd edn. CRC Press, Boca Raton
11. Wszolek M, Kupiec-Teahan B, Skov Guldager H et al (2006) Production of Kefir, Koumiss and other related products. Fermented Milks:174–216. https://doi.org/10.1002/9780470995501.ch8
12. FAO/WHO (2011) Codex Alimentarius Commission. Milk and milk products (CODEX STAN 243-2003), 2nd edn. World Health Organization (WHO) and Food and Agriculture Organization of the United Nations (FAO), Rome, pp 6–16

Chapter 16

Labneh, Ayran, Amasi, and Tarag

Usman Mir Khan, Anam Latif, and Rana Muhammad Aadil

Abstract

Fermented milk has been consumed throughout human history. Buttermilk, acidophilus milk, labneh, and yogurt are long-established fermented milk products well known to many people. The popularity of labneh, a concentrated fermented milk made by specific yogurt cultures named *Lactobacillus delbrueckii* subsp. *bulgaricus* and *Streptococcus thermophilus* have grown widely due to their nutritional advantages and storage qualities, leading to their increased economic significance. Ayran is a yogurt-based, salty beverage (southern Asian *lassi*) popular in Turkey, Azerbaijan, and Kazakhstan. It is made by a mixture of yogurt (made with yogurt cultures) with water and salt. Amasi, or sour milk, is a fermented food traditionally made by the *Zulus* of South Africa from raw bovine milk that naturally ferments at room temperature in clay pots and gourds used frequently. The lactic acid bacteria make up the majority of microflora present in the natural fermentation of amasi. Traditionally, tarag (alcoholic fermented milk of cows, goats, and yaks), produced by Mongolia's nomadic people and the northwest of China, is a unique naturally fermented alcoholic dairy beverage known for many ages. Significant research has to be done on the possibility of developing more ancient, fermented milk products in both traditional and commercial ways. These fermented milk beverage methods will prove helpful for researchers' findings to aid a significant number of importance in future dairy product development. There is also a need to follow safe and healthy fermented milk within standard operating procedures (SOPs) to have healthy and safe fermented milk products.

Key words Milk, Fermentation, Lactic acid bacteria, Fermented milk

1 Introduction: Traditional Fermented Milk-Based Beverages

One of the first preservation techniques used by humans was probably fermentation. Foods that have undergone controlled, deliberate microbial growth as well as enzymatic modifications of both their primary and secondary ingredients are referred to as fermented foods. The anaerobic breakdown of carbohydrates by bacteria is known as fermentation [1]. Probiotics, described as live microorganism and/or their components which when administrated that help the host's health, are naturally abundant in fermented dairy products. They help in managing obesity, dyslipidemia, or hypertension, and they may also have antioxidant, anti-inflammatory, antibacterial, anticancer, and immunomodulatory properties.

Adriano Gomes da Cruz et al. (eds.), *Dairy Foods Processing*, Methods and Protocols in Food Science,
https://doi.org/10.1007/978-1-0716-4144-6_16,

Microorganisms found in fermented foods extend shelf life and improve the texture, bio-functionality, flavor, and aroma of food [2].

Fermented milk has been consumed throughout human history. Many people were familiar with long-established fermented milk products such as yogurt, acidophilus milk, and labneh, but the fermentation process was not described scientifically. The fermentation process of dairy products and their beneficial bacteria have drawn a lot of attention over the years when it was realized how crucial it was for food to contain live bacteria for health benefits. Thus, Metchnikoff carried out the first scientific research on the advantages of fermented milk on human health at the turn of the early twentieth century [3]. The primary principle in the era of scientific research is that diet must satisfy a person's nutritional requirements while also serving as a disease-preventative. Currently, many items fall under the functional food spectrum, including cereals, spreads, dairy goods, baked goods, infant foods, meat goods, confectionery, snacks, ready meals, and various other alcoholic beverages [4]. Due to their (i) good capacity to add specific bioactive ingredients and nutrients, (ii) convenience in satisfying a variety of consumer demands, and (iii) ease of distribution and shelf stability with better-refrigerated storage conditions, dairy-based beverages, in particular, are currently active in functional foods category [5].

Based on these characteristics, there are four groups of fermented items, including:

1. Types that are slightly sour and have a pleasant diacetyl aroma.
2. Those fermented milk that are extremely sour and sour due to strong acid production.
3. Fermented milk products that have formation of both lactic acid and ethanol.
4. Fermented milk products with probiotics.

Depending on the type of milk used, starter cultures used, and preparation technique, each variety of fermented milk product has unique composition, flavor, and texture. These fermented milk products are detailed here with their specific characteristics (Table 1).

In the current chapter, the most recent reviews of dairy-fermented beverages of labneh, ayran, amasi, and tarag are shared in detail for a better understanding of their fermentation methods, origin, and product development. The popularity of labneh, a concentrated fermented milk product, has grown recently. It has higher nutritional advantages and storage qualities, leading to its increased economic significance. However, a lot of its consumer appeal depends on its sensory qualities, which in turn seem to be strongly influenced by how the material was processed. To produce

Table 1
Traditional fermented milk products with their origin, fermentation, and basic characteristics

Fermented milk products	Country/continent/area	Milk type	Starter cultures/ fermentation type	Basic characteristics	Form (drinkable or gel)	References
Yogurt	Various European regions and Middle East	All types, especially buffalo goat, cow, and sheep	*Lactobacillus delbrueckii* subsp. *bulgaricus* and *Streptococcus thermophilus*/ thermophilic and lactic acid type of fermentation	Lactic acid is created when the sugars in milk ferment, giving food its tart flavor and texture	Viscous	[8]
Labneh	Middle East/Libanon, Syria, and Jordan	Cow, goat, and sheep	*Lb. delbrueckii* subsp. *bulgaricus; S. thermophilus*/ thermophilic lactic acid fermentation	Yogurt that has been condensed into a ball and is stored in oil with more than 30% of total solids	Viscous	[9]
Kefir	Russia, Italy, Mongolia, Turkey, Tibet, Greece, Brazil, Hungary, Argentina, China, Europe, Asia, and Latin America	Cow, goat, sheep, and buffalo	Bacterial mixture, e.g., *Streptococcus*, *Leuconostoc*, *Lactobacillus*, and *Lactococcus* species, as well as yeasts such *as Galactomyces*, *Debaryomyces*, *Issatchenkia*, *Pichia*, *Saccharomyces*, *Kluyveromyces*, *Torulopsis*, *Kazachstania*, *Yarrowia*, and *Wickerhamomyces* species	Self-carbonated, thick, homogeneous, and creamy in consistency, with mildly alcoholic effects on health	Drinkable and viscous	[10]

(continued)

Table 1 (continued)

Fermented milk products	Country/continent/area	Milk type	Starter cultures/ fermentation type	Basic characteristics	Form (drinkable or gel)	References
Dahi (curd)	Asia India, and Bangladesh	Cow and buffalo	*Lactococcus lactis* subsp. *lactis*, *Lc. lactis* subsp. *cremoris*, *L. bifermentans*, *L. alimentarius*, *L. cascadei*, *S. thermophilus*, *L. helveticus*, *Lc. cremoris*, *L. delbrueckii* subsp. *bulgaricus*, *W. cibara*, *P. acidilactici*, *P. pentosaceus*, *W. paramesenteroides*, *Candida* spp., *L. fermentum*, and *Saccharomycopsis* spp.	Buttermilk and spontaneous fermentation	Drinkable	[11]
Yakult	Japan regions	Cow	*L. casei* subsp. *casei*/ spontaneous fermentation	Japanese one/fermented milk that has been sweetened; made using water, glucose-fructose syrup, skim milk, sugar, and other bacterial cultures	Drinkable	[12]
Kumis	Asia, Latin America, China, Kazakhstan, and Colombia	Ewe	*Saccharomyces* spp., *Lpb. plantarum*, *L. delbrueckii* subsp. *bulgaricus*, *L. buchneri*, *L. salivarius*, *L. helveticus*, and *L. acidophilus*	Spontaneous fermentation	Drinkable	[13]

Leben	Middle East (Jordan, Syria, and Libya)	Cow, goat, and sheep	*L. delbrueckii* subsp. *Bulgaricus* and *S. thermophiles*	Yogurt-like	Viscous	[7]
Ayran	Asia/Central Asia	Cow	*S. thermophilus*, *L. delbrueckii* subsp. *bulgaricus*, *L. helveticus*, and other yeasts	Yogurt-based beverage	Drinkable	[14]
Amasi	Africa/Zimbabwe and Republic of South Africa	Cow	*L. acidophilus*, *L. delbrueckii* subsp. *lactis*, *Lpb. plantarum*, *Lcb. casei* subsp. *pseudoplantarum*, *Lcb. casei* subsp. *casei*, *Lc. Diacetylactis*, *Lc. lactis*, *E. faecium*, *Lcb. casei subsp. pseudoplantarum*, *E. faecalis*, *Leuc. mesenteroides*, and *L. helveticus*	Spontaneous fermentation	Viscous	[15]
Tarag	Asia/Mongolia	Yak, cow, and goat	*S. thermophilus*, *L. delbrueckii* subsp. *bulgaricus*, *L. helveticus*, and yeasts	Spontaneous fermentation, acidic, and sour milk	Drinkable	[16]
Langfil	Europe/Sweden	Cow	Mesophilic LAB	Elastic texture with EPS	Viscous	[17]
Grassis	Central Asia/Nordic/Sudan	Camel	*Lc. plantarum*, *Leuconostoc* spp., *Lc. paracasei* subsp., *Lc. lactis*, and *Enterococcus* spp.	Produced by a semicontinuous/fed-batch fermenting method in large skin bags that retain plenty of previously soured foods. This food is consumed throughout Sudan	Viscous	[18]

(continued)

Table 1 (continued)

Fermented milk products	Country/continent/area	Milk type	Starter cultures/ fermentation type	Basic characteristics	Form (drinkable or gel)	References
Dadih	Asia/West Sumatra and Indonesia	Buffalo	Mesophilic LAB	Two days of fermentation at ambient temperature and spontaneously in a bamboo tube with natural LAB		[19]
Kurut	Asia and Middle East/China, Tibet, and Turkey	Yak and cow	*L. delbrueckii* subsp. Bulgaricus and *S. thermophiles*	Yogurt-like	Viscous	[20]
Khoormog	Asia/Mongolia	Camel	Mesophilic LAB and yeasts	Natural fermentation; a sour and alcoholic flavor	Drinkable	[21]
Viili	Europe/Finland	Cow	*Lc. lactis* subsp. *lactis biovar diacetylactis*, *G. candidus*, *Lc. Lactis* subsp. *cremoris*, *Lc. lactis* subsp. *lactis*, and *Leuc. mesenteroides* subsp. *cremoris*	Ropy/stringy texture	Drinkable	[9]
Buttermilk	Various Europe regions	Cow	Mesophilic LAB	A butter manufacturing byproduct	Drinkable	[22]
Acidophilus milk	America, Europe, Greece, and Scandinavia, Russia, Turkey, and North America	Cow	*L. acidophilus*	Therapeutic milk; fermented by *Lactobacillus* and *Lactococcus sps.*	Drinkable	[17]
Airag	Asia/China and Mongolia	Cow, ewe, and camel	LAB, *L. helveticus*, *Kl. marxianus*, *B. mongoliense*, *L. kefiranofaciens*, and yeasts	Spontaneous fermentation	Drinkable	[23]
Ayib	Central and East Africa	Goat	Mesophilic LAB; yeasts	Spontaneous fermentation	Drinkable	[16]

Chal	Asia/Iran	Camel	Mesophilic LAB	Spontaneous fermentation in skin bags, sometimes in bottles at normal room temperatures	Drinkable	[24]
Chaas/ mattha	Asia, specifically in India (northern regions)	Cow	Mesophilic LAB	A byproduct of desi butter, buttermilk is made using ice, salt, sugar, and artificial flavors	Drinkable	[25]
Chanklish	Middle East	Cow, goat, and sheep	*L. delbrueckii* subsp. *bulgaricus*; *S. thermophilus*	Condensed yogurt	Viscous	[9]
Aewsoo, huquan, and arera	Ethiopia/Africa	Cow	Mesophilic starters	Sour buttermilk	Drinkable	[26]
Chhurpi	Asia, Tibet, and Eastern Himalayas	Cow	*L. bifermentans*, *Ltb. curvatus*, *L. farciminis*, *Lcb. paracasei*, *L. hilgardii*, *Lb. kefir*, *L. alimentarius* *Leuc. mesenteroides*, *Lmb. fermentum*, *E. faecium*, *W. confuses*, and *Lpb. Plantarum*	It is a product of spontaneous fermentation used in soups/side dishes.	Viscous	[27]
Doogh	Asia (Afghanistan, Armenia, Azerbaijan, Iraq, Iran, and Syria)	Cow	*L. delbrueckii* subsp. *bulgaricus*; *S. thermophilus*	Yogurt-like drink	Drinkable	[28]
Chigee	Asia/Mongolia	Ewe	Mesophilic LAB and yeasts	Spontaneous fermentation: alcoholic/sour taste	Drinkable	[29]
Chhu	Asia/Tibet	Yak	*Candida* spp., *Saccharomycopsis* spp., *L. alimentarius*, *Lvb. brevis*, *L. farciminis*, *Lc. lactis*, *Lvb. Brevis*, and other yeasts	Product of spontaneous fermentation that resembles cheese and is used in soups	Viscous and coagulated	[25, 27]

(continued)

Table 1 (continued)

Fermented milk products	Country/continent/area	Milk type	Starter cultures/ fermentation type	Basic characteristics	Form (drinkable or gel)	References
Ergo	Africa/Ethiopia	Cow and goat	*L. delbrueckii* subsp. *bulgaricus*; *S. thermophiles*	Spontaneous fermentation and yogurt-like	Viscous	[30]
Kishk	Mideast and Africa, Iraq, Lebanon, Turkey, and Egypt	Cow, goat, sheep, camel, and buffalo	*L. delbrueckii* subsp. *bulgaricus*; *S. thermophiles*	Dried fermented milk/cereal mixture	Viscous and dried	[31]
Fillbunke	Europe/Finland	Cow	Mesophilic LAB	Elastic texture	Drinkable	[22]
Gariss	Africa/Sudan	Camel	*Lpb. plantarum*, *L. paracasei* subsp. *paracasei*, *Lc. lactis*, *L. fermentum*, *Leuconostoc* spp., *Enterococcus* spp., and yeasts	Spontaneous fermentation and hanging of product in bags made of goatskin leather	Drinkable	[32]
Felisouka	Europe/Poland	Cow	Mesophilic LAB and yeasts	Buttermilk with sugar added that has been fermented with *S. cerevisiae*	Drinkable	[9]
Frozen yogurt	Various Europe regions	Cow	*L. delbrueckii* subsp. *bulgaricus*; *S. thermophiles*	Low fat	Viscous and frozen	[33]
Mabisi	Africa/Zambia	Cow	Mesophilic LAB	Spontaneous fermentation	Drinkable	[34]
Krokmach	Europe/Bulgaria	Sheep	*S. thermophilus*, *Lc. Garviae* subsp. *garviae*, *L. delbrueckii* subsp. *bulgaricus*, *Torulaspora* spp., and *E. faecalis*	Yogurt-like; high-fat and high-salt contents	Viscous	[35]
Mish	Africa/Sudan	Cow	Mesophilic LAB	A one-month-long process of backsliding and hot fermentation of milk	Viscous	[36]

Kule naoto	Africa/Kenya and Tanzania	Cow	*Lpd. plantarum*, *L. fermentum*, *L. acidophilus*, and *L. paracasei* subsp. *paracasei*	The primary food source for the Maasai community is daily consumption. It is made by spontaneous fermentation	Drinkable	[37]
Lassi	Asia/India	Cow	Thermophilic LAB, *L. acidophilus*, and *S. thermophiles*	Buttermilk is a cool drink. Occasionally made with the addition of salt, ice, artificial flavors, and sugar	Drinkable	[38]
Maconi	Asia/Caucasian mountains	Cow	Mesophilic LAB and yeasts	Spontaneous fermentation and kefir-like	Drinkable	[14]
Meekiri	Asia/Sri Lanka	Buffalo	*Lpb. plantarum*, *Lmb. fermentum*, *Ltb. curvatus*, and *L. acidophilus*	Back-slopping and yogurt-like	Viscous	[39]
Matsoni	Europe/Georgia and Armenia	Cow, goat, sheep, and buffalo	*L. acidophilus*, *S. thermophilus*, *Lc. lactis* subsp. *cremoris*, *Lc. lactis*, *G. candidus*, *Candida sp.*, *Saccharomyces sp.*, *L. delbrueckii* subsp. *bulgaricus*, and *L. delbrueckii* subsp. *lactis*	Back-slopping fermentation	Viscous	[40]
Nono	Africa/Nigeria	Cow	Nonstandardized microflora	Spontaneous fermentation	Drinkable	[41]
Mursik	Africa/Kenya	Cow	*Lmb. fermentum*, *Lpb. plantarum*, *L. brevis*, and *Lcb casei*	Spontaneous fermentation	Drinkable	[42]
Mohi	Asia/Nepal	Cow	*L. alimentarius*, *Lc. lactis*, *Lc. cremoris*, and other yeasts	High-acid buttermilk is used as a refreshing beverage	Drinkable	[43]

(continued)

Table 1 (continued)

Fermented milk products	Country/continent/area	Milk type	Starter cultures/ fermentation type	Basic characteristics	Form (drinkable or gel)	References
Nunu	Africa/Ghana	Cow	*E. faecium*, *Lmb. fermentum*, *L. helveticus*, *Lpb. plantarum*, *E. italicus*, *Leuc. mesenteroides*, *W. confusion*, and other yeasts	Spontaneous fermentation and yogurt-like	Viscous	[44]
Philu	Asia/Bhutan and Eastern Himalayas	Cow and yak	*Lpb. plantarum Lcb. paracasei* subsp. *paracasei*, *Lc. lactis* subsp. *lactis*, *Lcb. paracasei* subsp. *pseudoplantarum*, *Lc. lactis* subsp. *cremoris*, *L. bifermentans*, *L. alimentarius*, *L. kefir*, *Lcb. paracasei* subsp. *paracasei*, and *Lb. hilgardii*	Back-slopping fermentation and solid cream	Viscous	[27]
Omashikwa	Africa/Namibia	Cow	Mesophilic LAB	Fermented buttermilk	Drinkable	[45]
Rayeb	Africa/Tunisia	Buffalo	*Leu. cremoris*, *S. acidomonas*, *E. durans*, *L. delbrueckii*, *L. helveticus*, *L. bulgaricus*, *L. acidophilus*, *S. thermophilus*, *A. viridans*, and *E. faecium*	Spontaneous fermentation of raw milk	Viscous	[46]
Podkvasa	Europe/Bulgaria	Sheep	Mesophilic LAB and yeasts	Spontaneous fermentation and kefir-like	Drinkable	[9]

Rob	Africa/Sudan	Cow, goat, and sheep	*S. thermophilus*, *Lmb. fermentum*, *L. bulgaricus*, *L. acidophilus*, *L. helveticus*, *Kl. lactis*, *Candida kefyr*, *S. globosus*, *S. exigua*, *S. cerevisiae*, *Kl. bulgaricus*, and *Lc. Lactis* subsp. *lactis*	Back-slopping fermentation	Viscous/water diluted	[47]
Ryazhenka	Europe/Russia	Cow	*L. delbrueckii* subsp. *bulgaricus*; *S. thermophilus*	Back-slopping fermentation	Drinkable	[48]
Roub	Africa/Sudan	Cow	Mesophilic LAB	Back-slopping fermentation	Drinkable	[49]
Shubat	Asia and Europe/ Kazakhstan, Uzbekistan, Russia, China, and Mongolia	Camel	*Lactobacillus* spp. and *Enterococcus* spp.	Spontaneous fermentation	Drinkable	[50]
Shyow	Asia/Sikkim, Bhutan, Tibet, and Labak	Yak	Mesophilic LAB	Thick gel	Viscous	[27]
Skyr	Iceland/Europe	Cow and sheep	*S. thermophilus*, mesophilic LAB, *Lactobacillus* spp., *L. delbrueckii* subsp. *Bulgaricus*, and yeasts could exist as pollutants.	Skim milk, rennin, and a back-slopping starter were added to make this product. Whey is removed by draining it through a cloth bag	Viscous	[9]
Semjolk	Europe/Sweden	Cow	Mesophilic LAB, such as *Leuconostoc* spp. and *Lactococcus* spp.	Homemade product	Drinkable	[22]
Shanklish	Middle East/Turkey, Syria, and Lebanon	Sheep, cow, and goat	*L. delbrueckii* subsp. *bulgaricus*; *S. thermophilus*	Spice-infused concen trated yogurt preserved in olive oil	Concentrated/ viscous	[51]
Snezhanka	Europe/Bulgaria	Cow, buffalo, and sheep	*L. delbrueckii* subsp. *bulgaricus*; *S. thermophilus*	Yogurt-like product with 6% added sugar	Viscous	[9]

(continued)

Table 1 (continued)

Fermented milk products	Country/continent/area	Milk type	Starter cultures/ fermentation type	Basic characteristics	Form (drinkable or gel)	References
Stagisto	Europe/Greece	Cow and sheep	*L. delbrueckii* subsp. *bulgaricus*; *S. thermophilus*	Concentrated yogurt	Viscous and concen trated	[8, 52]
Suusac	Africa/Kenya	Camel	Mesophilic LAB	Raw milk spontaneously ferments in gourds that have been washed and treated with smoke.	Drinkable	[53]
Tatmjolk	Europe/Finland, Sweden, and Norway	Cow	Mesophilic LAB; *Lactococcus* spp. and *Leuconostoc* spp. *as* EPS-producing	Homemade product	Drinkable	[22]
Somar	Asia/Tibet	Cow and yak	Mesophilic LAB	It is consumed by the Sherpas of the Himalayan mountains	Viscous	[43]
Torba	Middle East/Turkey	Cow, goat, and sheep	*L. delbrueckii* subsp. *bulgaricus*; *S. thermophilus*	Straining yogurt concentration in a specialized cloth bag	Viscous and concen trated	[54]
Tuzlu	Middle East/Turkey	Cow, goat, and sheep	*L. delbrueckii* subsp. *bulgaricus*; *S. thermophiles*	Boiled for 60 min; salted yogurt	Viscous	[55]
Trahanas	Greece and Cyprus/Europe	Goat and sheep	*Leu. cremoris*, *Lc. diacetylactis*, *Lcb. casei*, *L. lactis*, *L. acidophilus*, and *L. bulgaricus*	Wheat and fermented milk are combined to create the final product, which is then rolled into balls and dried in the sun. Boiled and then consumed	Viscous	[56]

Ymer	Europe/Denmark	Cow	Mesophilic LAB	Concentrated fermented milk product	Viscous and concen trated	[7, 52]
Zabuli	Middle East/Iran	Cow	*Lactobacillus*, *Pediococcus*, and *Streptococcus* species	Traditionally prepared with yogurt, wheat flour, salt, and flavorful local herbs and spices (cumin, coriander, dill, garlic, and turmeric)	Viscous and concen trated	[57]
Zhentitsa	East Carpathian Mountains/ Asia	Sheep/whey from sheep milk	Mesophilic LAB	Obtained during the making of Bundz cheese	Drinkable	[9]
Zabady	Africa/Egypt	Cow and buffalo	Thermophilic LAB	Similar to yogurt in consistency and firmness	Viscous	[58]

a product with the necessary level of quality attributes for the consumer, it should be feasible to adjust and regulate the process parameters using rheological data [6]. In Arab nations, labneh, a strained/concentrated yogurt, is frequently used in sandwiches. Optional additions include olives, cucumber slices, and various green herbs. It is possible to make it even thicker, roll it into balls, put them in olive oil, and then allow them to ferment for a few weeks. It is occasionally used to produce stuffing for various pies and kebabs by combining it with onions, meat, and nuts [7].

Ayran, a yogurt-based salty beverage, is made from a combination of salt and water in yogurt. It is popular in the Republic of Macedonia, Kazakhstan, Turkey, Azerbaijan, Iranian, Kyrgyzstan, and Bulgaria. The same beverage is known by several names in several countries, including "Tan" in Armenia, "Dough" in Iran, "Shenina" in Jordan, "Laban Ayran" in Syria and Lebanon, "Laban Arbil" in Iraq, and "Moru" in South India. Middle Easterners between Lebanon and Afghanistan enjoy a beverage called "Doogh" similar to ayran but contains herbs, mostly mint, and carbonation and typically mineral water. Yogurt and water are combined to make the famous beverage known as "Ayran" or "Airan." It is widely used throughout Central Asia, the Middle East, Armenia, Iran, Turkey, Azerbaijan, the Balkans, and Lebanon [59]. It is similar to Iranian "Doogh" and Armenian "Than," albeit "Doogh" occasionally has a natural carbonation. In Cyprus, it is referred to as "Ayrani," which started as a salt-based method of preserving yogurt. In addition, you can flavor it with garlic or substitute cucumber juice for some or all of the water. Ayran is sometimes seasoned with black pepper though regarded as uncommon in Bulgarian regions where typically served without salt. Another method popular in different regions is a combination of the ayran with finely chopped mint leaves. Extra salt is commonly added to Turkish restaurants in nations like Bosnia and Herzegovina to give the flavor of salt water in the ayran-made foods [60].

Amasi, or sour milk, is a fermented food traditionally made by the *Zulus* of South Africa from raw bovine milk naturally fermented at ±5 °C in clay pots and gourds. Amasi/hodzeko, a traditional fermented milk dish from Zimbabwe, is made by natural fermentation of raw cow milk. Yeasts and lactic acid bacteria comprise the majority of microflora present in the natural fermentation of amasi [61]. Another study on amasi assumes that because of bacteria already present in milk, the container and the air ferment the milk within 1–3 days, depending on the surrounding temperature. It is common practice to occasionally speed up the fermentation by combining fresh milk with the remaining sour milk from a previous batch [62]. Amasi has higher fat and protein but is usually consumed as a delightful beverage. Amasi is a traditional milk-fermented treat similar to other goods that have been enjoyed by many tribes as a meal or drink for centuries. It is one of the fermented milk products produced at home and thus generally less available because of higher cost. According to reports, amasi is used by several South African tribes for its health and nutritional benefits and favorable social, economic, and cultural consequences. Many tribes in most Southern Africa consume amasi along with phuto, or crumbled pap [63].

Traditional fermented milk products that include tarag (fermented milk of cows, goats, and yaks) and airag (fermented mare's milk) are produced by Mongolia's nomadic peoples and have been for generations. In Mongolia and northwest China, a unique naturally fermented dairy product produced by the fermentation (with rich microflora, particularly lactic acid bacteria) of cow milk is known as "Tarag." These areas have produced and used tarag for centuries [64]. Tarag was typically produced naturally, without any commercial starter cultures, and the natural cultures were then employed the following day to inoculate new milk. At least 5–8 days of fermentation would be required for tarag to reach the ideal levels of acidity, alcohol, and flavor at a temperature of 10–25 °C.

Along with the conventional tarag, a stable microbial phase was thereby spontaneously created, maintained, and passed down through the generations [65]. Usually 1–2 L of tarag were consumed on average per person per day. Tarag was often high in fat and protein. Tarag has gained significant importance in cold regions with higher population consumption, where the main focus is on the intake of high-quality protein [8]. In addition, tarag had a slightly greater calorie count because meals without enough energy prevent protein from being used to its maximum potential. Tarag is an ideal ingredient for human health due to its increased fat, crude protein, total solids, vitamins, and minerals. Tarag included yeast and big LAB. Tarag was shown to be nutritionally sound and demonstrated its potential health benefits [66].

Some fermented milk products can be converted to other dairy products. The draining of whey and sun-drying of sour milk that turns bitter after a few days and that has continued to ferment for some months (occasionally called qoormog or isgelen tarag) produces the cheese known as aarool/quruud. Unlike the other cheese of the same name made from tarag, cheese made from isgelen tarag is quite acidic. One more reason to do so is to produce isgelen tarag for use as a coagulant in acid-coagulated cheese (eedem) production [67]. In the Mongolian provinces (Töv and Vörqangay), sour milk (tarag) that has undergone additional alcohol fermentation is referred to as qoormog and tarag, respectively. In these regions, qoormog or tarag, made predominantly via alcoholic fermentation and the tarag made primarily through lactic acid fermentation, are produced in close succession [68].

1.1 Methods Related to the Development of Milk-Based Beverages

Indigenous microbes have gained a variety of approaches in food fermentation, such as spontaneous fermentation and preselection of starter culture with well-known features. The most prevalent microbes used in food fermentations worldwide are lactic acid bacteria (LAB). A few genera that make up LAB are *Lactococcus*, *Lactobacillus*, *Leuconostoc*, *Streptococcus*, *Bacillus* and *Pediococcus* species. Despite the LAB's diversity, many of the bacteria have been classified by the European Food Safety Authority and US Food and Drug Administration as having "qualified presumption of safety" (QPS) or as being "generally regarded as safe" (GRAS), respectively [69]. Therefore, the LAB fermentation process is a well-known method for producing such foods, which improves human health by modifying our gut microbiota. The production of fermented food heavily relies on dairy products (beverages and non-drinkable goods) containing conjugated linoleic acid (CLA), iso-flavones, minerals, other vitamins, and 3-phytosterols [70]. The market for dairy-based beverages is still very modest when compared to sales of yogurt and milk. In contrast, the market for functional dairy beverages is dominated by products that contain probiotics or prebiotics. But, nowadays, whey and soy-based functional beverages are becoming popular in addition to milk-based beverages. Thus, there is a need to implement such methods for the promotion of milk-based beverages, which are economical and commercially available in cost-efficient ways [71].

The modern approach, sometimes known as the mechanical method, and the old (manual) method are typically used to make labneh. While the mechanical method concentrates the particles and removes the whey using centrifugal and ultrafiltration techniques, the conventional method involves straining the set yogurt into cloth bags to remove the whey [72]. The old process is labor- and time-intensive, which has led to the development of the mechanized method. More people are now interested in labneh texture, particularly in concentration methods, due to the recent rise in labneh popularity in Europe. Reverse osmosis and ultrafiltration processes create the dynamic rheology of labneh [73].

In both cases, the set yogurt used to make the concentrated product is produced using heated milk and thermophilic starting cultures by the usual steps for manufacturing yogurt. The whey is removed from the product using the traditional method, which comprises straining the acidified milk through muslin cloth bags. Acidified skim milk is usually concentrated using a centrifugal separation technique (separator) in the mechanical method, and then cream (40% fat) is added and combined to create the finished product [74]. Labneh has a clear flavor, a creamy or white color, good spreadability, and soft and smooth texture with little syneresis, and a medium acidity [75]. Due to the syneresis process, its physical features are comparable to those of young cheeses and fermented milk with high water content, such as quark, boursin, and petit Suisse. It is considered a product that falls between these two groups of foods [76].

Labneh is thought to have nutritional and medicinal qualities comparable to or even superior to yogurt. Compared to regular yogurt, labneh contains 2.5 times more protein, 50% more minerals, and a significant amount more live microorganisms. Labneh is also more suitable for consumption by people who are lactose intolerant because its lactose concentration is low (around 2–5% depending on the type and region) due to its fermentation into lactic acid [77]. Labneh has a high total solids content making it a probiotics-friendly matrix since it provides protection when combined with probiotics [78]. In this procedure, yogurt is traditionally blended with water (30–50%), salt (0.5–1%), and yeast to make ayran, a fermented beverage. Because homemade ayran is created every day and eaten fresh, when it is provided at restaurants, buffets, and bakeries, there is no requirement for separate packaging [79].

Ayran can be produced commercially using two different processes. In traditional methods, it can be prepared by either diluting milk with water and then fermenting the result or it can be made by diluting yogurt with water. The steps involved in producing ayran industrially vary from traditional methods. In the first stage, raw milk's fat amount is standardized (0.15% for fat-free, 0.8% for half-fat, 1.5% for full-fat). In first step, standardized milk is diluted with water to a final solid content of 8%. Traditional and industrial methods require the proper homogenization and pasteurization of milk, but the industrial method does not require addition of water [80]. Ayran with good microbiological purity can be produced using pasteurization. Then pasteurized milk is inoculated with *S. thermophilus* and *L. delbrueckii* subsp. *bulgaricus* starter cultures of yogurt and incubated until 4.2–4.4 pH is achieved. The fermentation and acidity development are then stopped by cooling the fermented samples [56]. In step two, enough water is infused into fermented milk to increase the sample's overall solid concentration to 8%. The next second step is to add salt (0.5%). Ayran is then bottled in glass, polypropylene, or polystyrene when sold in dairy facilities. Various manufacturing processes impact acetaldehyde content; however, sensory analysis cannot pick up on these variations [81].

Amasi, also known as *mukaka wakakora* or *zifa*, is a popular fermented milk beverage in Zimbabwe. In the past, gathering raw and unpasteurized milk from source, bovine milk, into calabash or clay container that is usually sealable was the first step in preventing contamination by foreign contaminants. This procedure led to the initial development of *Amasi*, *Kule*, *naoto*, or *Kweionik* as fermented milk produced in Zimbabwe, Kenya, and Uganda, respectively [82]. Amasi is also symbolic in family business and rituals among Zulu-Xhosa groups in South Africa. The container is typically pre-smoked before milk is poured to stop mold growth or other germs. Then the jar is sealed and left to ferment for 2–3 days at room temperature. The fermentation process can be speed up to just a few days by leaving the milk outside in the direct heat of the sun [83]. It is customary to sample the milk after 24 h to see whether fermentation occurs. After 1 or 2 days, the milk will be divided into two parts: liquid portion and thick curd component. The watery or thin liquid, also known as the coagulant or amasi, is often extracted first [84]. When the desired properties are obtained, the thick section is frequently collected. If any amasi is left over, it can be added to raw milk and processed using back-slopping/same methods to serve as a starter culture for subsequent batches. In Fig. 2, the general processing stages are condensed. The fermentation process, which gives amasi its sour flavor, is a vital stage in the amasi production process. The thin liquid disposal is another crucial stage since it separates amasi from other closely related dairy-fermented products [65].

The tarag was produced using conventional techniques employing raw whole milk from goats of the Zang and Chaidamu breeds, two legally recognized domestic goat breeds for milk production in China. A large leather bag was filled with fresh, raw goat milk and carefully fastened with a leather string. Then to achieve the necessary levels of acidity, alcohol, and flavor, a natural fermentation process was carried out in a temperature range of 15–20 °C for at least 2 days. Along with lactic acid, tarag shares a trait with other traditional naturally fermented milk like kefir, koumiss, and kurut in the presence of alcohol [85].

Tarag, also known as airag in Inner Mongolia, is a sort of hard yogurt prepared from cow milk, occasionally from skim milk of sheep/goats. Lactobacillus kefiranofaciens dominate the microbiota of this product. Firstly, milk is heated and then a bacterial culture is added when it reaches the touch-cooling point (normal room temperature). The curd does not start to form for a couple of hours. Tarag has a pH range of 3.6–3.9 and comparatively high lactic acid content [86]. Tarag is often taken alone, occasionally with fruit. Additionally, using a homemade distiller, these alcoholic milk products are frequently turned into Mongolian milk liquor (arkhi). These fermented milk products from Mongolia are valuable sources of vitamins and minerals for nomadic nomads [87].

Dairy-based functional beverages are fast-growing segments of the functional food market, which includes dairy products as a substantial component. This chapter provides an overview of the most current studies on dairy-based fermented beverages with possible health advantages, any developments and recent trends in the field, and prospective areas for future research-based applications.

2 Materials

The materials and equipment used to produce labneh, ayran, amasi, and tarag must be hygienic and follow the procedures according to food safety legislation. The materials may vary from country or origin and depend on the procedures and precautions followed according to the country and consumer preferences.

2.1 Labneh

The following materials are required for the production of labneh in both traditional and industrial procedures. The standard operating procedure, food safety, and hygienic parameters must be followed for safe and healthy fermented food product development.

2.1.1 Milk

Buffalo and cow milk may be used for the production of labneh. The milk procurement from the dairy herd must be preferred and stored at 4 °C until usage. Whey protein or whole milk powder/ skim milk powder may be added for standardization of raw milk in preparation of labneh with standardized fat and SNF ratio. The milk used for the production of labneh in modern methods is usually standardized at 8% fat and 14% SNF content (dry matter content) (*see* **Note 1**).

2.1.2 Starter cultures

In the traditional method, yogurt with an unknown/mixed starter culture is used, while in industrial methods, three commercial starter cultures are usually utilized to make labneh. These cultures include the following LAB cultures, respectively:

I. *L. delbrueckii* subsp. *bulgaricus*
II. *S. thermophilus*
III. *L. acidophilus*
IV. *Bifidobacterium bifidum*

The starter cultures can be procured according to requirements, stored, and activated by the manufacturing requirements. The yogurt starter cultures (*S. thermophilus* and *L. delbrueckii* subsp. *bulgaricus*) were first individually activated in skim milk by inoculating skim milk with 2% lyophilized strains and incubating it at 37 °C for 24 h (*see* **Note 2**).

2.2 Ayran

The following materials are required to produce ayran in both traditional and industrial procedures. The standard operating procedure, food safety, and hygienic parameters must be followed for safe and healthy fermented food product development.

2.2.1 Milk

Buffalo and cow milk may be used for the production of ayran. The milk procurement from the dairy herd must be preferred and stored at 4 °C until usage. Whey protein or whole milk powder/skim milk powder may be added to standardize raw milk in preparation of labneh with standardized fat and SNF ratio. The milk used to produce labneh in industrial methods is usually standardized at 1.5% fat and 8% SNF (dry matter content) (*see* **Notes 3** and 7).

2.2.2 Starter cultures

In the traditional method, yogurt with an unknown/mixed starter culture is used. In industrial methods, three commercial starter cultures are usually utilized to make ayran. These cultures include the following LAB cultures, respectively:

I.
L. delbrueckii subsp. *bulgaricus*

II.
S. thermophilus (Y350B)

III.
L. acidophilus (LA 3)

IV. *B. bifidum* (SP 9)

The starter cultures can be procured according to requirements, stored, and activated by the manufacturing requirements. The yogurt starter cultures (*S. thermophilus* and *L. delbrueckii* subsp. *bulgaricus*) were first individually activated in skim milk by inoculating skim milk with 2% lyophilized strains and incubating it at 37 °C for 24 h (*see* **Note 2**).

2.3 Amasi

The following materials are required to produce amasi in both traditional and industrial procedures. The standard operating procedure, food safety, and hygienic parameters must be followed for safe and healthy fermented food product development.

2.3.1 Milk

Bovine milk is being used for the production of amasi. The milk procurement from the dairy herd must be preferred and stored at 4 °C until used. Whey protein or whole milk powder/skim milk powder may be added to standardize raw milk to prepare the amasi with standardized fat and SNF ratio. The milk used for the production of amasi usually must be whole milk with full cream for better quality and taste aspects (*see* **Notes 11** and **12**).

2.3.2 Starter cultures

In the traditional method, yogurt with an unknown/mixed starter culture is used. In industrial methods, three commercial starter cultures were used to manufacture amasi typically in the order of 10^8 CFU/ml. These cultures include the following LAB cultures, respectively:

I. *Lactobacillus* spp. (*L. casei*, *L. plantarum*, *L. paracasei*)

II. *Lactococcus* spp. (*Lc. lactis*, *Lc. lactis* subsp. *cremoris*)

III. *Leuconostoc* spp. (*Leu. pseudomesenteroides*)

IV. *Enterococcus* spp. (*E. faecalis*).

The starter cultures can be procured according to requirements, stored, and activated by the manufacturing requirements. The starter cultures were first individually activated in skim milk by inoculating skim milk with 2% lyophilized strains and incubating it at 37 °C for 24 h (*see* **Note 2**).

2.4 Tarag

The following materials are required for the production of tarag in both traditional and industrial procedures. The standard operating procedure, food safety, and hygienic parameters must be followed for safe and healthy fermented food product development.

2.4.1 Milk

Tarag is a hard type of yogurt made from cow milk and sometimes from the skim milk of sheep or goats. Traditionally, naturally fermented goat milk was first used for tarag production. But bovine milk is being used for the production of tarag. The procurement of milk from the dairy herd must be preferred, and it must be stored at 4 °C until raw milk for the development of tarag was standardized fat and SNF ratio. Whey protein or whole milk powder or skim milk powder may be added for standardization. The milk used to make tarag is typically standardized at 8% fat and 14% SNF content, but tarag can also be produced with skimmed or low-fat milk (*see* **Notes 1** and **24**).

2.4.2 Starter cultures

In the traditional method, yogurt with an unknown/mixed starter culture is used. In industrial methods, three commercial starter cultures are usually utilized to make tarag. These cultures include the following LAB cultures, respectively:

I. *L. delbrueckii* subsp. *bulgaricus*

II. *S. thermophilus*

III. *L. kefiranofaciens*

IV. *L. helveticus*

V. *L. fermentum*

After procurement, storage and activation of starter cultures must be done by the recommendation of the food manufacturer.

The starter cultures were first individually activated in skim milk by inoculating skim milk with 2% lyophilized strains and incubating it at 37 °C for 24 h (*see* **Note 2**).

3 Methods

The production of fermented milk must be carried out in hygienic equipment. All the procedures must be carried out at room temperature unless otherwise specified standard operating procedures (SOPs) must be followed.

3.1 Labneh

The schematic flow diagram of traditional and modern methods (with ultrafiltration technique) for the production of labneh is usually followed according to the consumer requirements or the industry's quality standards (Fig. 1).

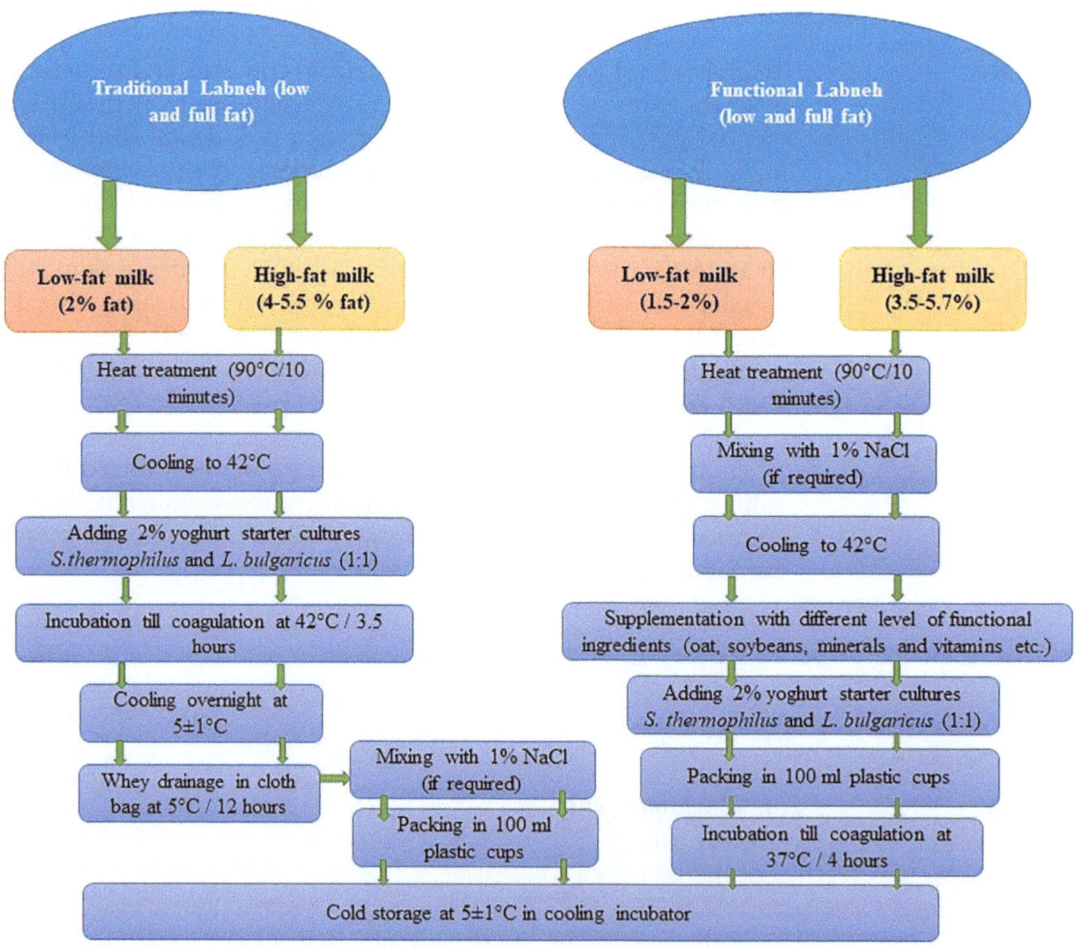

Fig. 1 Schematic flow diagram of manufacturing steps involved in labneh production

1. **In the traditional method**, labneh is produced by heating milk to medium heat till it's warm enough but not boiled then cooling it to room temperature.
2. After that, unknown/mixed starter culture or usually previously made yogurt is added and then cooling and hanging for water removal is done.
3. It can be washed with water for extra curd strength and salting according to the taste preferences.

1. **In industrial methods**, advanced methods must be followed according to SOPs or the low-/high-fat labneh requirement.
2. Raw bovine milk is first filtered before further processing to remove solid contaminants like white blood cells, dirt, and other debris from udder tissue.
3. The milk is then pasteurized at 85 °C for 30 min or 90 °C for 10 min to kill pathogenic microorganisms and cooled at 42 °C (*see* **Note 4**).
4. The milk is then inoculated with yogurt cultures of *S. thermophilus* and *L. delbrueckii* subsp. *bulgaricus* Lb-12 DRI-VAC at 1–2% rate.
5. *L. acidophillus* CH-1 may be added depending on the fermentation type and industrial standard processing requirements (can be obtained from the starter culture suppliers).
6. The starter cultures, 2% by the ratio of each mixture of cultures, are added to fresh buffalo milk and then incubated at 42 °C for 4 h.
7. The labneh is chilled to 10 °C overnight, mixed, and strained through cheesecloth by hanging the bag in a freezer at 5 °C to drain whey for 12 h.
8. The labneh can be mixed with 1% NaCl to enhance the salt content for flavor compensations or either way to protect from the spoilage bacteria as a preservative.
9. The labneh in modern industrial methods can be added with other functional ingredients such as oats, soybean oil, minerals, vitamins, etc., or beneficial LAB bacteria to enhance the functional characteristics of labneh to be used as a functional food (*see* **Note 5**).
10. If used, the plastic containers must be free from contamination and kept at 7 °C for 15 days. The cold storage at the retailer is usually done at 5 ± 1 °C in a cooling incubator/refrigerator temperature.
11. Labneh treatment samples are undergoing chemical and microbiological testing while still fresh and during storage.

12. The physicochemical parameters of labneh such as fat, protein, lactose, and total solids contents must be evaluated by the methods prescribed by AOAC [88].
13. The labneh must also be subjected to acidity, pH, ash, and microbiological analysis for quality evaluation.
14. The sensory evaluation must be done for consumer overall acceptability regarding the product's consumer preferences (*see* **Note 6**).

3.2 Ayran

The schematic flow diagram of traditional and industrial methods for the production of ayran is usually followed according to the consumer requirements or the industry's quality standards (Fig. 2).

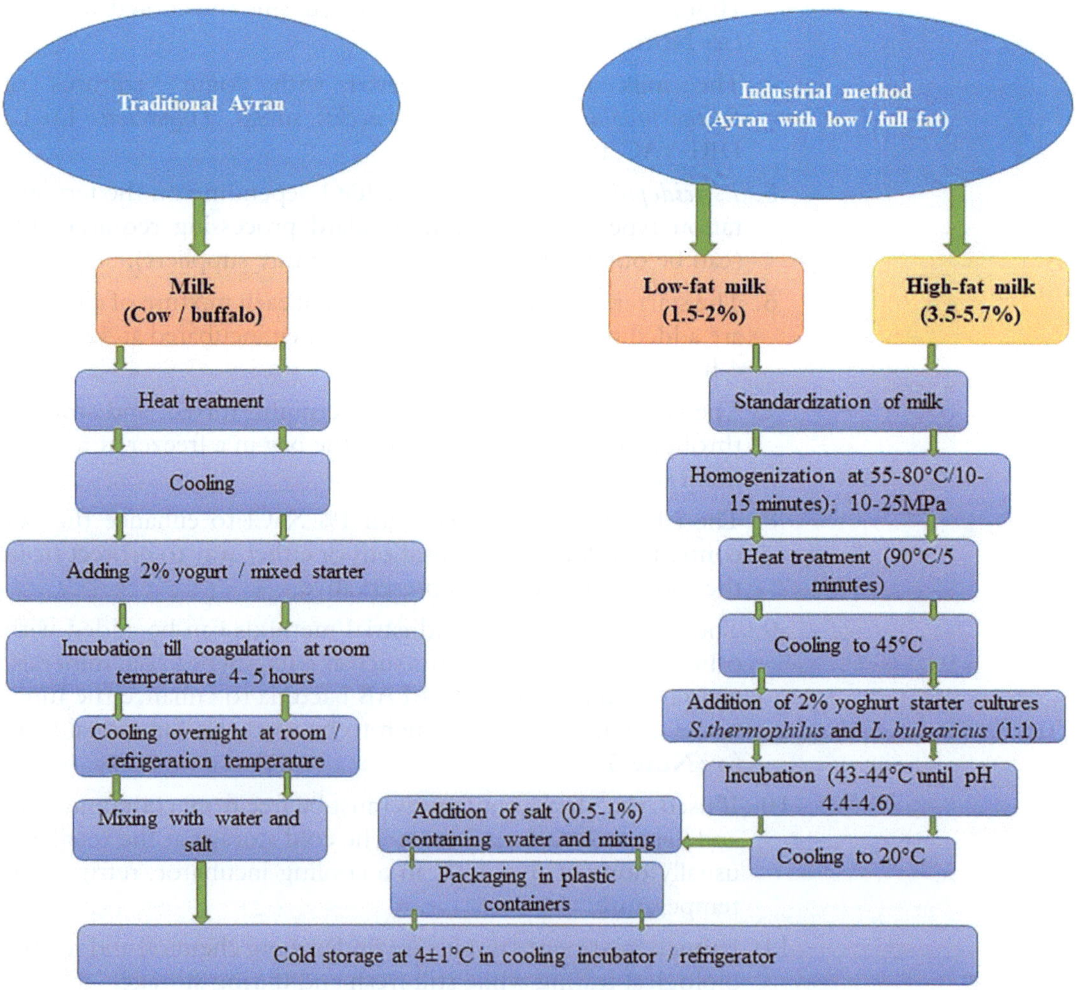

Fig. 2 Schematic flow diagram of manufacturing steps involved in ayran production

1. **In the traditional method**, ayran is produced by heating milk to medium heat until it feels warm but not boiled then cooling it to room temperature.
2. After that, unknown/mixed starter culture or usually previously made yogurt is added.
3. Later on, the yogurt is mixed with salt and water to have the final product.

1. **In industrial methods**, advanced methods must be followed according to SOPs or the low-/high-fat ayran requirement.
2. Raw bovine milk is first filtered before further processing to remove solid contaminants like white blood cells, dirt, and other debris from udder tissue (*see* **Note** 7).
3. After this, milk is standardized to 8% dry matter and 1.5% fat or high fat, depending on the industry's requirements.
4. The milk is then homogenized at 55–80 °C for 10–15 min at 10–25 MPa pressure. It varies depending on the milk type.
5. The milk is pasteurized at 85–90 °C for 5 min to kill pathogenic microorganisms and then cooled to 40–45 °C.
6. The milk is then inoculated with yogurt cultures of *S. thermophilus* and *L. delbrueckii* subsp. *bulgaricus* Lb-12 DRI-VAC at 2% rate.
7. *L. acidophillus* CH-1 may be added depending on the fermentation type and industrial standard processing requirements (can be obtained from the starter culture suppliers).
8. Add the starter cultures at 2% by the ratio of each mixture of culture in fresh buffalo milk and it must be incubated at 42 °C for 4 h.
9. Then incubation is done at 43–44 °C until the pH reaches 4.4–4.6, which is the fermentation process of the starter cultures.
10. Then the coagulum is cooled to 20 °C to stop the fermentation process.
11. Salty water is added to the coagulum to dilute it. The salt is usually added at 0.5–1% (*see* **Note 8**).
12. The final product has a pH range between 4.2 and 4.3.
13. The ayran is then packed depending on the packaging material and storage is done at 4–7 °C.
14. The cold storage at the retailer is usually done at 5 ± 1 °C in a cooling incubator/refrigerator temperature.

15. The composition of ayran differs and depends on milk type, fat removal efficiency, and dilatation rate in the final product after adding salty water (*see* **Note 9**).
16. The physicochemical parameters of ayran such as fat, protein, lactose, and total solids contents must be evaluated by the methods prescribed by AOAC [88].
17. The ayran must also be subjected to acidity, pH, ash, and microbiological analysis for quality evaluation.
18. The sensory evaluation must be done for consumer overall acceptability regarding the product's consumer preferences (*see* **Note 10**).
19. The schematic flow diagram of traditional and industrial methods for the production of ayran is usually followed according to the consumer requirements or the quality standards of the industry (Fig. 2).

3.3 Amasi

The schematic flow diagram of traditional and industrial methods for the production of amasi is usually followed according to the consumer requirements or the industry's quality standards (Fig. 3).

1. **In the traditional method**, fresh cow milk is allowed to naturally ferment in gourds (named "calabash," a wooden bucket shape like gourd vegetable) or earthenware (pots) or any other suitable containers at room temperature to make amasi (*see* **Notes 12** and **13**).
2. A clean calabash is smoked to remove possible contaminants.
3. Collection of raw bovine (cow/buffalo) milk poured into the pre-smoked calabash.
4. The fermentation of milk is carried out by microorganisms that present naturally in milk, container, and surrounding air within 1–3 days.
5. The fermentation depends on the ambient temperature condition of the surroundings of the pots (*see* **Note 13**).
6. After coagulation, a viscous product (amasi) is obtained and the liquid whey (wheylumlaza) is strained or drained/strained by making a plugged hole at the bottom of the container.
7. The remaining amasi is used as a starter culture for the next batch by adding raw milk and processing under similar conditions (known as back-slopping) (*see* **Note 14**).

On identification of the fermentation causing dominant bacterium within a diverse bacterial community, *Lc. lactis* subsp. *cremoris* was considered the fermentation-causing bacterium.

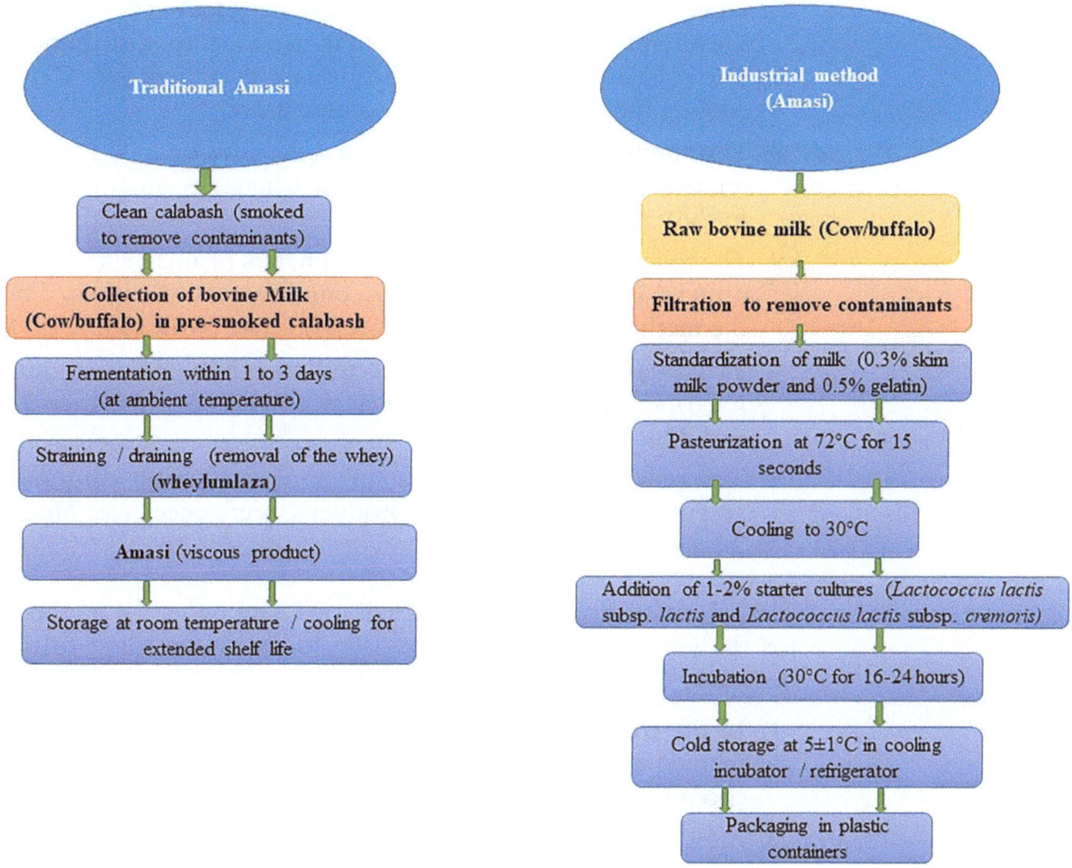

Fig. 3 Schematic flow diagram of manufacturing steps involved in amasi production

1. **In the industrial method**, advanced methods must be followed according to SOPs or the requirement of low-/high-fat amasi (*see* **Note 15**).
2. Raw bovine milk is first filtered before further processing to remove solid contaminants like white blood cells, dirt, and other debris from udder tissue (*see* **Notes** 7 and **14**).
3. The standardization of milk with adding about 0.3% skim milk powder and 0.5% gelatin, respectively.
4. The standardized milk is pasteurized at 72 °C for 15 s and cooled at 30 °C (*see* **Note 16**).
5. The milk is then inoculated with LAB starter cultures (*Lc. lactis* subsp. *lactis* and *Lc. lactis* subsp. *cremoris*) at 1–2% rate to give inoculum start up at 10^6 CFU/ml inoculum level (can be obtained from the starter culture suppliers).
6. The product is then incubated for 16–24 h at 30 °C. It is regarded as the final step of the fermentation process (*see* **Note 17**).

7. Once the required circumstances have been met, it is cooled at a low temperature of 5 ± 1 °C and then packed for distribution. The cold storage at the retailer is usually done at 5 ± 1 °C in a cooling incubator/refrigerator temperature.
8. The composition of amasi varies and depends on milk type, fat removal efficiency, and dilatation rate when salty water is added to the final product (*see* **Note 18**).
9. The following physicochemical parameters of amasi such as fat, protein, lactose, and total solids contents must be evaluated by the methods prescribed by AOAC [88] (*see* **Note 19**).
10. The amasi must also be subjected to acidity, pH, ash, and microbiological analysis for quality evaluation (*see* **Notes 20** and **21**).
11. The sensory evaluation must be done for overall consumer acceptability regarding the product's preferences (*see* **Notes 22** and **23**).

3.4 Tarag

The schematic flow diagram of traditional and industrial methods for the production of tarag is usually followed according to the consumer requirements or the industry's quality standards (Fig. 4).

1. **In the traditional method**, tarag is produced by heating milk to medium heat till it's warm enough but not boiled then cooling it to room temperature as easy to touch.
2. After that, unknown/mixed starter culture or usually previously made yogurt is added.
3. Traditionally, nomadic people use skimmed milk and add mixed/unknown strains of starter cultures for yogurt.
4. After coagulation, the coagulum is placed into a small cotton bag to extract water and used as a starter for following tarag fermentation.
5. The starter is added to milk at medium inoculation/warm temperature at 1:20 to 1:30 and then stirred to mix well.
6. After that, the milk temperature gradually drops and it starts to coagulate within 30 min.
7. The milk is fully coagulated into a soft tofu-like product (tarag) after 3–4 h of fermentation (*see* **Note 24**).
8. The fermentation usually continues until the following morning about 18–20 h.
9. This prolonged fermentation further drops the pH to 4–4.5. It sometimes takes at least 5 days of fermentation to achieve the desired alcoholic degree, acidity, and sensorial characteristics (*see* **Notes 25** and **26**).

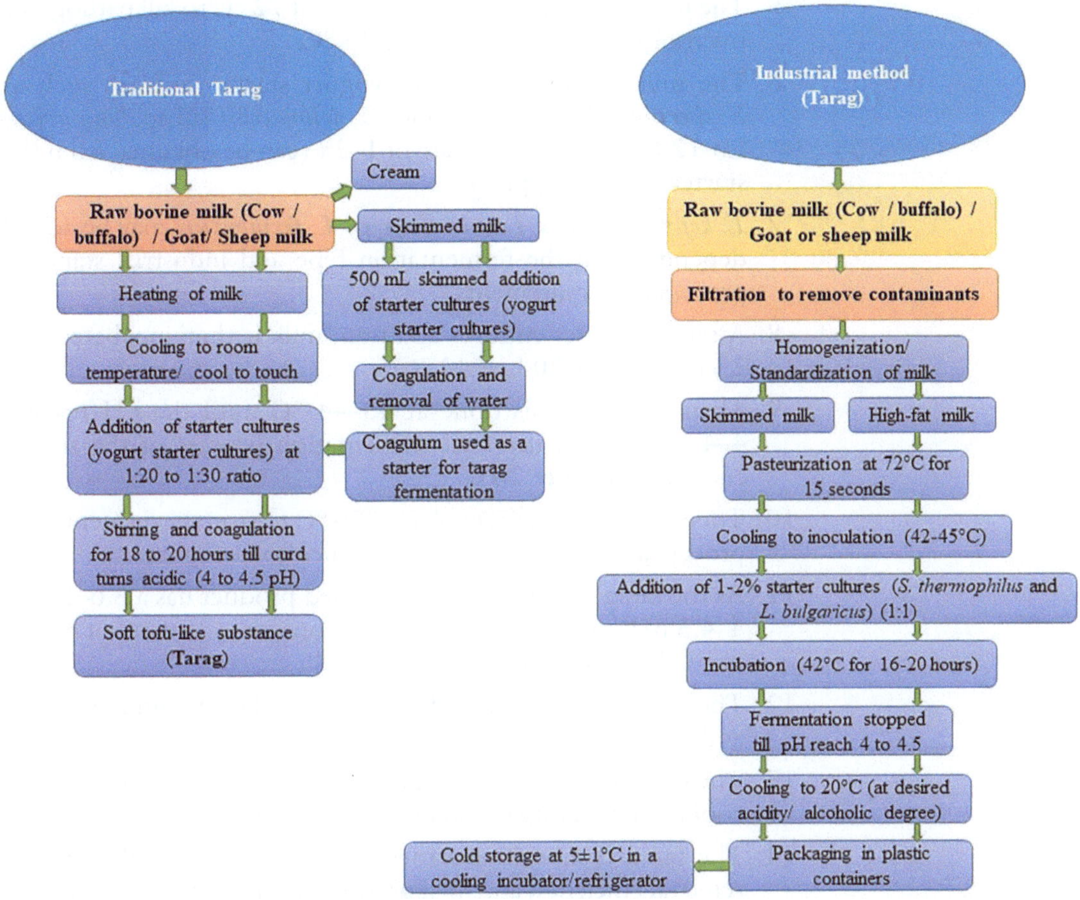

Fig. 4 Schematic flow diagram of manufacturing steps involved in tarag production

10. Tarag can be eaten directly or with some added fruits and in some cases further processed into fermented curdled milk and other dried forms (*see* **Note 25**).

1. **In industrial methods**, advanced methods must be followed according to SOPs or the low-/high-fat tarag requirement.
2. Firstly, milk is standardized to 8% dry matter and 1.5% fat (skimmed milk) or high fat (3–4% fat and 8010% dry matter) depending on requirements of dairy industry.
3. The milk may be then homogenized at 55–80 °C for 10–15 min at 10–25 MPa pressure. It varies depending on the milk type and industrial requirements.
4. The milk is subjected to cream separation for standardization or to make tarag from skimmed milk. It varies depending on the milk type and industrial requirements.

5. The milk is then pasteurized for 15 s at 72 °C to kill pathogenic microorganisms and cooled to 45 °C.
6. The milk is inoculated with yogurt starter cultures such as *S. thermophilus* (Y350B) and *L. delbrueckii* subsp. *bulgaricus* Lb-12 DRI-VAC at the rate of 1–2% (can be obtained from the starter culture suppliers).
7. *L. kefiranofaciens*, *L. helveticus*, or *L. fermentum* may be added depending on the fermentation type and industrial standard processing requirements.
8. After the starter cultures addition, milk is then incubated at a 2% rate for 16–20 h at 42 °C.
9. Then incubation is done at 43–44 °C until the pH reaches 4–4.5, which is the final fermentation process of the starter cultures (*see* **Note 26**).
10. Then the coagulum is cooled to 20 °C to stop the fermentation process at the desired acidity and alcoholic degree.
11. As the tarag is sour milk, the finished product has a 3.6–3.9 or 4–4.5 pH value depending on the consumer demand of higher or medium acidic tarag products.
12. The tarag is then packed depending on the packaging material and storage is done at 4–7 °C.
13. The cold storage at the retailer is usually done at 5 ± 1 °C in a cooling incubator/refrigerator temperature.
14. The tarag composition varies and depends on milk type, fat removal efficiency, and final product rate of dilatation with salty water addition (*see* **Note 27**).
15. The following physicochemical parameters of tarag, such as fat, protein, lactose, alcoholic degree, and total solids contents must be evaluated by the methods prescribed by AOAC [88].
16. The tarag must also be subjected to acidity, pH, ash, and microbiological analysis for quality evaluation (*see* **Note 28**).
17. The sensory evaluation must be done for overall consumer acceptability regarding the product's preferences (*see* **Note 29**).

4 Notes

1. The ultimate flavor of cultured milk products is the consequence of a combination of milk compositions present in a certain ratio. For instance, butter and buttermilk are both connected with the diacetyl flavor. It should be slightly acidic and prickly while drinking cultured milk. To fulfill these requirements, the lactic acid and carbohydrate content must

be monitored while the product is being made. When milk is properly cooled, acidification can be minimized, considering that the surplus carbon dioxide is eliminated by vacuum de-aeration or stirring at the end of fermentation [89]. To create a coagulum that may be easily blended into a smooth and viscous product, the whey proteins must be denatured. Heat treatments of 90 °C for 3 min or 85 °C for 30 min can achieve these requirements.

2. The Direct Vat Set (DVS) cultures of all the starter cultures can be procured from Christian Hansen (Hoersholm, Denmark) and DSM (Moorebank, NSW, Australia). If milk is of poor quality purity, sodium caseinate (90%) may be used. Glucose-δ-lactone (GDL), gellan, and phytosterols (β-sitosterol ≥70%) can be procured from Sigma-Aldrich (St. Louis, MO, USA). These stabilizing and emulsifying agents may also enhance the curd firmness and nutritional properties.
3. According to a recent study, Ayran typically comprises 5% total solids, 94% water, 1.7% protein, 1.2% fat, 1% ash, and 0.7% salt. Ayran is a different fermented milk beverage than yogurt beverages containing salt without any fruit flavoring [90]. At end of the storage period, viable yogurt bacteria count usually reaches 10^8 CFU/mL, which is typically enough to be effective as a nutritional supplement for humans. The product has acceptable sensory characteristics and can be kept at a refrigerator temperature for up to 15 days [91].
4. Thermal processing, which can destroy toxins and antinutritional components like lectins in red beans, is one of the most important methods for controlling, eliminating, or lowering infections down to manageable levels. The method for determining the necessary heat treatment for the two primary preservation processes (pasteurization and sterilization) depends on applying the thermal inactivation of microorganisms to new heat treatment techniques, such as electrical heating techniques and novel combined therapies [92].
5. Gelatin and skim milk powder are added to prevent syneresis and produce a smooth, homogenous food product. This boosts the nutritional content and functional aspects of fermented foods. Continuous cooling prevents pipelines and fillers' blockage by stopping the milk from hardening. Additionally, the use of many stabilizers in combination may be researched to offer an effect in acidified milk beverages at low levels without compromising flavor.
6. The shelf life of fermented food products is usually extended by fermentation process, which also improves milk taste and digestibility. These homemade goods undergo spontaneous fermentation by natural starter cultures from earlier batches with intricate microbial makeup.

7. Ayran has a low pH that makes it (like other acidic milk drinks) susceptible to textural instability during storage. The serum separation (≥30%) and low viscosity are the main textural flaws documented. Serum separation can be avoided if hydrocolloid stabilizers like pectin, guar gum, and gelatin are present [93].
8. Salt solutions with different concentrations were made by combining tap water and food-grade NaCl. The salt solutions were pasteurized by boiling for about 5 s, allowing them to cool to room temperature, and then repeating the process. They were then sufficiently mixed with yogurt to yield ayran.
9. In the preliminary trials, a variety of stabilizers, including k-carrageenan, low methoxyl pectin, Na alginate, locust bean gum (LBG), gelatin, agar agar, guar gum, carboxymethylcellulose, and high methoxyl pectin (HMP), were tested with the primary objective of preventing serum separation and the secondary objective of guaranteeing homogeneous texture form in ayran. However, carboxymethylcellulose caused the high separation of serum. The serum separation, gelation, and apparent aggregation were all caused by k-carrageenan. Agar agar, Na alginate, and low methoxyl pectin all contributed to serum separation and aggregation. The only stabilizers discovered effective for ayran were guar gum, LBG, gelatin, and HMP [94, 95].
10. However, it was discovered that the stabilizers' impact on the product's flavor limited the amount of stabilizers that could be employed. Even at low quantities, ayran was found to taste and smell strange when exposed to HMP and gelatin. Low quantities of gelatin and HMP did not prevent serum separation either. Guar gum prevented serum separation in ayran and had the highest apparent viscosity; however, due to the oily mouthfeel it produced, it was not preferred organoleptically. However, milk beverages with a high-fat content may be appropriate for guar gum. LBG at a dosage of 0.1–0.2% is suggested for acidic milk beverages to stabilize texture without altering product flavor [96]. The stabilizer solution was created by mixing the requisite amounts of stabilizer and salt in distilled water used to obtain the proper stabilizer concentration in the final 250 g ayran sample.
11. Amasi is now made from pasteurized milk under carefully controlled processing conditions to ensure food safety, hastening the fermentation process with the help of commercial starting cultures. Pasteurization renders all naturally present bacteria inactive, even harmful ones. The amasi product popularity and the city's growing population have caused it to spread from rural to urban regions [15].

12. Full-cream milk is used to make the fermented milk product amasi. The bacterial cultures present in homemade fermented amasi come from unpasteurized milk that was kept in clay or calabash containers used to store the milk. The primary tools utilized by the various tribes to create the homemade amasi were calabashes and clay pots [84]. Raw and fresh unpasteurized milk is placed in containers to ferment at normal room temperature for 2–3 days, or for a quicker reaction, it is placed immediately in the sun. The unsweetened curd with a smooth, yogurt-like texture can be characterized by fermentation.
13. Due to the numerous microbial activities in milk, which are present during fermentation and cause-specific chemical and physical changes during the process, amasi is produced and has specified features. The lactic acid bacteria (LAB) that give amasi its acidic qualities give it a fresh sour smell. Its pH ranges from 3.6 to 4.2, and its consistency is slightly thicker than normal yogurt [97].
14. The whitish thick liquid/coagulum (amasi) is usually extracted after the watery thin liquid. The thick curd mass part is frequently collected when the desired properties are obtained. The remaining amasi could be used as a starter culture for subsequent batches by adding raw milk and processing in similar methods (back-slopping). The sour flavor of amasi is the result of the fermentation process, which is a crucial stage in amasi processing. Another critical step is discarding the thin liquid since it distinguishes amasi from other nearly related dairy-fermented products [98].
15. The commercialization technique for amasi differs slightly from locally produced amasi since it uses microbial starting cultures to speed up the fermentation process. Other components, including skim milk and gelatin, are also added to stabilize the food. Because of laws governing safety of food aspects, raw milk is first pasteurized before fermentation to kill potentially harmful microbes. The starter culture that was added causes the flavor, lactic acid, and aroma components to grow [99].
16. The nutritional makeup of amasi varies greatly and is affected by type of milk used, storage conditions, and the temperature and duration of the fermentation process. The literature on the nutritional makeup of amasi includes various micronutrients, including minerals and vitamins. Carbohydrates ($\geq$6%) make up majority of the chemical composition, followed by lipids (4.06%) and proteins (3.87%) [100].
17. During milk fermentation, the acid coagulation and proteolysis processes make the proteins more easily digestible. Lactose in milk is converted to acid during fermentation, lowering pH

and thickening the end product. The enzyme β-galactosidase hydrolyzes around 30% of lactose, converting it to glucose and galactose [101]. Amasi may therefore be a good choice for lactose-intolerant consumers.

18. It is commonly known that amasi contains several vitamins when it is fermented and made into food. The preparation method, milk type, milk microbes, and method of preparation influence the quantity of vitamins. Aside from that, it was mentioned that amasi is a suitable source of nutrition for including in food-based dietary requirements because it is abundant in minerals like calcium, iron, and zinc [34].
19. The literature is lacking in information on health benefits of amasi. Direct evidence for health benefits of amasi came from a South African food allergy study that examined regular consumption of fermented milk and discovered that it was protective against asthma, allergic rhinitis, and atopic dermatitis issues in urban children but not rural children. A recent editorial published in the South African Journal of Child Health suggested that the probiotic component of amasi may contribute to the medicine's efficacy in treating pediatric diarrhea. The South African Department of Health for Children (*Road to Health* book) suggested amasi (also known as "maas") as a weaning food for kids who aren't breastfed beginning at the age of 12 months [102].
20. The organic acids produced during fermentation of amasi cause milk pH to decrease to 4.29. This pH may cause the milk to coagulate by preventing the growth of such bacteria which frequently may lead to food degradation. Amasi has a longer shelf life than raw milk as it can be kept for more than 21 days at low temperatures of 4–5 °C. Amasi has low pH which helps it to have positive effects on satiety, glycemic response, and appetite control by delaying stomach emptying [103].
21. When amasi is fermented, the conversion of lactose to lactic acid lowers the lactose level compared to fresh milk, and lactose-intolerant people may usually tolerate this reduced lactose content [104, 105].
22. According to the World Bank, the socioeconomic development of countries depends increasingly on animal-based products like dairy products. Such dairy products, especially amasi, should be a part of a healthy diet because they contain crucial nutrients and microbes that are considered good for human digestion and overall health. However, there are underlying problems with product safety and processing that call for appropriate and prompt measures to protect the populace [106]. Despite this amasi popularity and widespread

consumption in Southern Africa, there are still gaps in its nutritional makeup and health advantages. A thorough understanding of its microbiota is also necessary even if little research exists.

23. The development of "omics" techniques, such as metabolomics, foodomics, genomics, transcriptomics, metagenomics, volatilomics, and proteomics has made it possible to understand better the nutritional components, functionalities, and metabolic interactions underlying molecular mechanisms in fermented foods, also other potential health benefits regarding amasi [107]. These techniques are useful in the commercial as well as advanced industrial methods for better health advantages of the amasi products.
24. The ancient naturally fermented goat milk dish called tarag was a mainstay of the Mongolian diet due to alcohol production in natural ways. Per person per day, 1–2 L of tarag were consumed on average. The product was created using raw whole milk from goats of the Zang and Chaidamu types, two domestic goat species officially recognized in China for producing milk [108, 109].
25. China traditionally consumes tarag, a traditional naturally fermented alcoholic beverage derived from goat milk. Each person in the Mongolian population consumes an average of 1–2 L per day. The absence of nutritious composition is because it is typically manufactured in homes rather than factories [110].
26. Tarag goods had high levels of yeast and LAB. It should be included in the category of products from yeast lactic fermentations if the yeast is one of the natural microflora of the milk and the item contains alcohol. Hence, a starter for tarag is mild, homogenized acid curd and lubricated from earlier preparation along with low water content and 4.5 pH [111].
27. Tarag was often found to be high in fat and protein. The findings support tarag significance for the region population consumption of high-quality protein. In addition, tarag had a slightly greater calorie count because meals without enough energy prevent protein from being used to its maximum potential [112, 113].
28. The higher amounts of protein and fat improved the ability to clot and coagulate, resulting in a coagulum that was dense and homogeneous [114], thus making it acceptable to consumer demands according to the rich taste and flavor.
29. According to one study, milk coagulation characteristics were mostly influenced by its protein and fat content. The rich flavor and consistency of tarag, which is comparable to unsweetened commercial yogurt or fresh cheese, were also influenced by this feature [115].

5 Recommendations

1. There are significant researches that has to be done on the possibility of developing more ancient fermented milk products in both traditional and commercial ways. Also, the addition of probiotics into the production of well-liked traditional fermented dairy products can be a future approach to have specific benefits by using genetically modified starter culture strains.
2. It is becoming more and more important to develop convenient dry, cultured, fermented food products with live bacteria using different methods, such as microencapsulation and to fortify such powders or concentrated products with vitamins and minerals.
3. The production of appropriate mixtures comprising beneficial components such as dietary fiber, organic vitamins, and folic acid opens up new study opportunities in this rapidly increasing fermented food production field.
4. There will undoubtedly be more potential for growth in the fermented beverage business due to the development of various health beverages and dietetic preparations as value-added food products based on conventional fermented methods.
5. In addition to the overall nutrient increase gained, more research is required to determine the medicinal advantages of viable and nonviable starter cultures available in fermented beverages available in various ancient regions of the world.
6. There is also a need to use a good systematic preventative technique for risk management in the dairy production sector. Given the safety precautions and hygienic protocols for food products, these procedures must be followed properly through SOPs accordingly.
7. Nearly all dairy products share common manufacturing stages, these fermented milk beverage methods will prove helpful for researchers' findings to aid a significant number of importance in future dairy product development. This will ultimately provide better knowledge for safe fermented food product development with advancements in health aspects for the end consumers.

Acknowledgments

This work was supported by the National Institute of Food Science and Technology, Faculty of Food Nutrition and Home Sciences, University of Agriculture, Faisalabad, Pakistan.

References

1. Salminen S, Collado MC, Endo A et al (2021) The International Scientific Association of Probiotics and Prebiotics (ISAPP) consensus statement on the definition and scope of postbiotics. Nat Rev Gastroenterol Hepatol 18: 649–667. https://doi.org/10.1038/s41575-021-00440-6
2. Hill D, Sugrue I, Arendt E et al (2017) Recent advances in microbial fermentation for dairy and health. F1000Research 6:751. https://doi.org/10.12688/f1000research.10896.1
3. García-Burgos M, Moreno-Fernández J, Alférez MJM et al (2020) New perspectives in fermented dairy products and their health relevance. J Funct Foods 72:104059. https://doi.org/10.1016/j.jff.2020.104059
4. Van Kleef E, Dagevos H (2015) The growing role of front-of-pack nutrition profile labeling: a consumer perspective on key issues and controversies. Crit Rev Food Sci Nutr 55:291–303. https://doi.org/10.1080/10408398.2011.653018
5. Li D, Wang P, Luo Y et al (2017) Health benefits of anthocyanins and molecular mechanisms: update from recent decade. Crit Rev Food Sci Nutr 57:1729–1741. https://doi.org/10.1080/10408398.2015.1030064
6. Kalam Saleena LA, Phing PL, Gan R-Y et al (2023) Fermented dairy products from Middle Eastern and Northern African (MENA) countries: Insight on production and physiochemical characteristics. Int Dairy J 141: 105614. https://doi.org/10.1016/j.idairyj.2023.105614
7. Tamime AY, Hickey M, Muir DD (2014) Strained fermented milks – a review of existing legislative provisions, survey of nutritional labelling of commercial products in selected markets and terminology of products in some selected countries. Int J Dairy Technol 67: 305–333. https://doi.org/10.1111/1471-0307.12147
8. Tamime AY, Robinson RK (2007) Tamime and Robinson's yoghurt: science and technology. Elsevier, London
9. Baschali A, Tsakalidou E, Kyriacou A et al (2017) Traditional low-alcoholic and non-alcoholic fermented beverages consumed in European countries: a neglected food group. Nutr Res Rev 30:1–24
10. Farag MA, Jomaa SA, Abd El-Wahed AR, El-Seedi H (2020) The many faces of kefir fermented dairy products: quality characteristics, flavour chemistry, nutritional value, health benefits, and safety. Nutrients 12:346
11. Dehkordi FS, Yazdani F, Mozafari J, Valizadeh Y (2014) Virulence factors, serogroups and antimicrobial resistance properties of Escherichia coli strains in fermented dairy products. BMC Res Notes 7:1–8
12. Matsuoka K, Uemura Y, Kanai T et al (2018) Efficacy of Bifidobacterium breve fermented milk in maintaining remission of ulcerative colitis. Dig Dis Sci 63:1910–1919
13. Dimov SG (2022) The unusual microbiota of the traditional Bulgarian dairy product Krokmach – a pilot metagenomics study. Int J Dairy Technol 75:139–149
14. Kocabaş H, Ergin F, Aktar T, Küçükçetin A (2022) Effect of lactose hydrolysis and salt content on the physicochemical, microbiological, and sensory properties of ayran. Int Dairy J 129:105360
15. Maleke MS, Adefisoye MA, Doorsamy W, Adebo OA (2021) Processing, nutritional composition and microbiology of amasi: a Southern African fermented milk product. Sci Afr 12:e00795
16. Konuspayeva G, Baubekova A, Akhmetsadykova S, Faye B (2023) Traditional dairy fermented products in Central Asia. Int Dairy J 137:105514
17. Gebreselassie N, Abrahamsen RK, Beyene F, Narvhus JA (2012) A survey on spontaneously fermented buttermilk in Northern Ethiopia. Afr J Food Sci Technol 3:78–89
18. Abdelgadir W, Nielsen DS, Hamad S, Jakobsen M (2008) A traditional Sudanese fermented camel's milk product, Gariss, as a habitat of Streptococcus infantarius subsp. infantarius. Int J Food Microbiol 127:215–219
19. Harun-ur-Rashid M, Togo K, Ueda M, Miyamoto T (2007) Identification and characterization of dominant lactic acid bacteria isolated from traditional fermented milk Dahi in Bangladesh. World J Microbiol Biotechnol 23:125–133
20. Dimassi O, Iskandarani Y, Afram M et al (2020) Production and physicochemical properties of labneh anbaris, a traditional fermented cheese like product, in Lebanon. Int J Environ Agric Biotechnol 5:509–516
21. Gran HM, Gadaga HT, Narvhus JA (2003) Utilization of various starter cultures in the production of Amasi, a Zimbabwean naturally fermented raw milk product. Int J Food Microbiol 88:19–28

22. Yam BAZ, Khomeiri M, Mahounak AS, Jafari SM (2014) Hygienic quality of camel milk and fermented camel milk (chal) in Golestan Province. Iran J Microbiol Res 4:98–103
23. Guo L, Xu W, Li C et al (2021) Comparative study of physicochemical composition and microbial community of Khoormog, Chigee, and Airag, traditionally fermented dairy products from Xilin Gol in China. Food Sci Nutr 9:1564–1573. https://doi.org/10.1002/fsn3.2131
24. Hingmire SR, Lembhe AF, Zanjad PN et al (2009) Production and quality evaluation of instant lassi. Int J Dairy Technol 62:80–84
25. Dewan S, Tamang JP (2006) Microbial and analytical characterization of Chhu-A traditional fermented milk product of the Sikkim Himalayas. J Sci Ind Res 65(9):747–752
26. Sudun, Wulijideligen, Arakawa K et al (2013) Interaction between lactic acid bacteria and yeasts in airag, an alcoholic fermented milk. Anim Sci J 84:66–74
27. Güler Z (2007) Levels of 24 minerals in local goat milk, its strained yoghurt and salted yoghurt (tuzlu yoğurt). Small Rumin Res 71:130–137
28. Admasu MA, Cione E, Aquaro S (2016) Microbiological characteristics and physicochemical parameters of fermented milk product ergo-A traditional yogurt product of Ethiopia. Food Sci Qual Manag 49:42–45
29. Abesinghe A, Priyashantha H, Prasanna PHP et al (2020) Inclusion of probiotics into fermented buffalo (Bubalus bubalis) milk: an overview of challenges and opportunities. Fermentation 6:121
30. Bintsis T, Papademas P (2022) The evolution of fermented milks, from artisanal to industrial products: a critical review. Fermentation 8:679
31. Kondybayev A, Loiseau G, Achir N et al (2021) Fermented mare milk product (Qymyz, Koumiss). Int Dairy J 119:105065
32. Arrizza S, Ledda A, Sarra PG, Dellaglio F (1983) Identification of lactic acid bacteria in "Gioddu". Sci e Tech Latt 34:87–102
33. Shori AB (2012) Comparative study of chemical composition, isolation and identification of micro-flora in traditional fermented camel milk products: Gariss, Suusac, and Shubat. J Saudi Soc Agric Sci 11:79–88
34. Moonga HB, Schoustra SE, Linnemann AR et al (2021) Influence of fermentation temperature on microbial community composition and physicochemical properties of mabisi, a traditionally fermented milk. LWT 136:110350
35. Mathara JM, Schillinger U, Kutima PM et al (2004) Isolation, identification and characterisation of the dominant microorganisms of kule naoto: the Maasai traditional fermented milk in Kenya. Int J Food Microbiol 94:269–278
36. Sulieman AME, Abd Elgadir HO, Elkhalifa AE (2011) Chemical and microbiological characteristics of fermented milk product, Mish. Int J Food Sci Nutr Eng 1:1–4
37. Liu WJ, Sun ZH, Zhang YB et al (2012) A survey of the bacterial composition of kurut from Tibet using a culture-independent approach. J Dairy Sci 95:1064–1072
38. Bagal SG, Chavan KD, Kulkarni MB (2007) Studies on preparation of lassi from high acid cow milk. J Dair Foods Home Sci 26:80–84
39. El Zubeir IEM, Abdalla WM, El Owni OAO (2005) Chemical composition of fermented milk (roub and mish) in Sudan. Food Control 16:633–637
40. Priyashantha H, Ranadheera CS, Rasika DMD, Vidanarachchi JK (2021) Traditional Sri Lankan fermented buffalo (Bubalus bubalis) milk gel (Meekiri): technology, microbiology and quality characteristics. J Ethn Foods 8:1–17
41. Akabanda F, Owusu-Kwarteng J, Glover RLK, Tano-Debrah K (2010) Microbiological characteristics of Ghanaian traditional fermented milk product, Nunu. Nat Sci 8:178–187
42. Fagbemigun O, Cho G-S, Rösch N et al (2021) Isolation and characterization of potential starter cultures from the nigerian fermented milk product nono. Microorganisms 9:640
43. Elemanova R, Musulmanova M, Ozbekova Z et al (2022) Rheological, microbiological and sensory properties of fermented khainak milk fermented with different starter cultures. Int Dairy J 134:105453
44. Akabanda F, Owusu-Kwarteng J, Tano-Debrah K et al (2013) Taxonomic and molecular characterization of lactic acid bacteria and yeasts in nunu, a Ghanaian fermented milk product. Food Microbiol 34:277–283
45. Abd El Gawad IA, Abd El Fatah AM, Al Rubayyi KA (2010) Identification and characterization of dominant lactic acid bacteria isolated from traditional rayeb milk in Egypt. J Am Sci 6:728–735
46. Abdelgadir WS, Ahmed TK, Dirar HA (1998) The traditional fermented milk products of The Sudan. Int J Food Microbiol 44:1–13
47. Abdalla MOM, Hussain SIK (2010) Enumeration and identification of microflora in Roub,

a Sudanese traditional fermented dairy product. Br J Dairy Sci 1:30–33

48. Nehme L, Salameh C, Tabet E et al (2019) Innovative improvement of Shanklish cheese production in Lebanon. Int Dairy J 90:23–27
49. Kochetkova TV, Grabarnik IP, Klyukina AA et al (2022) Microbial communities of artisanal fermented milk products from Russia. Microorganisms 10:2140
50. Fondén R, Leporanta K, Svensson U (2006) Nordic/Scandinavian fermented milk products. In: Fermented milks. Blackwell Science Ltd, Oxford, pp 156–173
51. Akhmetsadykova S, Baubekova A, Konuspayeva G et al (2015) Lactic acid bacteria biodiversity in raw and fermented camel milk. Afr J Food Sci Technol 6(3):84–88
52. Odunfa SA (1988) African fermented foods: from art to science. MIRCEN J Appl Microbiol Biotechnol 4:259–273
53. Şanal H, Güler Z (2010) Changes in non-essential element concentrations during torba yoghurt production. Akad Gıda 8:6–12
54. O'Callaghan YC, Shevade AV, Guinee TP et al (2019) Comparison of the nutritional composition of experimental fermented milk: wheat bulgur blends and commercially available kishk and tarhana products. Food Chem 278:110–118
55. Kahala M, Mäki M, Lehtovaara A et al (2008) Characterization of starter lactic acid bacteria from the Finnish fermented milk product viili. J Appl Microbiol 105:1929–1938
56. Kabak B, Dobson ADW (2011) An introduction to the traditional fermented foods and beverages of Turkey. Crit Rev Food Sci Nutr 51:248–260
57. Vasiee A, Falah F, Mortazavi SA (2022) Evaluation of probiotic potential of autochthonous lactobacilli strains isolated from Zabuli yellow kashk, an Iranian dairy product. J Appl Microbiol 133:3201–3214
58. Mashak Z, Sodagari H, Mashak B, Niknafs S (2014) Chemical and microbial properties of two Iranian traditional fermented cereal-dairy based foods: Kashk-e Zard and Tarkhineh. Int J Biosci 4:124–133
59. Yıldız F (2016) 1: Overview of yogurt and other fermented dairy products. In: Development and manufacture of yogurt and other functional dairy products. CRC Press, Boca Raton, pp 1–45
60. Aladjadjiyan A, Zheleva I, Kartalska Y (2016) Traditional Bulgarian dairy food. In: Traditional Foods: General and Consumer Aspects. Springer, USA, pp 115–122
61. Mokoena MP, Mutanda T, Olaniran AO (2016) Perspectives on the probiotic potential of lactic acid bacteria from African traditional fermented foods and beverages. Food Nutr Res 60:29630
62. Gadaga TH, Lehohla M, Ntuli V (2013) Traditional fermented foods of Lesotho. J Microbiol Biotechnol Food Sci 2:2387–2391
63. Kroll, F. (2016) Foodways of the Poor in South Africa How Value-Chain Consolidation, Poverty and Cultures of Consumption Feed Each Other, Institute for Poverty, Land and Agrarian Studies (PLAAS). Retrieved from https://coilink.org/20.500.12592/rrmpqh on 27 Aug 2024. COI: 20.500.12592/rrmpqh
64. Sun Z, Liu W, Bao Q et al (2014) Investigation of bacterial and fungal diversity in tarag using high-throughput sequencing. J Dairy Sci 97:6085–6096
65. Tamang JP, Cotter PD, Endo A et al (2020) Fermented foods in a global age: East meets West. Compr Rev Food Sci Food Saf 19:184–217
66. Muthukumaran MS, Mudgil P, Baba WN et al (2023) A comprehensive review on health benefits, nutritional composition and processed products of camel milk. Food Rev Int 39:3080–3116
67. Hirata M, Hirata M (2020) Milk culture of North Asia. In: Milk culture Eurasia constructing a hypothesis monogenesis–bipolarization. Springer, Singapore, pp 107–147
68. Azam M, Mohsin M, Ijaz H et al (2017) Lactic acid bacteria in traditional fermented Asian foods. Pak J Pharm Sci 30:1803–1814
69. Alonso S, Carmen Castro M, Berdasco M et al (2019) Isolation and partial characterization of lactic acid bacteria from the gut microbiota of marine fishes for potential application as probiotics in aquaculture. Probiotics Antimicrob Proteins 11:569–579
70. Bimbo F, Bonanno A, Viscecchia R (2016) Do health claims add value? The role of functionality, effectiveness and brand. Eur Rev Agric Econ 43:761–780
71. Hati S, Das S, Mandal S (2019) Technological advancement of functional fermented dairy beverages. In: Engineering tools in the beverage industry. Elsevier, Duxford, pp 101–136
72. Farkye NY (2017) Quark, quark-like products, and concentrated yogurts. In: Cheese: Chemistry, Physics and Microbiology. Elsevier Inc., USA, pp 1103–1110
73. Minz PS, Singh RRB (2016) Modernization of manufacturing process for traditional

Indian dairy products. In: Modrnization of Traditional Food Processs and Products. Springer, USA, pp 161–174

74. Chandan RC (2017) An overview of yogurt production and composition. In: Yogurt in Health and Disease Prevention. CRC Press, USA, pp 31–47
75. Jørgensen CE, Abrahamsen RK, Rukke E-O et al (2019) Processing of high-protein yoghurt – a review. Int Dairy J 88:42–59
76. da Cruz AG, Silva MC, Pimentel TC et al (2023) Probiotic foods and beverages: technologies and protocols. Springer Nature, Berlin
77. Desai NT, Shepard L, Drake MA (2013) Sensory properties and drivers of liking for Greek yogurts. J Dairy Sci 96:7454–7466
78. El-Sayed SM, El-Sayed HS (2021) Antimicrobial nanoemulsion formulation based on thyme (Thymus vulgaris) essential oil for UF labneh preservation. J Mater Res Technol 10: 1029–1041
79. Akarca G, Ozkan M, Ozcan T (2022) The impact of combination of solution plasma processing and pulsed electric field on the viability of probiotic bacteria, microbial growth, and structure of yoghurt drink. J Food Process Preserv 46:e16616
80. Salameh C, Banon S, Hosri C, Scher J (2016) An overview of recent studies on the main traditional fermented milks and white cheeses in the Mediterranean region. Food Rev Int 32:256–279
81. Kumar A, Hussain SA, Raju PN et al (2017) Packaging material type affects the quality characteristics of Aloe-probiotic lassi during storage. Food Biosci 19:34–41
82. Pinto MGV, Franz CMAP, Schillinger U, Holzapfel WH (2006) Lactobacillus spp. with in vitro probiotic properties from human faeces and traditional fermented products. Int J Food Microbiol 109:205–214
83. Groenenboom AE, Shindano J, Cheepa N et al (2020) Microbial population dynamics during traditional production of Mabisi, a spontaneous fermented milk product from Zambia: A field trial. World J Microbiol Biotechnol 36:1–14
84. Maleke MS, Adefisoye MA, Doorsamy W, Adebo OA (2021) Processing, nutritional composition and microbiology of amasi: A Southern African fermented milk product. Sci Afr 12:1–8
85. Nedović V, Gibson B, Mantzouridou TF et al (2015) Aroma formation by immobilized yeast cells in fermentation processes. Yeast 32:173–216
86. Choi S-H (2016) Characterization of airag collected in Ulaanbaatar, Mongolia with emphasis on isolated lactic acid bacteria. J Anim Sci Technol 58:1–10
87. Buell PD, Anderson EN, de Pablo Moya M, Oskenbay M (2020) Food by country. In: Crossroads of cuisine. Brill, Leiden, pp 204–286
88. Paez V, Barrett WB, Deng X et al (2016) AOAC SMPR® 2016.002. J AOAC Int 99: 1122–1124
89. Sharma R, Garg P, Kumar P et al (2020) Microbial fermentation and its role in quality improvement of fermented foods. Fermentation 6:106
90. Meybodi NM, Nasab SS, Khorshidian N, Mortazavian AM (2021) Probiotic beverages: health benefits and current trends in the Middle East. In: Probiotic beverages. Elsevier, London, pp 99–126
91. Marco ML, Sanders ME, Gänzle M et al (2021) The International Scientific Association for Probiotics and Prebiotics (ISAPP) consensus statement on fermented foods. Nat Rev Gastroenterol Hepatol 18:196–208. https://doi.org/10.1038/s41575-020-00390-5
92. Deak T, Mohácsi-Farkas C (2023) Thermal treatment. In: Food safety management. Elsevier, London, pp 405–419
93. Yousefi M, Jafari SM (2019) Recent advances in application of different hydrocolloids in dairy products to improve their techno-functional properties. Trends Food Sci Technol 88:468–483
94. Koksoy A, Kilic M (2004) Use of hydrocolloids in textural stabilization of a yoghurt drink, ayran. Food Hydrocoll 18:593–600
95. Manzoor M, Singh J, Bandral JD et al (2020) Food hydrocolloids: functional, nutraceutical and novel applications for delivery of bioactive compounds. Int J Biol Macromol 165:554–567
96. Gyawali R, Ibrahim SA (2016) Effects of hydrocolloids and processing conditions on acid whey production with reference to Greek yogurt. Trends Food Sci Technol 56: 61–76
97. Moonga HB, Schoustra SE, Linnemann AR et al (2022) Towards valorisation of indigenous traditional fermented milk: mabisi as a model. Curr Opin Food Sci 46:100835
98. Moyane JN, Jideani AIO (2013) The physicochemical and sensory evaluation of commercial sour milk (amasi) products. Afr J Food Sci 7:56–62

99. Osvik RD, Sperstad S, Breines EM, et al (2013) Bacterial diversity of amasi, a South African fermented milk product, determined by clone library and denaturing gradient gel electrophoresis analysis. Afr J MicrobRes 7:1–13
100. Moonga HB, Schoustra SE, Linnemann AR et al (2019) The art of mabisi production: a traditional fermented milk. PLoS One 14: e0213541
101. Agyei D, Owusu-Kwarteng J, Akabanda F, Akomea-Frempong S (2020) Indigenous African fermented dairy products: Processing technology, microbiology and health benefits. Crit Rev Food Sci Nutr 60:991–1006
102. Karenzi E, Mashaku A, Nshimiyimana AM et al (2013) Kivuguto traditional fermented milk and the dairy industry in Rwanda. A review. Biotechnol Agron Soc Environ:17, 383–391
103. Dlamini BC, Buys EM (2009) Adaptation of Escherichia coli O157: H7 to acid in traditional and commercial goat milk amasi. Food Microbiol 26:58–64
104. Terpou A, Gialleli A-I, Bekatorou A et al (2017) Sour milk production by wheat bran supported probiotic biocatalyst as starter culture. Food Bioprod Process 101:184–192
105. Dapčević-Hadnađev T, Tomić J, Škrobot D et al (2022) Processing strategies to improve the breadmaking potential of whole-grain wheat and non-wheat flours. Discov Food 2: 11
106. Maleke MS (2021) Effect of fermentation conditions (time and temperature) on the physicochemical, bacteria community and nutritional composition of amasi (a Southern African fermented dairy product). University of Johannesburg (South Africa), Johannesburg
107. Bintsis T (2023) Dairy fermentation. Fermentation 9:252
108. Tamime AY, Wszolek M, Božanić R, Özer B (2011) Popular ovine and caprine fermented milks. Small Rumin Res 101:2–16
109. de Melo PGV, de Carvalho Neto DP, Maske BL et al (2022) An updated review on bacterial community composition of traditional fermented milk products: what next-generation sequencing has revealed so far? Crit Rev Food Sci Nutr 62:1870–1889
110. Widyastuti Y, Febrisiantosa A, Tidona F (2021) Health-promoting properties of lactobacilli in fermented dairy products. Front Microbiol 12:673890
111. Oki K, Dugersuren J, Demberel S, Watanabe K (2014) Pyrosequencing analysis of the microbial diversity of airag, khoormog and tarag, traditional fermented dairy products of Mongolia. Biosci Microbiota Food Health 33:53–64
112. Zhang H, Chen X, Dan T, Dong J (2014) Traditional Chinese fermented dairy foods. In: Lact Acid Bacteria: Fundamentals and Practices, Springer Dordrech, Netherlands, pp 493–535
113. Macori G, Cotter PD (2018) Novel insights into the microbiology of fermented dairy foods. Curr Opin Biotechnol 49:172–178
114. Yu J, Wang WH, Menghe BLG et al (2011) Diversity of lactic acid bacteria associated with traditional fermented dairy products in Mongolia. J Dairy Sci 94:3229–3241
115. Jatmiko YD, Howarth GS, Barton MD (2018) Naturally fermented milk and its therapeutic potential in the treatment of inflammatory intestinal disorders. In: AIP conference proceedings. AIP Publishing, Melville

Chapter 17

Non-fermented Dairy Beverages

Junio César Jacinto de Paula, Ana Flávia Coelho Pacheco, Renata Golin Bueno Costa, Denise Sobral, and Paulo Henrique Costa Paiva

Abstract

A dairy beverage is a product from a mixture of milk and whey, with or without adding food products or substances. Because they are economically viable, non-fermented dairy drinks are an alternative to other traditional dairy products and allow the utilization of whey from cheese making or the permeate resulting from the membrane concentration process. Technologically, using whey in manufacturing beverages can be a viable strategy due to its nutritional composition and the possibility of reducing the environmental pollution generated by its inappropriate disposal. Various non-fermented drinks can be made to popularize access to nutritious and environmentally appropriate food. This chapter discusses the main methods of making non-fermented drinks.

Key words Whey, Technology, Nutritional, Dairy drink, Milk beverage

1 Introduction

In recent years, dairy beverages have gained market share due to their variability, versatility, nutritional quality, sensory attributes, and their more affordable price when compared to other traditional dairy products [1]. In addition, increased knowledge of the health benefits of whey, mainly related to its biological properties, has increased interest in whey beverages [2].

In Brazil, Normative Instruction no. 16 of 23 August 2005, which approves the Technical Regulation on the Identity and Quality of Dairy Beverages, defines it as the dairy product resulting from a mixture of milk (fresh, pasteurized, sterilized, UHT, reconstituted, concentrated, powdered, whole, semi-skimmed or partially skimmed, and skimmed) and whey (liquid, concentrated, and powdered), whether or not added with food products or substances. The dairy base must represent at least 51% of the total ingredients of the product [3].

Adriano Gomes da Cruz et al. (eds.), *Dairy Foods Processing*, Methods and Protocols in Food Science,
https://doi.org/10.1007/978-1-0716-4144-6_17,

In the case of non-fermented dairy drinks, there must be no addition of microorganism cultures or fermented milk products. From an industrial point of view, the processing technology for unfermented milk drinks is simple and low cost [4]. Basically, whey is mixed with milk and then optional ingredients are added, such as sucrose (usually in concentrations of 5–8%) or sweeteners, flavorings, and colorings (which can be natural and bioactive pigments, such as β-carotene and lutein); flavoring ingredients, such as cocoa or fruit-based preparations (usually in concentrations of 3–5%); and thickeners and/or stabilizers, to maintain viscosity and stability to avoid phase separation of the drink during its shelf life [3]. We have seen an increase in launches of lactose-free and carbonated products, the addition of prebiotics and probiotics, sugar reduction, and the addition of proteins (for athletes), vitamins, and minerals. In this context, research into new products is fundamental to serving consumers and increasing the competitiveness of industries [5].

The main quality characteristics of dairy drinks can vary according to the raw material used (milk and whey) and the subsequent ingredients and various manufacturing conditions (e.g., standardization, homogenization, heat treatment, cooling). This chapter aims to describe a general protocol for producing unfermented milk drinks on a laboratory and pilot scale.

2 Material

Use whey and fresh milk of good microbiological quality, with a maximum acidity of 13°D, obtained under hygienic conditions and without additives such as colorings and sodium nitrate when making the cheese. The fat content of whey varies according to the cheese produced, but it can be skimmed to utilize the fat in other products and make a skimmed drink (0% fat). The milk should preferably have a standardized fat content between 0% (for skimmed drinks, 0% fat) and 3%. Only use food-grade ingredients stored under the conditions described by the manufacturer/supplier. Prepare all beverage ingredients using quality products authorized by the competent authorities. Prepare and store all additives and ingredients at room temperature unless otherwise stated.

2.1 Ingredients

1. Whey: This should be used immediately after the cheeses have been made. Give preference to whey without added additives or whey powder suitable for human consumption (*see* **Note 1**).
2. Milk: Whole milk, skimmed milk, skimmed milk powder, and concentrated skimmed milk can be used.

3. Dairy products: Whey protein concentrate (WPC), whey protein isolate (WPI), and milk protein concentrate (MPC) can be used to fortify milk nonfat solids (MSNF; mainly lactose, protein, and minerals) to comply with legislation or to increase the protein content in products for athletes. Milk cream can be used to standardize the fat content of product.
4. Flavorings: Normally used to impart aroma and flavor to improve product acceptance, various natural flavors, fruit juices, fruit concentrates, chocolate powder, honey, sugars, and other ingredients can be added, which are distributed homogeneously and determine the final flavor of product.
5. Sweeteners: Sweeteners and sugar can be added if necessary: sucrose, glucose syrup, sucralose, aspartame, among others.
6. Stabilizers: In whey drinks, the solids content is reduced due to the addition of whey, which dilutes the protein concentration, so it is necessary to use stabilizers and/or thickeners to maintain or increase the viscosity and stability of the product, avoiding phase separation during shelf life (*see* **Note 4**). The most commonly used additives are gelatins, gums, and modified starches. In the case of acidified protein drinks, pectin is necessary to stabilize the system.
7. Acidulants: In the case of acidified drinks, different types of acids can be used, such as citric, fumaric, acetic, lactic, and a mixture of these (*see* **Note 5**).

3 Methods

Carry out all procedures at room temperature, unless otherwise indicated. For the product to be successfully manufactured, certain precautions must be followed. The product must be made with pasteurized milk and by good hygiene practices, which involve cleaning and sanitizing the utensils, environment, and handlers to avoid contamination and health problems. It is essential to wear a mask and cap to prevent contamination. Wearing ornaments (watches, earrings, among others) that could fall into the product and create a health risk or microbial contamination is not recommended. Next, we will present the steps involved in making a non-fermented dairy drink on a laboratory or pilot scale:

3.1 Preparation of Milk Drink

Brazilian legislation [3] states that dairy drinks without additions, where the dairy base represents 100% (one hundred percent) weight/weight (w/w) of the total ingredients of product, contain a minimum protein content of 1.7% and a minimum fat content of 2%. For milk drinks with added food products or substances and vegetable fat, where the milk base represents 51% (one hundred percent) weight/weight (w/w) of the total ingredients of product, it contains a minimum protein content of 1% and a minimum fat content of 2%. Generally, commercial drinks include fat content in a proportion of approximately 0–3%. The dairy beverage production process on a laboratory or pilot scale is summarized in Fig. 1. Three

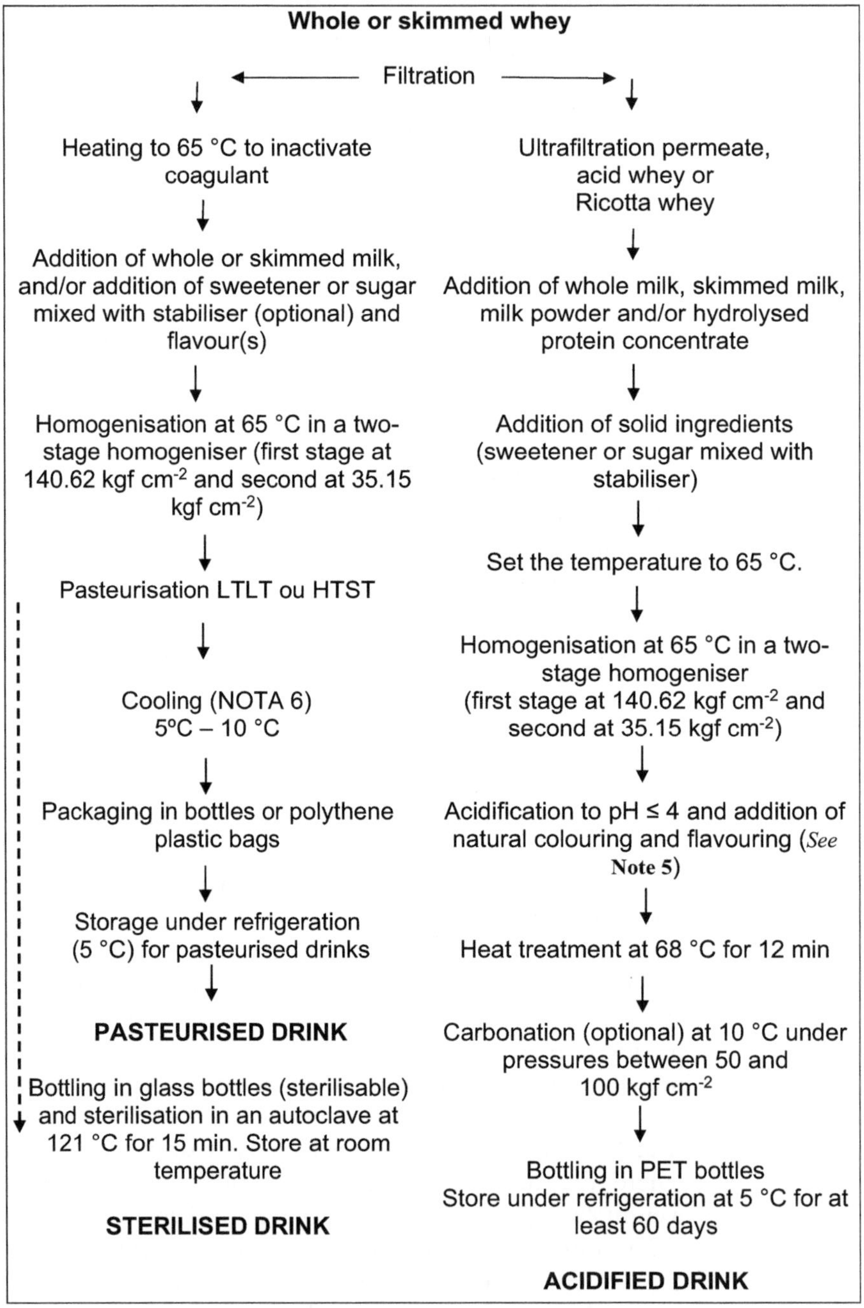

Fig. 1 Non-fermented dairy drink manufacturing processes

classes of methods for producing representative dairy beverages are described in the following text.

3.1.1 Pasteurized Milk Drink

Pasteurized milk drinks are much simpler and quicker to prepare than sterilized or acidified ones as they do not require sterilization or acidification. This type of drink is considered to be similar to milk without the addition of stabilizers and flavorings, with the advantage of being less expensive, but it generally has a lower protein and fat content [6].

Cheese whey should be fresh, with an acidity of up to 13 °D and free of colorings and sodium nitrate, obtained under good hygienic conditions. Give preference to freshly obtained whey, i.e., the drink should be produced immediately after the cheese has been made. If this is impossible, cool it directly to 5 °C and store the whey under refrigeration at 5 °C. The amount of whey should be based on the desired composition of the final product and compliance with legislation, especially about the protein content of the drink. Remember that the whey must always be strained with a suitable utensil to remove any small residues of cheese mass that may be present in the product.

Heat the whey to a temperature of 65 °C before mixing in the other ingredients (*see* **Note 7**). Then add good quality milk to the whey container, which has been heated to 65 °C. Then add the ingredients (sweetener or sugar, chocolate, or fruit juice and/or flavoring); the sugar needs to be mixed with the stabilizer. Otherwise, it will be necessary to use suitable equipment, such as a homogenizer or tri-blender, to disperse the stabilizer in the drink.

In addition to dispersing the food ingredients, the homogenization stage aims to reduce the size of milk fat globules, increasing the stability of final product. Homogenization occurs at 65 °C in a two-stage homogenizer, the first at a pressure of 140.62 kgf cm^{-2} and the second at 35.15 kgf cm^{-2} (*see* **Note 3**).

Heat treatment can be carried out by heating the mixture of milk, whey, and stabilizer under stirring at a temperature of 65 °C for 30 min, known as low-temperature long-time (LTLT), or in a plate pasteurizer, known as ultra high temperature or ultra-high temperature (HTST). The main objectives of heat treatment for beverage pasteurization are to eliminate pathogenic microorganisms and denature whey proteins.

Cool the mixture in an ice water bath to a temperature of between 5 °C and 10 °C. It is important that this cooling is done as quickly as possible to avoid contaminating the product. In the industry, the milk drink production tank has an ice water cooling device in the tank jacket. Filling must be done hygienically, in suitable polythene or PET bottles. The bottles must be previously sanitized in a solution of between 100 and 200 ppm active chlorine. Optionally, it can be filled into 1 L polyethylene bags with heat sealing.

Although the process is simple, as it is a pasteurized drink, the shelf life of product is short, approximately 3–5 days, depending on good manufacturing practices. It must be kept and sold at temperatures not exceeding 10 °C.

3.1.2 Sterilized Milk Drink

The basic difference between pasteurized and sterilized milk drinks is precisely the type of heat treatment used. In this case, it is filled into sterilizable glass bottles and sterilized in an autoclave at 121 °C for 15 min (*see* **Note 11**). This type of product lasts approximately 60–90 days at room temperature. In this case, it is necessary to use sodium citrate as a protein thermal stabilizer to prevent phase separation with protein sedimentation (*see* **Note 8**).

3.1.3 Acidified Milk Drink

Acidified milk drinks are produced in different parts of the world and come in different variations, such as milk and fruit drinks, yogurt drinks, buttermilk drinks, whey drinks, and kefir, among others. Stabilizers are added to these drinks to prevent the formation of sediment and phase separation (*see* **Note 2**). Caseins at milk pH are present in micelles, and repulsive steric interactions stabilize these protein particles. When the pH of milk is lowered, many phenomena take place that modify the stability of casein micelles. At pH 4.6, calcium is soluble [7, 8] and casein aggregation is irreversible, forming a three-dimensional network of aggregated filaments. When the acidified milk drink is processed, the physical properties of the protein network are modified by heat treatment, which contributes to the acceleration of protein aggregation process, and homogenization destroys the gel, producing caseinate fragments. Under these conditions, sedimentation can only be prevented by adding a stabilizer [8, 9].

In the case of acidified milk drinks, which are usually prepared by direct acidification with fruit juices and/or the addition of weak acids, the pH ranges from 3.4 to 4.6. During acidification, the micelles aggregate at a pH around the isoelectric point (pH 4.6) mainly due to the collapse of the extended layer formed by the kappa-casein on the micelle surface [9]. At a pH below the isoelectric point, milk proteins that are positively charged and negatively charged stabilizers such as pectin are commonly used to return the system to a stable state without phase separation. When a stabilizer such as pectin is added before acidification, it adsorbs onto the surface of casein micelles by electrostatic interactions. It prevents their aggregation and phase separation in the product [10]. The time-temperature binomial of heat treatment, which denatures whey proteins in milk and allows them to bind to casein by disulfide bridges before acidification, as well as the homogenization process, can significantly improve the pectin stabilization system in the acidified drink or milk [11]. The acidified drink production process is described in stages in the following text.

Different types of whey can also be used to make acidified milk drinks (acid whey, ricotta whey, and membrane concentration permeate).

1. Fresh whey, preheated to 65 °C to inactivate the coagulant, with a maximum acidity of 12 °Dornic, must be filtered through a desiccator filter to ensure that no cheese particles are incorporated into the membrane concentration permeate; acid whey or ricotta whey can be used.
2. Whole or skimmed milk is added to the whey, adjusting the volume according to the protein content in the final drink (*see* **Note 9**), as established by the Technical Regulation on the Identity and Quality of Milk Drinks [3].
3. The solid ingredients, including sweetener or sugar, sodium citrate, and stabilizer, must be mixed beforehand and always added while stirring to facilitate dispersion. The temperature of the mixture should then be adjusted to 65 °C.
4. The mixture is homogenized in a two-stage homogenizer, with pressures of 140 kgf cm^{-2} in the first and 35 kgf cm^{-2} in the second.
5. Acidification is carried out with citric acid until a pH close to 4 is reached. Flavoring is carried out with citrus flavors, which go best with this type of drink due to its low pH. Natural lemon flavor seems to be the most suitable.
6. In the pasteurization stage, the drink must be heat-treated at 68 °C for 12 min and then quickly cooled to 10 °C (*see* **Note 6**).
7. Optionally, the drink can be cold carbonated (up to 10 °C) under 50 and 100 kgf cm^{-2} pressures. After 10 min in contact with CO_2 under pressure, the drink is filled cold into PET bottles, previously cleaned and sanitized in a solution containing 200 mg L^{-1} of active chlorine.
8. Bottled drinks can be stored under refrigeration at 5 °C. Shelf life is 60 days or more after manufacture, under refrigeration, and the product can be transported at room temperature (*see* **Note 10**).

4 Notes

1. Using whey and fresh milk of good microbiological quality is essential when making unfermented milk drinks. Whey with acidity above 13°D can precipitate during thermal processing and other organoleptic changes.

2. Stabilizers and thickeners stabilize the protein gel by strengthening existing interactions, preventing syneresis, and increasing the viscosity of drink.
3. High homogenization pressures can be used for greater firmness and viscosity due to an increase in surface area caused by forming smaller fat particles and increasing the interaction between protein and stabilizer.
4. Stabilizers and thickeners, if used, need to be mixed with sugar before addition to improve dispersion in drink. These additives require heat treatment and homogenization to enhance their action and interaction with the beverage components, mainly to avoid phase separation during shelf life.
5. For acidified drinks, using citric acid is better in sensory terms.
6. Rapid cooling is more suitable for improving product stability.
7. Heating the whey to 65 °C before adding the milk is necessary to eliminate coagulant residues and prevent the drink from forming clots or lumps, a defect.
8. In the case of sterilized milk drinks, a darkening of the product due to the Maillard reaction is to be expected. In this case, chocolate is more suitable to improve the product appearance.
9. If protein concentrate is used to standardize the protein content, adding milk may not be necessary.
10. The processing recommended for acidified drinks uses microbiological barrier technology (low pH, heat treatment, and refrigeration) and guarantees "commercial sterility," significantly increasing the product's shelf life.
11. For UHT, or ultrahigh temperature, drinks, the product is subjected to a sterilizer for 2 to 4 seconds at a temperature between 130 °C and 150 °C, immediately cooled to a temperature below 32 °C, filled under aseptic conditions in carton packaging Tetra Pak aseptic.

References

1. Cappato LP, Ferreira MVS, Moraes J et al (2018) Whey acerola-flavoured drink submitted ohmic heating: bioactive compounds, antioxidant capacity, thermal behavior, water mobility, fatty acid profile and volatile compounds. Food Chem 263:81–88
2. Oliveira GAR, Guimarães JT, Ramos GLPA et al (2022) Benefits of thermosonication in orange juice whey drink processing. Innov Food Sci Emerg Technol 75:Article 102876
3. Brasil. Ministério da Agricultura, Pecuária e Abastecimento. Regulamento Técnico de Identidade e Qualidade de Bebida Láctea (2005) Instrução Normativa n.16, de 23 de agosto de 2005. Diário Oficial da União, Brasília
4. Pacheco AFC, Souza LB, Paula JCJ et al (2022) Bebidas lácteas: produção, tendências e desafios. Boletim de Extensão. Universidade Federal de Viçosa, Viçosa – MG. Divisão de Extensão
5. Paula JCJ, Pinto MS, Cortez MAS et al (2017) Bebidas lácteas fermentadas: tecnologia, padrões e tendências. Rev Ing Tecn 35:70–73
6. Paula JCJ, Almeida F, Sobral D et al (2010) Tecnologia de fabricação de bebida láctea

fermentada e não fermentada. Epamig, Belo Horizonte

7. Eleya OMM, Banon SD, Hardy J (1995) A comparative study of pH and temperature effects on the acidic coagulation of milks from cows, goats, and sheeps. J Dairy Sci 78:2675–2682
8. Laurent MA, Boulenguer P (2003) Stabilization mechanism of acid dairy drinks (ADD) induced by pectin. Food Hydrocoll 17(4): 445–454
9. Tamime AY, Robinson RK (1999) Yoghurt: science and technology, 2nd edn. CRC Press, Boca Raton
10. Nakamura A, Yoshida R, Maeda H, Corredig M (2006) The stabilizing behaviour of soybean soluble polysaccharide and pectin in acidified milk beverages. Int Dairy J 16:361–369
11. Sedlmeyer F, Brack M, Rademacher B, Kulozik U (2004) Effect of protein composition and homogenisation on the stability of acidified milk drinks. Int Dairy J 14:331–336

Chapter 18

Cream and Butter

Hannah Caroline Santos Araujo, Rafael Donizete Dutra Sandes, Mônica Silva de Jesus, Narendra Narain, Adriano Gomes da Cruz, and Maria Terezinha Santos Leite Neta

Abstract

Butter and cream are important products from dairy industry. The production of these dairy products has been done by the humanity for years. They are products with high fat content, which are very susceptible to reactions such as lipid oxidation. Thus, knowing and standardizing the production stages of each of these products is extremely important to obtain dairy products of high nutritional and sensorial quality that satisfy the consumer. In this chapter, we will present about the protocols of obtaining cream and butter.

Key words Butter, Cream, Process, Milk fat, Sour cream, Sweet cream

1 Introduction

Cream and butter are one of most common dairy products, could be found in every market, and are very appreciated. They could be used as ingredients in dishes or eaten individually, being very important nutritionally and functionally. Cream is the fat of milk after being separated by centrifuging process and it is the precursor of butter. Butter is a solid emulsion composed of 80–82% of milk fat, air, and water that contains also bioactive compounds, such as tocopherols, carotenoids, and vitamins [1].

Cream is obtained from the milk fat, and it is an important constituent for aroma and flavor of dairy products. It was formerly obtained by separation gravity, method that did not provide a satisfactory yield [2]. Currently, cream is obtained from the separation of milk fat with centrifuges that use the centrifugal force generated by the rotation of the device to separate the fat from the milk, thus allowing a good yield in obtaining the product. For cream, two parameters are very important. One of them is the content of milk fat that should be in the range of 35–40%. And the other parameter is pH; the cream should have pH of 6.4–6.7. If

Adriano Gomes da Cruz et al. (eds.), *Dairy Foods Processing*, Methods and Protocols in Food Science,
https://doi.org/10.1007/978-1-0716-4144-6_18,

the pH of the cream is too low (below pH 6.4), it is indicative of the presence of undesired bacteria [3].

Butter is a multiphase water-in-oil-type plastic emulsion consisting of crystalline fat, aqueous dispersed phase, and fat globules (FGs) and must contain >80% milk fat and <16% water [4]. And it is one of the most popular fat-rich dairy products that is produced that can play an important role in human nutrition and has a unique taste and composition [5, 6]. Butter can be obtained from pasteurized cream or fermented pasteurized cream with lactic acid bacteria submitted to a high-speed churning that will induce a phase inversion from an oil-water emulsion to a water-in-oil emulsion [3].

Butter and cream are products that have a high fat content. Due to that, they are susceptible to some undesirable reactions, such as rancidity caused by peroxidation of their lipid fraction and hydrolytic rancidity [3, 4]. The majority composition of the milk's fat is of saturated fatty acids around 70%, monounsaturated fatty acids around 25%, and finally polyunsaturated fatty acids with an amount of approximately 5% [2]. Due to this rich lipid content, it is important to know very well each step of the obtaining process of cream and butter to avoid these reactions in final product.

Processing conditions of cream and butter, such as cooling rate, crystallization temperature, ripening rate and efficiency, as well as the quality of the milk, are important to produce a high-quality product in the final process. It is not possibly making quality butter if you do not have quality cream [5]. Therefore, the objective of this chapter is to present the processing steps of manufacturing of cream and butter.

2 Cream Obtaining Process

The manufacturing of cream is similar around the world. The initial step consists in the best choice of feedstock, in this case milk. The quality of raw milk is crucial to obtain a quality cream. After the selection of feedstock, the process for obtaining cream is basically concentrated in four steps: separation, standardization of fat content, and heat treatment. Other steps will be included depending on what type of cream you want to obtain at the end of the process. The cream flowchart is shown in Fig. 1.

2.1 Fat Separation and Standardization

In this step, the fat is separated from the milk through a centrifugal force. After removal of all fat from the milk, the cream obtained is mixed with the skimmed milk at the centrifuge outlet. Control valves and density and flowmeters control all the process in a fully automated system.

The temperature at which standardization occurs is usually above 40 °C. For this reason, this step must occur quickly, and then the cream must proceed to heat treatment. The purpose of

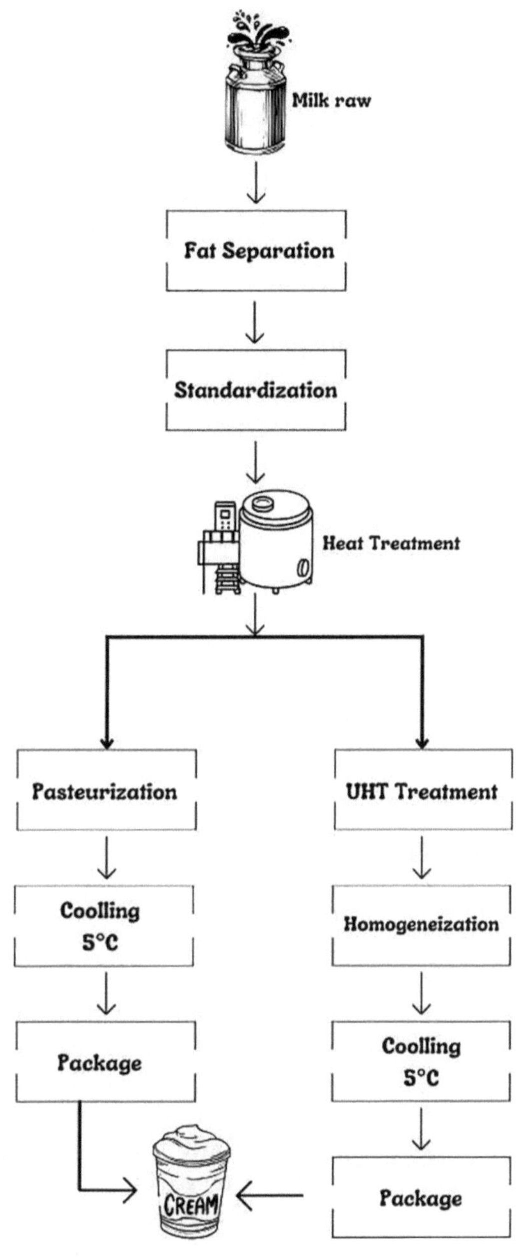

Fig. 1 Cream production flowchart

quickly carrying out standardization is to prevent microbial growth in the cream.

2.2 Heat Treatment

As demonstrated in the flowchart in Fig. 1. Cream must undergo heat treatment to inactivate enzymes and microorganisms. The treatment applied to the cream will be modified according to the process to which it will be subjected to manufacture a product. For production of butter, it is recommended that pasteurization be applied at a combination of time and temperature of 75 °C/15 s. It is also recommended that the fat content in the cream to be pasteurized is at least 20% (*see* **Note 1**).

For the cream that will be used in other preparations besides butter, the heat treatment commonly used is UHT treatment where the working range can be 130–150 °C for a few seconds. One example is the temperature/time binomial of 142 °C/4 s. In all forms of heat treatment, the objective is the destruction of pathogenic microorganisms, as well as inactivating enzymes, especially peroxidase. If the heat treatment applied is very intense and hindered a long period of time, problems will occur in the final product, such as a cooked flavor.

2.3 Homogenization

Homogenization aims to reduce the size of the fat globule to prevent cream separation. This step process is normally applied to creams that receive UHT treatment and are marketed for culinary purposes. Creams that will be used to other purposes such as whipped cream and butter should not be homogenized. For the whipped cream, small fat globules impair the formation of foam that will help stabilize the whipped cream. In addition, for the butter, the small particles will harm the phase inversion (oil-water emulsion→water-oil emulsion) that occurs in the butter.

3 Butter Obtaining Process

Butter is the most popular product manufactured from milk fat due to sensory characteristics and nutritional value. It contains at least 80% of fat and a maximum of 16% of water [7]. It is obtained after churning cream that will promote a phase inversion that will turn an oil-water emulsion to water-oil emulsion, that will be the final product, knows butter. Butter can be obtained from two different types of cream: sweet cream (cream without formed acid, cooled immediately after pasteurization to temperatures between 4 and 6 °C) (*see* **Note 2**) or sour cream (cream cooled to temperatures between 15 and 20 °C and inoculate with 2–4% starter culture).

The process to obtaining butter requires a lot of attention as any error during processing can lead to defects in flavor, texture, and aroma in the final product, displeasing the consumer. One of the most common problems in butter production are lipolysis and oxidation of the fatty acids that will release free fatty acids that will impact the flavor, color, and texture and reduce the quality and nutritional value of butter, creating serious problems and economic losses in the dairy and food distribution industries [8]. The butter production flowchart is shown below in Fig. 2.

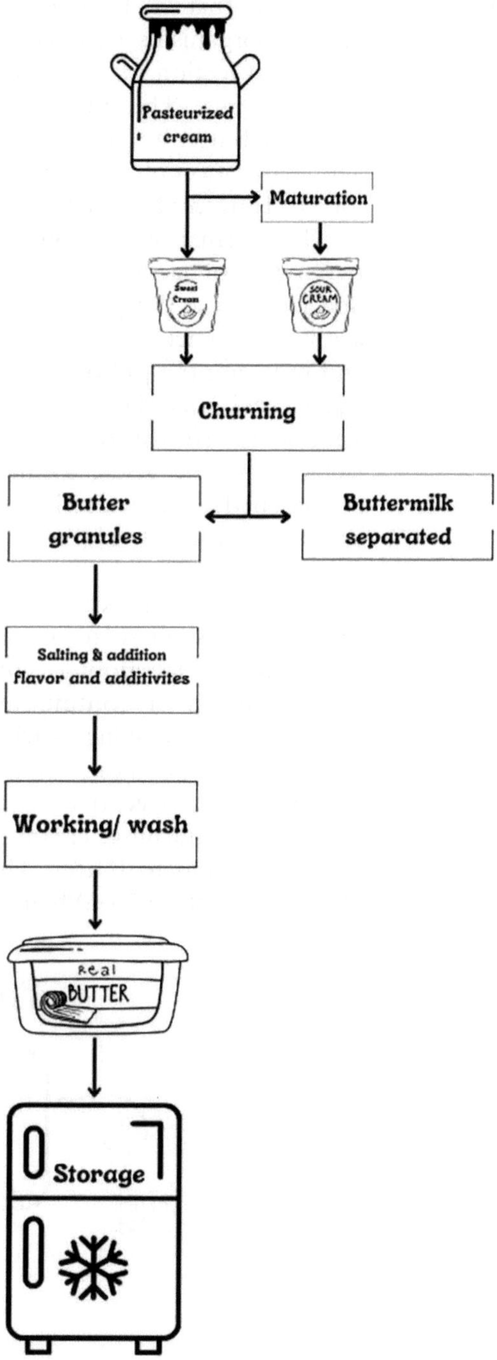

Fig. 2 Butter production flowchart

3.1 Maturation (Optional)

The cream that will be used to produce butter may receive a culture of microorganisms that will promote the formation of diacetyl, lactic, and acetic acid, providing the formation of a soft and pleasant butter aroma. The addition of acid-producing microorganisms gives the butter a strong aroma and it also improves the fat yield. The maturation phase occurs in a temperature range of 10–15 °C for 10–15 h (*see* **Note 3**).

The formation of aroma and flavor of butter happens due to two types of lactic bacteria, considered acidifying or flavoring. Species such as *Streptococcus cremoris* and *Streptococcus* belong to acidifiers that act on lactose, promoting the formation of lactic acid and reducing the pH of the cream to values between 4.4 and 4.6. Lactic acid bacteria such as *Streptococcus lactis sub diacetylactis*, *Leuconostoc citrovorum*, *and Leuconostoc paracitrovorum* act on lactic and citric acids, forming compounds such as acetic acid, acetaldehyde, and diacetyl that are responsible for the characteristic aroma of butter [3, 4] (*see* **Notes 4** and **5**).

3.2 Churning

Churning is the process of converting the fat in cream from a discontinuous phase to a continuous phase, culminating in the conversion of fat globules into fat granules, which can occur in either a batch or continuous process (continuous process uses modern butter-making machines). This step occurs traditionally in a cylindrical, conical, cubical, or tetrahedral churn with adjustable speed [9] (*see* **Note 6**).

During churning, cream is being transformed from an oil-in-water emulsion to butter. This event is called "phase inversion" (Fig. 3) (*see* **Note 7**). When the cream is churned, air bubbles are

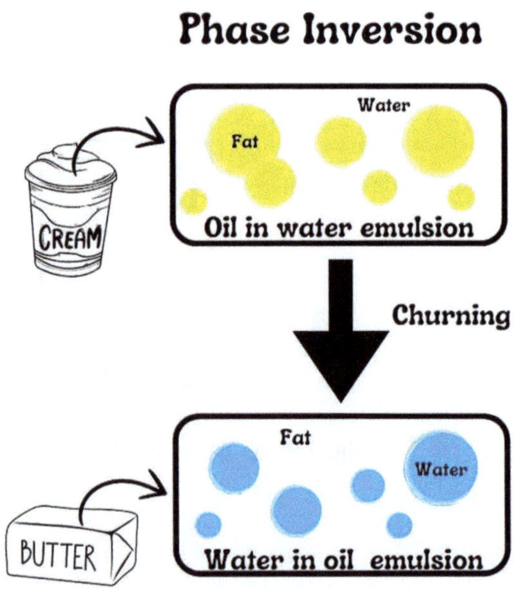

Fig. 3 Graphical diagram of phase inversion during butter production

formed, due to continuous shear, which results in the collapse of fat globules. This way, the fat globules coalesce, aggregate, come together, and form solid butter granules/clumps. As this process continues, a more solid structure is formed made up of these butter granules, while all the buttermilk is being released and drained from the churn [10].

The churning stage must be stopped when the fat granules attain the desired size (diameter of 3–5 mm). In the churning stage, it is possible to calculate yield (churning recovery), which is a measure of how much fat in the cream has been converted to butter. It is calculated in terms of how much fat remains in the buttermilk expressed as a percentage of the total fat in the cream. For example, a churning recovery of 0.50 means that 0.5% of the cream fat has remained in the buttermilk and that 99.5% has been turned into butter. Churning yield is considered acceptable if the value is less than 0.70 [3].

After removing the buttermilk at the end of churning, the butter grains can be washed with water at refrigeration temperature to remove the little buttermilk that remains in the butter grains (*see* **Note 8**). It is recommended to wash the butter grains two to three times or until the water runs clear. It is important to highlight that the water used must be of good quality to avoid microbiological contamination [3, 10].

3.3 Working

This step is responsible to remove the moisture between butter grains, resulting in a homogeneous and compact mass and providing greater buttermilk removal. At this stage, salt and other additives can be added to the butter, if necessary, and working promotes a uniform distribution of these ingredients. It is important to promote good mixing to avoid the presence of large drops of water that can cause changes in the flavor and aroma of the butter, because of an oxidation process, in addition to reducing its shelf life (*see* **Note 9**).

3.4 Packaging and Storage

The final steps are packaging and storage of butter. Different materials can be used to package butter such as proof paper, vegetable parchment, laminated foil, lacquered tins, or plastic tubs. However, it is important that everyone has the same objective, which is to protect the butter does not lose moisture or absorb a foreign flavor during storage (*see* **Note 10**).

After packaging, butter must be stored at temperatures of 5 °C. Storage is a very important step in manufacturing butter since problems at this stage can affect the quality of the final product, such as its spreadability. Therefore, it is always recommended to keep the butter in lower temperatures to improve the keeping quality and reduce the risk of package distortion during handling [3].

4 Notes

1. Heat treatment should be strong enough to result in a negative peroxidase test but not so intense as to cause defects such as cooked flavor. If the temperature during heat treatment was above ideal, it is recommended to apply vacuum aeration. This step heated the cream to 78 °C and after it is pumped by a vacuum chamber where the pressure corresponds to a boiling temperature of 62 °C. The reduced pressure causes an escape of the aromatic compounds in the form of gas when the cream is flash-cooled. After this treatment, the cream is returned to the heat exchanger for pasteurization and cooling and then continues to the ripening tank.
2. Sweet cream butter is the predominant butter type in the United States, while in Europe, butter is obtained from cultured cream or sour cream. It is recommended that the sweet cream after pasteurization be refrigerated for 4–6 °C and remain at that temperature for 8–10 h. This condition will allow the partial formation of the fat crystals (aging).
3. When the fermentation process is complete, it is recommended to lower the temperature to 8–11 °C, which is the churning temperature range.
4. The "aging" treatment is important for cream, with or without microorganisms. Aging is a processing step that approaches to modify the butter consistency (Wright and Marangoni, 2006).
5. Skim milk is the substrate mostly used as growth medium for starter cultures as it is easier to detect taste defects in skim milk cultures.
6. In the industry are used churns of 8.000 to 12.000 L capacity or more, and it is recommended that the churns be filled to 40–50% to allow space for foaming.
7. The recommended velocity of the churns is between 20 and 30 rpm, varying according to the equipment. The speed used must be lower than that necessary to overcome the force of gravity. Very high speeds cause the cream to stick to the walls of the equipment, preventing the formation of quality butter, while very low speeds cause an extension of time, damaging the final quality of the butter.
8. If used a good quality cream and all process was well handled, it is possible to use chilled buttermilk to wash the butter granules.
9. To ensure the effectiveness of water removal, it is necessary to ensure that the butter grains attain a certain level of firmness; otherwise, it will not be possible to remove or incorporate water/moisture.

10. One alternative to increase shelf life of butter is infuse nitrogen into the butter to help the butter to keep being soft and spreadable even at refrigeration temperatures.

Acknowledgments

The authors thanks Conselho Nacional de Desenvolvimento Científico e Tecnológico (CNPq) and the Coordenação de Aperfeiçoamento de Pessoal de Nível Superior (CAPES), Brazil, for their fellowships.

References

1. Taylan O, Cebi N, Yilmaz MT (2020) Detection of lard in butter using Raman spectroscopy combined with chemometrics. Food Chem 332:127344. https://doi.org/10.1016/j.foodchem.2020.127344
2. Soundous I, Amaury G, Juliette D, Sybille D, Aurelie L, Viviane P, Marianne S (2019) Overview of the local production process of raw milk butter in Wallonia (Belgium). Int J Dairy Technol 72:466–471. https://doi.org/10.1111/1471-0307.12608
3. Butter and dairy spreads. In: Dairy process handbook (2015) Tetra Park, 486 pages ISBN: 9789176111321
4. Seo CW (2023) Physicochemical changes during cream fermentation and quality characteristics of lactic butter produced using cultured creams with different fermentation levels. Int Dairy J 145:105730. https://doi.org/10.1016/j.idairyj.2023.105730
5. Ceylan O, Ozcan T (2020) Effect of the cream cooling temperature and acidification method on the crystallization and textural properties of butter. LWT Food Sci Technol 132:109806. https://doi.org/10.1016/j.lwt.2020.1098066
6. Tahmas-Kahyaoğlu D, Cakmakci S, Hayaloglu AA (2022) Changes during storage in volatile compounds of butter produced using cow, sheep or goat's milk. Small Rumin Res 211:106691. https://doi.org/10.1016/j.smallrumres.2022.106691
7. Lee J, Martini S (2018) Effect of cream aging temperature and agitation on butter properties. J Dairy Sci 101:7724–7735. https://doi.org/10.3168/jds.2018-14725
8. Mendez-Cid F, Centeno JA, Martinez S, Carballo J (2017) Changes in the chemical and physical characteristics of cow's milk butter during storage: effects of temperature and addition of salt. J Food Compos Anal 63:121–132. https://doi.org/10.1016/j.jfca.2017.07.032
9. Funahashi F, Horiuchi J (2008) Characteristics of the churning process in continuous butter manufacture and modelling using an artificial neural network. Int Dairy J 18:323–328. https://doi.org/10.1016/j.idairyj.2007.08.001
10. Prudencio ES, Amboni RDM, Freire CBF et al (2017) Manteiga e Crème de leite. In: Processamento de Produtos lácteos: queijos, leites fermentados, bebidas lácteas, sorvete, manteiga, crème de leite, doce de leite, soro em pó e lácteos funcionais, vol 3. Elsevier, Rio de Janeiro. ISBN:978-85-652-8085-2
11. Wright AJ, Marangoni AG (2006) Crystallization and rheological properties of milk fat. In Fox PF, McSweeney PLH (eds), Advanced Dairy Chemistry, vol 2. Lipids (3rd edn., pp. 245e291). Boston, MA, USA: Springer. e-ISBN: 0-387-28813-9

Chapter 19

Ice Cream and Frozen Yogurt

Eliara Acipreste Hudson, Letícia Bruni de Souza, Jaqueline de Paula Rezende, and Ana Clarissa dos Santos Pires

Abstract

Frozen desserts are emulsion-based products that undergo the freezing process while being aerated and are consumed in their frozen state. Notably, the popularity of frozen desserts has seen a remarkable increase in recent years. Among the various frozen desserts available, ice cream and frozen yogurt are prominent choices. Ice cream is a complex multiphase system composed of ice crystals, air cells, and fat globules embedded in a highly viscous, freeze-concentrated matrix phase. The production of frozen yogurt closely resembles that of ice cream, with the key distinction in the fermentation of the mix using yogurt cultures before freezing. This fermentation imparts distinctive flavors and textures to the final product. The ingredients used in ice cream and frozen yogurt production can vary to achieve the desired physical, chemical, and sensory properties. High-quality ingredients are crucial for maintaining the quality and flavor of these products. The production of these frozen desserts involves several critical steps, each playing a crucial role in determining the ultimate product's texture, structure, flavor, and overall quality. This chapter provides a comprehensive guide to the processing of ice cream and frozen yogurt, encompassing essential steps, technical considerations, and practical insights. It is tailored for food scientists, technologists, and professionals engaged in producing frozen desserts.

Key words Frozen dessert technology, Dairy processing protocol, High-quality dairy products, Dairy-based frozen products, Processing challenges

1 Introduction

Frozen desserts, like ice cream and frozen yogurt, are intricate colloidal systems that undergo simultaneous freezing and aeration and are ideally consumed while they are in the frozen state. They are typically presented as scooped servings or individual items, occasionally on sticks, and often accompanied by various confectionery items [1].

In general, frozen desserts encompass different ingredient types: lipids (either from dairy or nondairy sources), nonfat milk solids (a primary protein source that can be replaced with plant-based protein for dairy-free alternatives), sweetening agents, and

Adriano Gomes da Cruz et al. (eds.), *Dairy Foods Processing*, Methods and Protocols in Food Science,
https://doi.org/10.1007/978-1-0716-4144-6_19,

water, as the major ingredients. Stabilizers, emulsifying agents, and diverse flavorings and colorants are employed as minor ingredients [2]. In addition, other optional ingredients can be used, like pieces of fruit, chocolates, and nuts, among others. However, it is important to emphasize that some aspects set frozen desserts, like ice cream and frozen yogurt apart.

Ice cream is the most globally popular and consumed frozen dessert, and its classification can vary according to each country's specific regulations. Typically, categorization considers the product's concentration of lipids and total solids. However, ice cream is an exceptionally versatile product, providing consumers a broad spectrum of options. These range from premium offerings with high-fat content to low-fat and nonfat products. Moreover, choices have reduced sugar content or completely sugar-free varieties [3]. Furthermore, ice cream can be categorized as soft-serve or hard-type ice cream. The soft-serve products are frozen at the point of sale using a mix that has been produced and distributed. They are consumed while still in a soft-frozen state shortly after preparation, without undergoing any hardening process. On the other hand, in hard-type ice cream manufacture, the freezing procedure persists in containers without stirring until the temperature drops to −18 °C (0 °F) or lower, preferably falling within the range of −25 to −30 °C [1].

In the context of frozen yogurt, it represents a fermented frozen dairy product that mirrors the physical characteristics of ice cream mentioned earlier while maintaining the sensory and nutritional attributes linked to fermented milk products [4]. Concerning texture, ice cream is known for its rich and dense consistency, primarily due to the inclusion of cream and emulsifiers. In contrast, frozen yogurt presents a lighter, more aerated texture due to the fermentation process. In the flavor domain, ice cream presents a broad spectrum of alternatives, encompassing both traditional selections and innovative mixes. Conversely, frozen yogurt delivers a more subdued, slightly acidic flavor profile, owing to the presence of yogurt constituents [4, 5]. Both products can be segmented into hard-type items that undergo an additional freezing stage following the dynamic freezing process and soft-serve products that are consumed immediately after dynamic freezing without undergoing any additional hardening phase.

The popularity of frozen desserts has experienced substantial growth over the years, driven by several factors. Although often considered a treat, these sweet and cold foods can contribute to a balanced diet in several ways. It is important to note that ice cream and frozen yogurt contain proteins, an essential nutrient for growth, tissue repair, and muscle strengthening. Additionally, milk and its derivatives are common ingredients in these products, making them a source of calcium, which is crucial for bone and dental health. Additionally, frozen desserts can provide energy due to their caloric content, which can be beneficial for a quick energy boost when needed [6]. The global appreciation for frozen desserts arises from their delightful fusion of flavors and textures. The versatility of these frozen treats contributes to their universal popularity. Ice cream and frozen yogurt are available in many flavors and toppings, enabling consumers to tailor their choices to suit their preferences. This customization appeals to individuals of all ages and backgrounds, rendering these desserts suitable for diverse palates. Furthermore, frozen desserts are enjoyed year-round. While they provide refreshing relief during the hot summer months, they also bring comfort and joy in the colder seasons. Seasonal flavors and holiday-themed variations render them a favored choice during special occasions and celebrations [7].

The global frozen dessert market was valued at US$91.62 billion in 2021 and is expected to grow 5.4% until 2030, reaching more than US$145 billion. The ice cream segment accounted for the largest share of revenue in 2021, and this is attributed to the increasing demand for ice cream on a large scale due to its growing popularity and acceptance among all age groups. Factors such as low-fat content, flavor variety, and sugar-free options, among others, are driving the growth of the ice cream market [8].

Making ice cream and frozen yogurt presents challenges, especially when ensuring they are truly enjoyable for the consumer. To achieve the desired sensory characteristics, it is crucial to adhere to rigorous quality control measures, maintain hygienic production standards, and carefully manage ingredient selection, mixing, pasteurization, and freezing processes. A great challenge is getting the texture right, making it smooth and creamy. Getting the right texture means controlling the air added during freezing, also known as overrun. The right air balance makes the dessert creamy, affecting how it tastes in your mouth. It is also necessary that the ice crystals are not too large as large crystals can make the dessert gritty. To face these challenges, much research is done to improve the freezing process. Scientists and food experts have experimented with different freezing methods. Currently, the dynamic freezing process, in which the ice cream or frozen yogurt ingredients are churned simultaneously as they freeze, has proven to be the most efficient method [9, 10]. But it is not just about freezing. Addressing these challenges requires research, development, and rigorous quality control at all stages of production, from ingredient selection to storage and distribution of the final products. Furthermore, considering the evolving consumer preferences for natural ingredients and sustainable practices, it is prudent for practitioners in the frozen dessert industry to adapt and innovate their production methods in line with the ever-changing landscape. With these protocols in mind, the potential for creating exceptional frozen treats that delight consumers while meeting stringent quality and safety standards remains within reach.

2 Materials

2.1 Ingredients

The ingredients used in ice cream and frozen yogurt production can vary according to the physicochemical and sensory needs desired for the product. High-quality ingredients are essential for ice cream and frozen yogurt quality and flavor maintenance. Thus, the ingredients commonly used are:

- Milk (*see* **Note 1**).
- Milk cream or butter (*see* **Note 2**).
- Active yogurt starter cultures in the case of frozen yogurt (*see* **Note 3).**
- Sugars (*see* **Note 4**).
- Stabilizers and emulsifiers (*see* **Note 5**).
- Water (*see* **Note 6**).
- Coloring and flavoring (*see* **Note 7**).
- Optional ingredients, such as fruit pieces, nuts, chocolate, or other toppings (*see* **Note 8**).

2.2 Equipment

Achieving excellence in ice cream and frozen yogurt processing involves adhering to good manufacturing practices, which include preventive maintenance and thorough equipment sanitization. The following is a list of essential equipment required for the production of ice cream and frozen yogurt:

- Doser/balance.
- Mixer/blender.
- Mixing tanks with stirrers.
- Pasteurizer/heat exchanger.
- Homogenizer or high-pressure pumps (*see* **Note 9**).
- Yogurt fermentation tank in the case of frozen yogurt (*see* **Note 10**).
- Maturation tanks with stirrers.
- Scraped surface heat exchanger.
- Freezer for hardening.
- Automatic filling and sealing machines (*see* **Note 11**).

3 Methods

3.1 Ice Cream

3.1.1 Flowchart

The production of ice cream involves several critical steps, each of which contributes to the final product's texture, structure, flavor, and quality. Figure 1 shows the ice cream manufacturing flowchart.

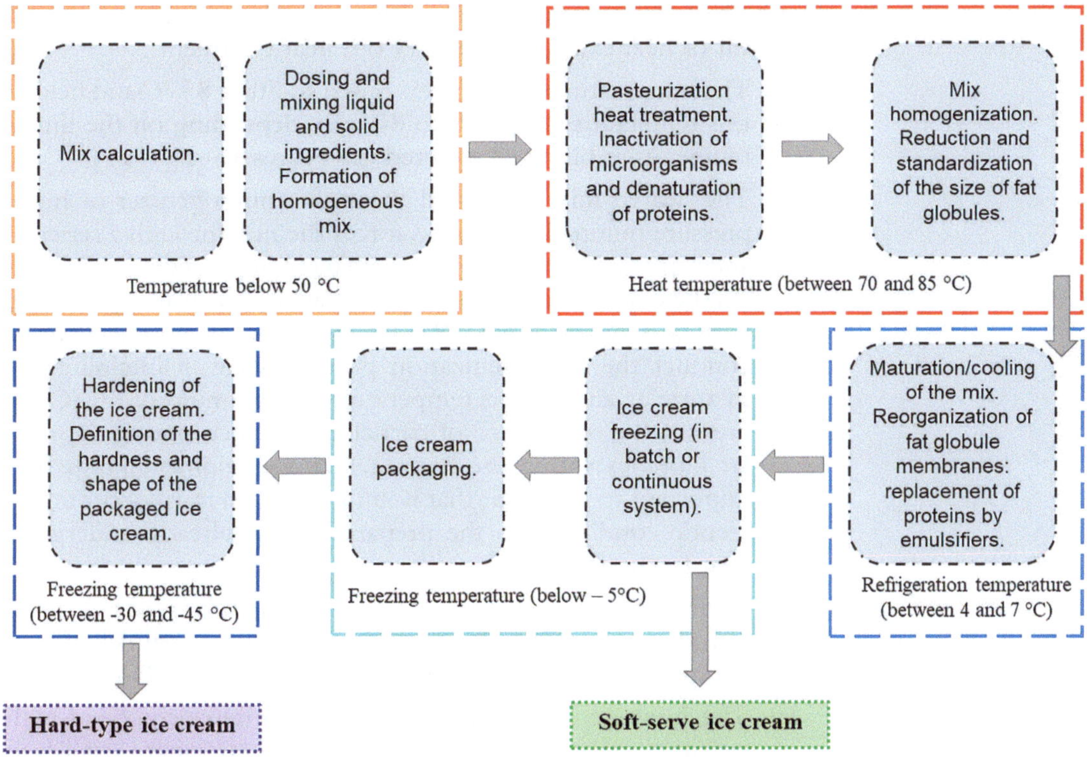

Fig. 1 Flowchart with ice cream manufacturing steps

The following is a more detailed description of the individual steps required for ice cream production:

3.1.2 Preparation of the Mix

1. Calculate the formulation considering the available ingredients and the product's storage conditions. Traditional milk-based ice cream formulations contain, on average, 10 to 16% sugar, 8 to 16% fat, 9 to 11% milk solids nonfat, and 0.5% of stabilizers and emulsifiers.
2. Mixing tanks, typically stainless steel, equipped with agitators and a heating system, are used to prepare the ice cream mix. Based on the quantities defined from the mixing calculations, start measuring the liquid ingredients (milk, cream) and adding them to the mixing tank (*see* **Note 12** and **Note 13**).
3. Mix the previously weighed dry ingredients (sugar, stabilizers, powdered flavorings) in a separate container.
4. Slowly add the dry ingredients to the mixing tank containing the liquid ingredients while agitating continuously. This helps prevent lumps and ensures even distribution (*see* **Note 14**) [11].
5. Mix until all dry ingredients are fully dissolved and a homogeneous ice cream base is formed. This base is often called the "mix."

6. After, the homogeneous mix is pumped into a pasteurization vat or heat exchanger to carry out heat treatment.
7. The temperature of the mix is raised to 70 to 85 °C and held at this temperature for 25 s to 30 min, depending on the time/temperature binomial adopted (*see* **Notes 15–17**) [11].
8. The heated mix is pumped through a homogenizer or high-pressure pump. This process forces the mix through a series of small nozzles at high pressure, breaking down milk fat globules (in the case of dairy fat-based ice creams) and ensuring uniform particle size (*see* **Note 18**). Furthermore, it is imperative to conduct the homogenization process while maintaining the mixture at an elevated temperature (a minimum of 50 °C) to prevent the occurrence of partial coalescence among the milk fat globules within the homogenizing equipment [12]. It is important to highlight that homogenization must occur under aseptic conditions as the preparation has already undergone heat treatment.
9. After pasteurization/homogenization, the mix must be cooled quickly at 4 to 7 °C. The result is a smoother, creamier texture and better fat dispersion throughout the mix (*see* **Note 17**).
10. The homogenized mix is transferred to tanks equipped with agitators, where it is maintained at 4 to 7 °C for at least 4 h to maturation (aging) of the mix (generally, maturation occurs overnight) (*see* **Note 19**).

3.1.3 Freezing Operations

1. The matured cold mix (around 4 °C) is pumped into the freezing chamber of a batch or continuous freezer. In a batch freezer, the mix is churned as it freezes, while in a continuous freezer, it is frozen while being pumped through a freezing cylinder (*see* **Note 20**). The freezing process reduces the mix's temperature to around −5 °C.
2. Air bubbles are introduced into the mix during fast freezing, expanding its volume by approximately 30% to 40% in batch-produced ice cream and 100% in the case of continuous freezer production, a parameter technically referred to as "overrun." Additionally, this overrun contributes to creating a smooth, creamy texture in ice cream. At the end of this step, it exits the freezer as soft-serve ice cream and is ready for this consumption. The fat particles form a three-dimensional fat network that adsorbs at the interfaces of the air bubbles, increasing the stabilization of the foam (*see* **Note 21**).
3. After the churning/freezing step, optional ingredients (e.g., chocolate, biscuits, and fruits) may be added to the ice cream (*see* **Note 22**). For hard-type ice cream, the freshly churned ice cream is packaged into pre-labeled containers or cartons. Automatic filling and sealing machines ensure accurate portions and

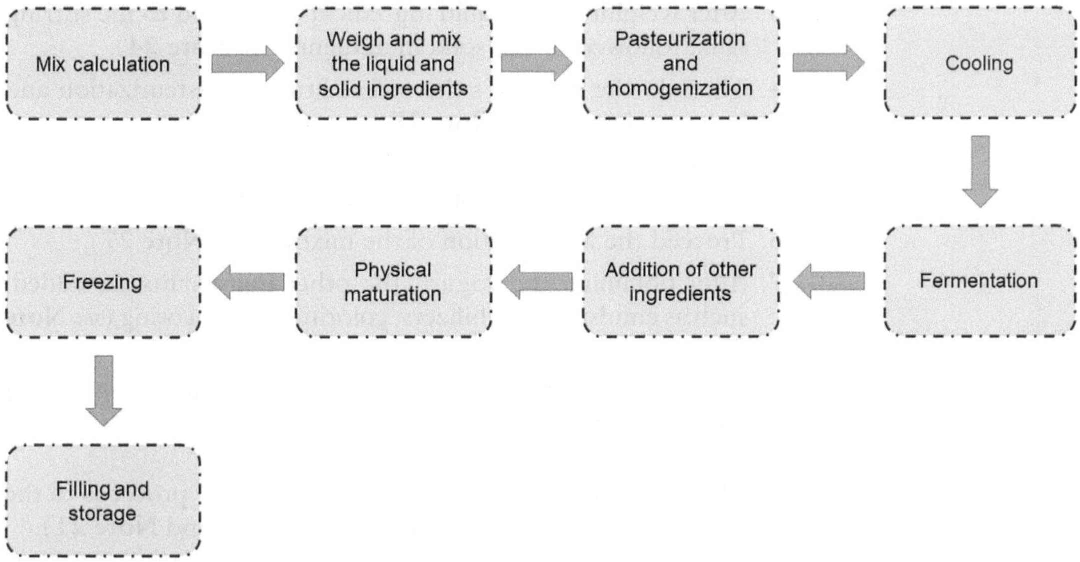

Fig. 2 Frozen yogurt preparation flowchart

lids or seals are applied immediately to maintain product freshness. Some packaging machines include date coding and labeling functions to meet regulatory requirements.

4. After packaging, the fresh ice cream (around −5 °C) is transferred to a hardening room or freezer at temperatures between −30 and − 45 °C. Thus, the ice cream ideally hardens, reaching service (around −18 °C) and storage (around −25 °C) temperatures. Hardening typically takes 2 to 4 h, when the ice cream becomes firm and stable for long-term storage (*see* **Note 23**).

3.2 Frozen Yogurt

3.2.1 Flowchart

The production of frozen yogurt closely resembles traditional ice cream manufacturing. Nevertheless, the mixture must undergo fermentation under precise conditions required for yogurt starter culture (Fig. 2). Throughout all the steps of the process, it is imperative to guarantee the viability of bacteria in the final product until consumption.

The following is a more detailed description of the individual steps required for frozen yogurt production.

3.2.2 Preparation of the Mix

1. Calculate the formulation considering the ingredients available and the products' storage conditions. Frozen yogurt formulation has around 5% fat and 12% nonfat solids (approximately 2.7% of milk proteins and 9% sugars).
2. For frozen yogurt production, first weigh the ingredients, milk, and sugars that will be utilized according to the defined formulation.

3. After weighing, the liquid ingredients are added to the stirring tank, followed by the solid ingredients (*see* **Note 24**).
4. Then do the mixture's thermal treatment/pasteurization and homogenization (*see* **Note 25**).
5. Afterward, cool the mixture for culture inoculation and incubation in the yogurt fabrication tank (*see* **Note 26**).
6. Proceed the fermentation of the mixture (*see* **Note 27**).
7. After obtaining the yogurt, the other ingredients are added, such as emulsifiers, stabilizers, coloring, and flavoring (*see* **Note 28**).
8. The mixture is added to the maturation tanks for its aging process (*see* **Note 29**).

3.2.3 Freezing Operations

1. After physical maturation (aging), the mixture proceeds to the scraped surface heat exchanger (*see* **Note 20** and **Note 21**).
2. Afterward, the frozen yogurt can be served (if it is a soft-service type), or it is packaged and sent for final hardening and storage in the case of hard-type frozen yogurt production (*see* **Note 23**).

4 Notes

1. The choice of milk plays a crucial role in determining ice cream and frozen yogurt's texture and fat content. Skim or low-fat milk is typically selected to create healthier options, while whole milk is chosen for a creamier and more indulgent version. Additionally, skimmed milk powder is frequently incorporated to enhance the texture by contributing essential milk solids to the product. The skimmed milk powder is preferred because it contains more protein (~35%, m/m) than whole milk (~25%, m/m); besides, it is not suitable for fat oxidation. The creation of a stable emulsion that contributes to the texture and body of ice cream is achieved through the adsorption of proteins to fat globules and air bubble interfaces. Other dairy protein sources, including condensed milk and concentrated or isolated whey proteins, may be employed. Nondairy protein sources—such as legumes, grains, and nuts—can also be included [1]; however, it is important to consider that other proteins may not offer the same physicochemical properties as dairy proteins.
2. Milk cream (36%, m/m of fat or more) or unsalted butter are the main sources of fat specifically for ice cream, contributing to a softer texture, improving its body, increasing its viscosity, lowering the cold sensation, raising fusion resistance, assisting

with emulsion stability, and being a conduit for flavor. Furthermore, other fat sources, including butter oil and plant-based sources, such as palm oil, coconut oil, and cocoa, can be used. It is important to highlight that the type of fat being used influences the product's final characteristics. In general, when choosing the fat source, it should be considered the fat crystallization tax, crystal structure, fusion point, and flavor [1].

3. Yogurt starter cultures are indispensable for initiating yogurt fermentation, contributing to the creamy texture characteristic of frozen yogurt. The traditional culture is composed of *Lactobacillus delbrueckii* subsp. *bulgaricus* and *Streptococcus thermophilus.* Probiotics such as *Lactobacillus acidophilus*, *Bifidobacterium*, and *Lacticaseibacillus casei* can also be added for health benefits. Frozen yogurt and ice cream are considered excellent matrices for carrying probiotic bacteria due to preserving bacterial viability achieved through freezing.
4. Simple sugars, such as monosaccharides and disaccharides, constitute the primary carbohydrate source in ice cream and frozen yogurt. They can be introduced directly or indirectly as lactose is incorporated when dairy ingredients are included in the formulation. Added sugars, such as sucrose, are responsible for granting sweetness while also contributing to lowering its freezing point. Other sources can be used, such as fructose, inverted sugar, glucose, corn syrup, and maltose [3]. Milk has lactose, a natural sweetener with less relative sweetness than sucrose, contributing to antifreeze power. The antifreeze power is the capacity of water-soluble ingredients, such as sugars in this case, to lower the freezing point of water in the mixture. The relative sweetness is the capacity of sugars to impart a sweet flavor to the ice cream. For those properties, sucrose is used as a reference, which has 100 antifreeze power and 100 relative sweetness. Glucose, for instance, has 190 antifreeze power and 74 relative sweetness. To calculate the antifreeze power of a given sugar, for example, glucose, the molar mass of the reference sugar (i.e., sucrose; 342 g/mol) is divided by the molar mass of the other sugar (glucose; 180 g/mol) and multiplied by 100. So $(342/180) \times 100 = 90$. This means that glucose has 1.9 times more antifreezing power than sucrose. Sweeteners can be used as substitutes for sugars. Alternatively, ice cream and frozen yogurt with low sugar content can be produced. However, the antifreeze power, relative sweetness, solubility, stability, and ice cream flavor should be analyzed.
5. Guar gum, alginate, carrageenan, pectin, xanthan gum, and gelatin are commonly used for stabilizers. The goal of adding stabilizers is to grant a uniform texture, avoid lactose crystallization, reduce ice crystal development, raise melt resistance, and improve air incorporation. Emulsifiers are responsible for

improving fat networks and granting melting resistance. Regarding frozen yogurt, its low pH value must be considered when choosing the stabilizer as some of them are not functional in acidic media, such as CMC (carboxymethylcellulose). Guar and xanthan gums are the stabilizers most used in producing frozen yogurt. Mono- and diglycerides and polysorbates are the most used as emulsifiers for ice cream.

6. Water can dissolve solid ingredients, such as powdered milk. For instance, the water may be added directly or indirectly when fluid milk is used.

7. Coloring and flavoring give the ice cream and frozen yogurt color and flavor. They can be added through artificial or natural means using extracts, fruit pieces, or other spices. Incorporating natural flavorings, such as vanilla, or fruit extracts, such as strawberry pulp, can enhance the flavor profile of frozen yogurt. Nowadays, there is an increasing demand for natural ingredients. Thus, anthocyanins, betalains, and beta-carotene have been used as colorants for ice cream and frozen yogurt.

8. Adding optional ingredients such as fresh sliced fruits, pieces of chocolate, nuts, granola, cookies, and others constitutes a strategy destined to enrich the frozen yogurt and ice cream's flavor and texture. These ingredients are added after the freezing and before the hardening steps.

9. The homogenizer is dispensable when the liquid ingredients are already homogenized.

10. It is used for producing frozen yogurt. A yogurt fermentation tank is a stainless-steel tank with an agitator, double jacket, inlets for hot and cold water, an analog thermometer, and a side outlet. The steps of mixing, pasteurization, cooling to 42 °C, inoculation, incubation, and cooling to 4 °C can be carried out in this equipment if available. If it is impossible to acquire this equipment, the mentioned steps can be carried out in individual stainless-steel tanks/containers with a separate refrigerator and incubator.

11. These machines are used for hard-type ice cream and frozen yogurt.

12. It is recommended to heat liquid ingredients (between 30 and 50 °C), especially if they include milk or cream, to help dissolve the solid ingredients more effectively. Ingredients like sugar and stabilizers are more soluble at higher temperatures, and heating ensures they are evenly distributed throughout the mix. This results in a smoother and more homogeneous ice cream base. When adding ingredients such as stabilizers and emulsifiers, it is worth observing the manufacturer's instructions regarding the appropriate temperature for their solubilization [11].

13. The dosage of liquid ingredients can be done manually using buckets with volumetric graduations. In larger, more modern ice cream production facilities, automatic dosing machines are often integrated into larger automated production lines, allowing for seamless and efficient production processes. These devices are designed to provide highly accurate and consistent dosing of liquid ingredients. This ensures greater accuracy in the composition of the mixtures and, consequently, better standardization between batches of the final product.
14. Equipment known as powder or ingredient feeders are commonly used in the food industry to accurately introduce dry or solid ingredients into a mixing tank containing liquid ingredients. The meter begins with a funnel-shaped hopper or container at the top, where the solid ingredients are loaded. Below the hopper, there is a metering mechanism (such as an auger or screw conveyor) designed to control the flow of solid ingredients that will be dispensed into a pipe attached to the inlet line of the mixing tank.
15. Each ice cream manufacturer's choice of the time/temperature binomial must be based on federal regulations for pasteurization standards. The binomial adopted must be appropriate to the requirements of the health authorities responsible for both the manufacturing region and the product market region.
16. The objective of the pasteurization goes beyond ensuring the microbiological safety of the mixture through the destruction of spoilage and pathogenic bacteria to safe levels. Subjecting the mixture to high temperatures also promotes the denaturation of the proteins, improving the texture and flavor of the ice cream and increasing the uniformity of the mixture due to the complete hydration of the dry ingredients. In batch-type processes, apply longer heat treatment times, such as 30 min, combined with lower temperatures (around 70 °C) (LTLT, low-temperature x long time), and the mixture is continuously agitated to ensure even heating and to prevent scorching. In continuous processes, heat treatment times are much shorter, in the order of seconds, using higher temperatures (around 85 °C) (HTST, high-temperature x short time). Therefore, batch processes promote protein denaturation and hydration of solids to a greater extent, when compared to the continuous process [11].
17. Ice cream production can occur through a continuous or batch process. In the batch process, the pasteurization, homogenization, and cooling steps are carried out sequentially independently. In this case, pasteurization occurs in double-walled (jacketed) tanks with a temperature control system. On the other hand, in the continuous production process, a plate

heat exchanger-type pasteurizer is generally used, and the homogenizing equipment is installed along the pasteurization line between the heating and cooling sections. Therefore, the flow of the mixture occurs in the following direction: mixing tank ➡ regeneration section ➡ heating section ➡ homogenizer ➡ cooling section [13].

18. Homogenization results in a notable increase in the quantity of fat globules, leading to a significant expansion of the fatty/aqueous phase interfacial area, ranging from four to ten times the original size. To mitigate the interfacial tension between fatty and aqueous phases, proteins adsorb at this interface, enhancing the fat globule stability in suspension by leveraging steric and electrostatic repulsions promoted by the protein chains [12]. This stability holds particular importance up to the churning step. However, it becomes essential for partial coalescence of fat globules to occur during churning. Consequently, it is crucial to allow for an adequate aging period in the mixture to facilitate the rearrangement of the fat globule membranes. This process is marked by the partial replacement of proteins with emulsifying molecules, leading to the formation of less stable fatty structures. Furthermore, subjecting the mixture to low temperatures during aging is essential to solidify part of the lipids, forming partially crystallized fat globules. Therefore, the maturation time depends on the source of fat used in the formulation. For example, milk lipids are more saturated than palm lipids so that a milk fat-based mixture would take longer to age.

19. Thermosensitive ingredients, such as aromas and colorings, must be added at the beginning of the maturation step to avoid their thermal degradation in the heat treatment step.

20. The scraped surface heat exchanger-type freezer has advantages for use in the ice cream churning stage. The design of this equipment allows the exchange of energy in the form of heat more efficiently, promoting rapid freezing of the mixture and, consequently, the formation of numerous small ice crystals (20 to 30 μm in diameter) (Fig. 3) [14]. Smaller ice crystals contribute to a smoother and creamier texture in the ice cream. This fine texture is highly desirable among consumers, providing a pleasant mouthfeel.

21. The churning process, which occurs in the dynamic freezing step, induces multiple collisions among the fat globules, favoring the occurrence of partial coalescence between them. This phenomenon gives rise to the formation of a three-dimensional fatty network-like structure. Air is also incorporated into this system (overrun) as the mixture is beaten and frozen simultaneously. Thus, the fatty network adsorbs at the interface of the air bubbles, increasing the kinetic stability of the foam. At this

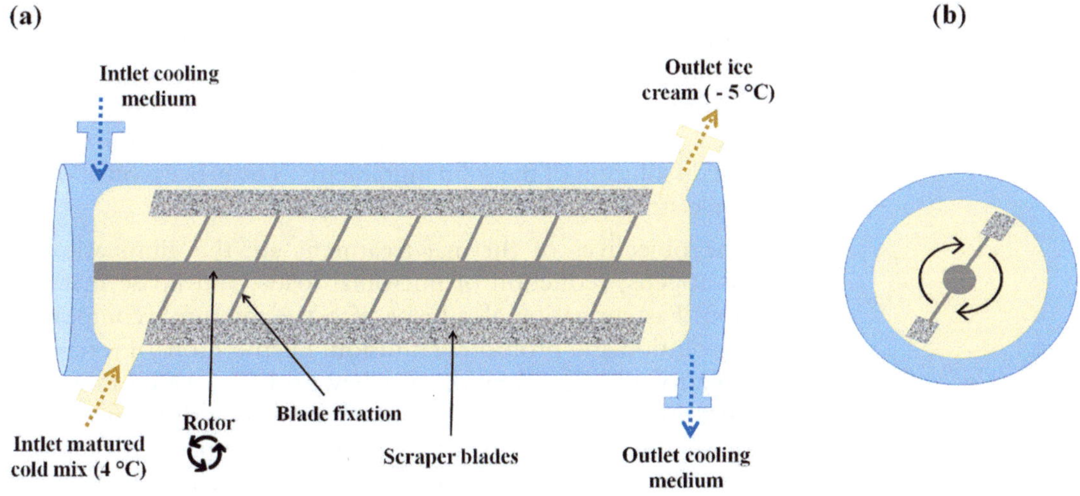

Fig. 3 (**a**) Schematic of a scraped surface heat exchanger (SSHE) used in ice cream manufacturing. (**b**) Representation of the SSHE cross-section

point, it is worth highlighting the importance of choosing both the lipid sources of the formulation and the aging conditions so that partially crystallized fatty structures are formed at the temperature at which the beating step occurs (around 5 °C). The stabilization of air bubbles by the fat network is essential to define the melting properties of the ice cream, such as the melting time [15, 16].

22. Ingredients that must remain in the form of larger particles (such as biscuits, nuts, chocolate flakes, and plums) can be added when packaging, when leaving the freezer. When convenient, these ingredients can also be included before the beating step. If using scraped surface heat exchangers, attention must be paid to the hardness of the ingredients to avoid damaging the scraper parts of the equipment. Furthermore, particles may have their size reduced, compromising the product's texture and the consumer's sensory experience.

23. The time spent in the hardening stage depends on the size and type of packaging, ice cream/frozen yogurt composition, and overrun, which will define the heat conduction through the product. The time must be sufficient for the frozen desserts to acquire the shape of the packaging and reach the appropriate hardness. The hardening step is not required if a soft-serve frozen yogurt system is chosen. Hardening is obligatory to produce hard-type frozen yogurt. It involves static freezing (in tunnels or freezers with forced cold air circulation between −15 and − 25 °C) of the product already packaged for retail sale. This stage is responsible for giving frozen ice cream/ yogurt its smooth and consistent texture, preserving its shape and preventing unwanted ice crystal growth.

24. In frozen yogurt production, the solid ingredients are sugar and powdered milk, and the liquid ingredients are milk or some liquid sugar. The other ingredients (emulsifiers, coloring, and flavorings) are added when the yogurt is ready. When adding the liquid ingredients, heat them to a 45 °C temperature to aid the dissolution of the solid ingredients. The process must occur while stirring to avoid creating lumps.
25. The objectives of thermal treatment are the elimination of pathogens, reduction of microbial loads, and lipase inactivation. Pasteurization also helps with the mixture's component dissolution and promotes fat fusion. Furthermore, it is responsible for the denaturation of whey proteins, which raises the final product's viscosity. This pasteurization is done in a plated or shell-tube heat exchanger. It can be done in a binomial time/temperature of 90 °C/5 min or equivalent. The homogenization, in turn, is responsible for emulsion formation and fat globule size reduction. After homogenization, there is a new fat globule membrane formation in which proteins adsorb to lower interfacial tension between fatty and aqueous phases.
26. The mixture is cooled to 42 °C, which is an optimal temperature for the growth of many microorganisms used in fermentation.
27. The fermentation occurs until it reaches a pH of 4.6 and an acidity higher than 0.6% of lactic acid. Protein aggregation and gelification occur at this point, resulting in the formation of yogurt. Fermentation time can vary according to temperature, culture type, and total solid concentration. Therefore, pH and acidity should be thoroughly measured during fermentation.
28. Upon stirring, the rest of the ingredients are added until 42 °C to solubilize them. As the yogurt has already passed through the pasteurization process, the ingredients must be added aseptically to guarantee the product's microbiological quality. Thus, once the yogurt is produced, it will not be subjected to thermal treatment to keep bacterial viability [1, 3].
29. Aging occurs at temperatures below 4 °C for at least 12 hours. During this phase, proteins are replaced by emulsifiers on the membrane of fat globules to prepare for fat destabilization in the subsequent step, which involves simultaneous churning and dynamic freezing. Furthermore, aging promotes fat crystallization, a critical step for the partial coalescence process and network formation. Throughout aging, proteins and stabilizers hydrate, leading to an increase in the mixture's viscosity.

Acknowledgments

The authors thank Coordenação de Aperfeiçoamento de Pessoal de Nível Superior - CAPES (Finance Code 001), Conselho Nacional de Desenvolvimento Científico e Tecnológico - CNPq (311788/2021-1, 406919/2021-6, and 403724/2023-6), and Fundação de Amparo à Pesquisa do Estado de Minas Gerais—FAPEMIG (APQ-02024-21, RED-00186-22, and APQ-00823-22).

References

1. Goff HD. The Structure and Properties of Ice Cream and Frozen Desserts. Encyclopedia of Food Chemistry [Internet]. 2019 Jan 1 [cited 2019 Sep 10];47–54. Available from: https://www.sciencedirect.com/science/article/pii/B9780081005965217034
2. Clarke C (2012) The science of ice cream. 2a. RSCPublishing, Cambridge, pp 1–527
3. Goff HD, Hartel RW (2013) Mix ingredients. In: Goff HD, Hartel RW (eds) Ice cream. 7a. Springer, New York, pp 45–85
4. Adil S, Jana AH, Mehta BM, Chandgude PB (2021) Ice cream and frozen yogurt - a suitable carrier for probiotics: a review. Agric Rev Of
5. Daw E, Hartel RW (2015) Fat destabilization and melt-down of ice creams with increased protein content. Int Dairy J 1(43):33–41
6. Hartel RW, Rankin SA, Bradley RL (2017) A 100-year review: milestones in the development of frozen desserts. J Dairy Sci 100(12)
7. Sipple LR, Racette CM, Schiano AN, Drake MA (2022) Consumer perception of ice cream and frozen desserts in the "better-for-you" category. J Dairy Sci 105(1)
8. Polaris Market Reasearch [Internet] (2023) Frozen dessert Market share, size, trends, industry analysis report, by product (Ice-cream, frozen yogurts, confectionary & candies, others); by distribution channel; by region; segment forecast. In: Frozen desserts Market, pp 2022–2030
9. Goff HD, Hartel RW (2013) Ice cream structure. Ice Cream [Internet]. [cited 2023 Jun 18];313–52. Available from: https://link.springer.com/chapter/10.1007/978-1-4614-6096-1_11
10. Bolliger S, Goff HD, Tharp BW. Correlation between colloidal properties of ice cream mix and ice cream. Int Dairy J [Internet] 2000 Jan 1 [cited 2019 Sep 10];10(4):303–9. Available from: https://www.sciencedirect.com/science/article/pii/S0958694600000443#BIB2
11. Goff HD, Hartel RW (2013) Ice cream [Internet]. Springer US, Boston, MA. Available from: https://link.springer.com/10.1007/978-1-4614-6096-1
12. Wang Y, Liu M, Qu X, Wang S, Ma Z, Zhang R et al (2019) Changes in the fat globule membrane protein components of pasteurized milk caused by different homogenization conditions determined using a label-free proteomic approach. LWT [Internet] 115:108430. Available from: http://www.sciencedirect.com/science/article/pii/S0023643819307728
13. Clarke C (2012) The science of ice cream, 2nd edn. The Royal Society of Chemistry, Cambridge, UK, p 253
14. Rao CS, Hartel RW (2006) Scraped surface heat exchangers. Crit Rev Food Sci Nutr [Internet] 46(3):207–219. Available from: http://www.tandfonline.com/doi/abs/10.1080/10408390500315561
15. Liu X, Sala G, Scholten E (2023) Structural and functional differences between ice crystal-dominated and fat network-dominated ice cream. Food Hydrocoll 138:108466
16. Liu X, Sala G, Scholten E (2022) Effect of fat aggregate size and percentage on the melting properties of ice cream. Food Res Int 160:111709

Chapter 20

Dulce De Leche

Analía Rodríguez, Adriana Gámbaro, Luis A. Panizzolo, and Sergio D. Rozycki

Abstract

Dulce de leche [milk caramel] is one of the most exquisite milk derivatives with particular color, taste, and consistency that differentiate it from other dairy products. These attributes are developed during the manufacturing process through nonenzymatic browning reactions (Maillard reaction and caramelization), where the Maillard reaction is the main responsible. The heating processing (at temperatures above 100 °C) and concentrating milk with ingredients, especially sucrose (partially or not substituted by monosaccharides) and pH regulation, favors these reactions. The gelation of milk proteins also occurs and partially causes the development of the *dulce de leche* texture where the Maillard reaction would be affecting it by cross-linking reactions. Various processing variables are involved in the manufacturing of *dulce de leche*, such as temperature, type and concentration of carbohydrates, type, and concentration of amino acids, time, pH, water activity, and metal cations. There are many varieties of *dulce de leche*, such as *familiar* [classic or *homestyle*], *repostero* [for baking], *pastelero* [for patisserie], or *heladero* [for ice-cream making]. The above has led to different manufacturing processes and formulations, with modifications depending on the country and the selected methodology, which also depends on the production volumes, the company's investment capacity, the costs, and the desired profit. Besides similarities among the manufacturing protocols, significant disparities exist as well and this chapter intends to clarify and to relate them to the distinctive attributes of the several styles of *dulces de leche*.

Key words Dulce de leche, Doce de leite, Concentrated dairy products, Bakery, Manjar blanco, Arequipe, Confiture de lait

1 Introduction

Dulce de leche [milk caramel] is one of the most exquisite milk derivatives due to its particular color, taste, and consistency, which strikes the difference in comparison to other dairy products. *Dulce de leche* is distributed and consumed in all Ibero-American countries and nowadays is being distributed in other regions such as Europe, Africa, and Asia.

Every October 11, the "*Día Mundial del Dulce de leche*" [World *Dulce de Leche* Day] has been celebrated since 1995. In

Adriano Gomes da Cruz et al. (eds.), *Dairy Foods Processing*, Methods and Protocols in Food Science,
https://doi.org/10.1007/978-1-0716-4144-6_20,

2022, the "*Consejo Mundial del Dulce de leche*" [World *Dulce de Leche* Council] was appointed with the participation of international members.

Dulce de leche is known by different names: *dulce de leche* in Argentina, Uruguay (*see* **Note 1**), Bolivia, Costa Rica, Cuba, Spain, Paraguay, Puerto Rico, and Dominican Republic; *manjar* in Chile and Ecuador; *manjar blanco* in Colombia, Bolivia, Ecuador, Panama, and Peru; *arequipe* in Colombia and Venezuela; *cajeta* in Mexico; *bollo de leche* in Nicaragua; *bienmesabe* in Panama; and *doce de leite* in Brazil and Portugal. There is also a similar product in France, typical in Breton, Norman, and Savoyard cuisines, known as *confiture de lait* [1, 2].

It is a product obtained by heating and concentrating milk, and adding ingredients, especially sucrose (partially or not substituted by monosaccharides), under regular or reduced pressure [3, 4].

The valued sensory attributes of *dulce de leche* (color, taste, texture) are developed based on nonenzymatic browning reactions (Maillard reaction and caramelization), where the Maillard reaction is the main responsible (*see* **Notes 2** and **3**). Due to the manufacturing process, with heat processing at temperatures above 100 °C and concomitant concentration, these reactions are favored by the formation of brown pigments and volatile compounds contributing to the distinctive flavor and aroma [5–8]. The gelation of milk proteins also occurs, and partially causes the development of the *dulce de leche* texture [7, 9], where the Maillard reaction would be affecting it by cross-linking reactions [10, 11].

Various elements participate in the initiation and development of the Maillard reaction, and they can be considered as processing variables in the manufacturing of *dulce de leche*, such as temperature, type and concentration of carbohydrates, type, and concentration of amino acids, time, pH, water activity, and metal cations, where the temperature is a strong and influent variable [6–8].

Dulce de leche manufacturing, which began at a culinary level and then scaled to an industrial level, has several manufacturing alternatives with different methodologies and technology. The common variable among them is heat processing, which consists of two essential stages, together or separately: concentration of solids and color development (Maillard reaction) to obtain the desired attributes of *dulce de leche* based on the target market preferences. The process of concentrating solids must result in maximum humidity of 30% while also achieving a protein conformational change and interaction to obtain the distinctive consistency, texture, and rheology. In addition, the Maillard reaction must occur to an exact extent because insufficient development as well as overdevelopment in the reaction result in a product without the desired attributes of *dulce de leche* regarding color, flavor, and aroma. The manufacturing technologies include the traditional technology in an open *paila* as well as the technologies using concentration by evaporation, which use evaporators working under pressures lower than the atmospheric pressure (boiling temperature below 100 °C), with continuous manufacturing systems, or systems combining both technologies (mixed/semi-batch). Any modifications to manufacturing and materials used are based on the style of *dulce de leche* to be produced, the production volumes, the company's investment capacity, the costs, and the desired profit.

There are many varieties of *dulce de leche* based on the way of using and consuming it, such as, among others, *familiar* [classic or *homestyle*], *repostero* [for baking], *pastelero* [for patisserie], or *heladero* [for ice-cream making] [2]. The above has led to different manufacturing processes and formulations, which have various modifications depending on the country and the selected methodology.

Although there are similarities among the manufacturing protocols used in different countries and companies, significant disparities exist as well. Therefore, this chapter intends to clarify such disparities and to relate them to the distinctive attributes of the several styles of *dulces de leche*.

2 Materials

2.1 Manufacturing of Dulce de Leche

Dulce de leche is not included in the Codex Alimentarius, but it is regulated by the Mercosur.

MERCOSUR/GMC/RES. no. 137/96: "*Dulce de Leche* is the product derived from milk or reconstituted milk, with or without the addition of other food substances, obtained by the concentration and action of heat, under normal or reduced pressure, with or without the addition of milk solids and/or cream, and with the addition of sucrose (partially or not substituted by monosaccharides and/or other disaccharides)."

The name *dulce de leche* is reserved for the product in which the dairy base does not contain nondairy fat and/or proteins.

The required ingredients are milk and/or reconstituted milk.

The optional ingredients are cream; dairy solids; monosaccharides and disaccharides substituting the sucrose at a maximum of 40% w/w; starch or modified starches in a maximum ratio of 0.5 g/100 ml in milk; and cocoa, chocolate, coconut, almonds, peanuts, dried fruits, grains, and/or other foodstuffs isolated or mixed in a ratio between 5% and 30% w/w of the final product [12].

2.2 Materials

2.2.1 Raw Materials

In the manufacturing of *dulce de leche*, often named "traditional" manufacturing, an initial mixture of whole fluid milk and sucrose (white sugar) is formed (*see* **Note 4**).

Whole fluid milk (generally adjusted to a 3% fat content) is used and sucrose is added up to 30% relative to the whole fluid milk, according to the MERCOSUR Regulation [12].

The whole fluid milk, and other raw materials, could be optionally replaced, in total or in part, by skimmed fluid milk and cream or butter. Whole or skimmed powdered milk, milk whey, whey proteins and/or liquid and/or powdered milk permeate, glucose, and/or fructose (up to 10% according to information collected from the industry), among other materials, could also be used in the nonconventional manufacturing of *dulce de leche*.

2.2.2 Additives

Generally, sodium bicarbonate is added to the initial mixture as a neutralizing agent, in a 0.1% ratio relative to the whole fluid milk. For the same purpose, calcium hydroxide can be used in combination or as a replacement, in a ratio of half the amount of calcium hydroxide to sodium bicarbonate, both food grade (*see* **Note 5**).

Potassium sorbate may be added to the cooled *dulce de leche* as a preservative (against fungi and yeasts) up to 600 ppm (mg/kg) expressed as sorbic acid, and a hydroalcoholic solution of the same preservative, in different concentrations, and even saturated (58.2% w/w at 20 °C), may be optionally sprayed on the product surface (already packaged). The surface can be sprayed with a natamycin solution, at a rate of 1–2 mg/dm^2, for the same purpose. Sometimes, calcium lactate is also added as a texturizing agent and sorbitol is added as a wetting agent (< 5%) [12].

Additives, such as anti-foaming agents (glyceryl monostearate, 0.025–0.05%) and stabilizing agents for the protein system (sodium tripolyphosphate, 0.01–0.02%/sodium citrate, 0.1–0.2%), are often used in *dulces de leche* for industrial uses.

2.3 Equipment

Different equipment and processing conditions are used depending on the manufacturing method (*see* **Notes 6** and 7).

In the ***batch method***, a *paila dulcera* [sweet mixing vat] is used. This equipment is made of stainless steel (formerly, the interior was coated with a thin layer of copper), cylindrical in its upper part and hemispherical in the lower part (Fig. 1). It has a double jacket where steam from a kettle can enter as a heating medium to achieve partial evaporation of the water in the initial mixture, thereby concentrating the total solids (TS) up to 70 or 75%, depending on the style of *dulce de leche*. After the *dulce de leche* has been produced, tap or well water at room temperature can be poured into the double jacket as a cooling medium to reach the packaging temperature (65–75 °C). It also comprises an anchor-like agitator that has the same shape as the lower part of the *paila* (hemispherical) and has hard plastic scraper blades with springs that force them to adhere and strongly press to the inner surface of the *paila*, thereby vigorously scraping to avoid the material from sticking and the presence of burnt taste and/or smell.

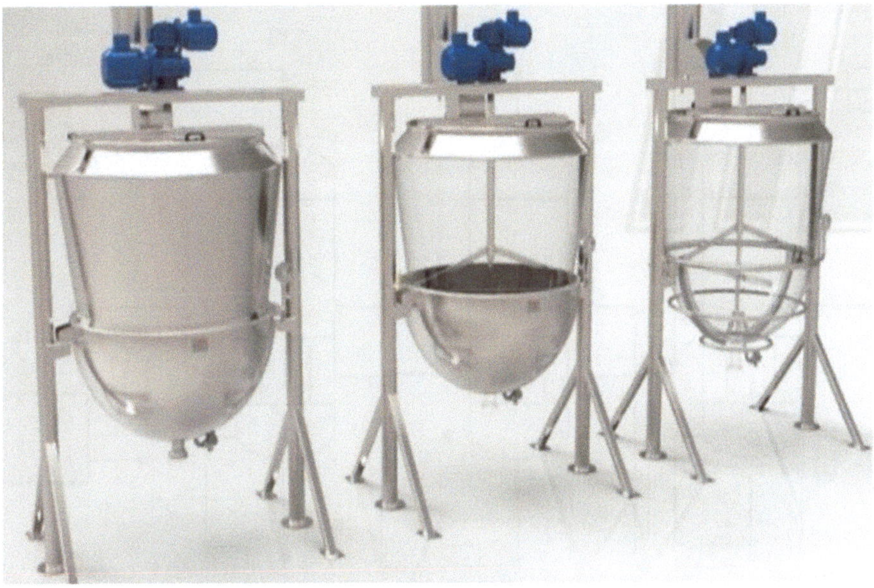

Fig. 1 *Paila dulcera* equipment

In the ***semi-batch method***, the initial mixture is previously concentrated in a "multiple-effect evaporator" (different "evaporation bodies" or "evaporators"), usually three effects or "triple effect" (Fig. 2), and the concentration (manufacturing) ends in a *paila* (Fig. 3).

The evaporators used are "falling film" evaporators (Fig. 4), where the initial mixture falls like a film inside the tubes of each "effect" at a considerable rate, thereby exchanging heat and concentrating, while the heating steam surrounds these tubes on the outside. In the "first effect," the steam comes directly from the steam-generating kettle.

In the "second effect," the heating steam is the steam produced in the "first effect" ("previous effect") and it provides enough energy for evaporation (and concentration), and so on. This occurs because such "second effect" is at a lower pressure than the "first effect," and therefore, the water evaporates at a lower temperature than in the "first effect." This results in great energy efficiency compared to the evaporation efficiency (kg of evaporated water/ hour), in addition to being compact equipment that occupies little volume.

It is possible to reach a concentration level (50–55% TS) higher than normal (48–50% for whole powdered milk and skimmed powdered milk) because the mixture being concentrated has a high concentration of sugars (approximately 44–45%), which are molecules of lower size and molecular weight compared to the remaining molecules (caseins and fat) that provide lower viscosity. The optimal number of "effects" is between 3 and 4 (better cost-

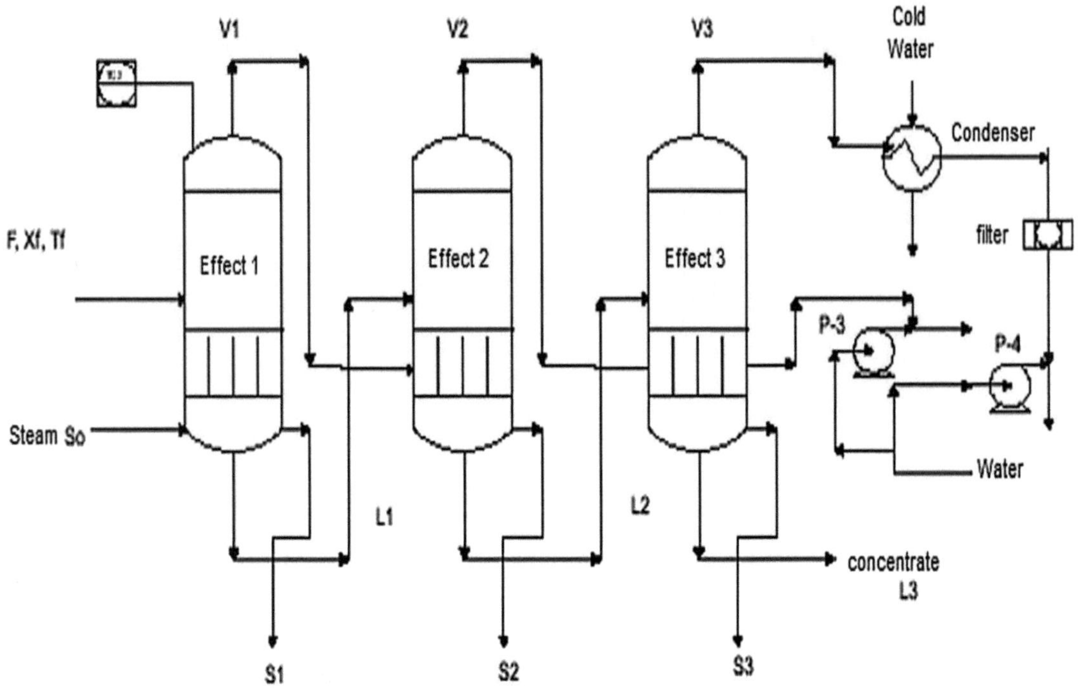

Fig. 2 "Triple effect evaporator" equipment

benefit ratio between the initial investment cost, which increases as the number of "effects" increases, and the system operation cost, which decreases as the number of "effects" decreases, because of the aforementioned energy efficiency).

In the ***continuous method***, there are two possible operative dispositions ("arrays"):

(a) The initial mixture is first concentrated as much as possible in an evaporator system (50–55% TS). Then it is "colored," that is, it develops the distinctive dark brown color of most commercially available *dulces de leche*, mainly due to Maillard reactions, and finally, the concentration is completed to its final value (70 or 75% TS, depending on the style of *dulce de leche*) in a scraped film evaporator.

(b) First, the initial mixture is "colored" and then the concentration is performed in the two stages described above, first in the evaporators and finally in the scraped film evaporator (Fig. 5).

Generally, the scraped film evaporator is a tubular heat exchanger, wherein the product circulates through the central tube, where there are tubular blades that scrape the inner wall such that the product does not stick or burn, surrounded by a tubular jacket through which the kettle heating steam circulates. Optionally, but less frequently, a plate heat exchanger can be used.

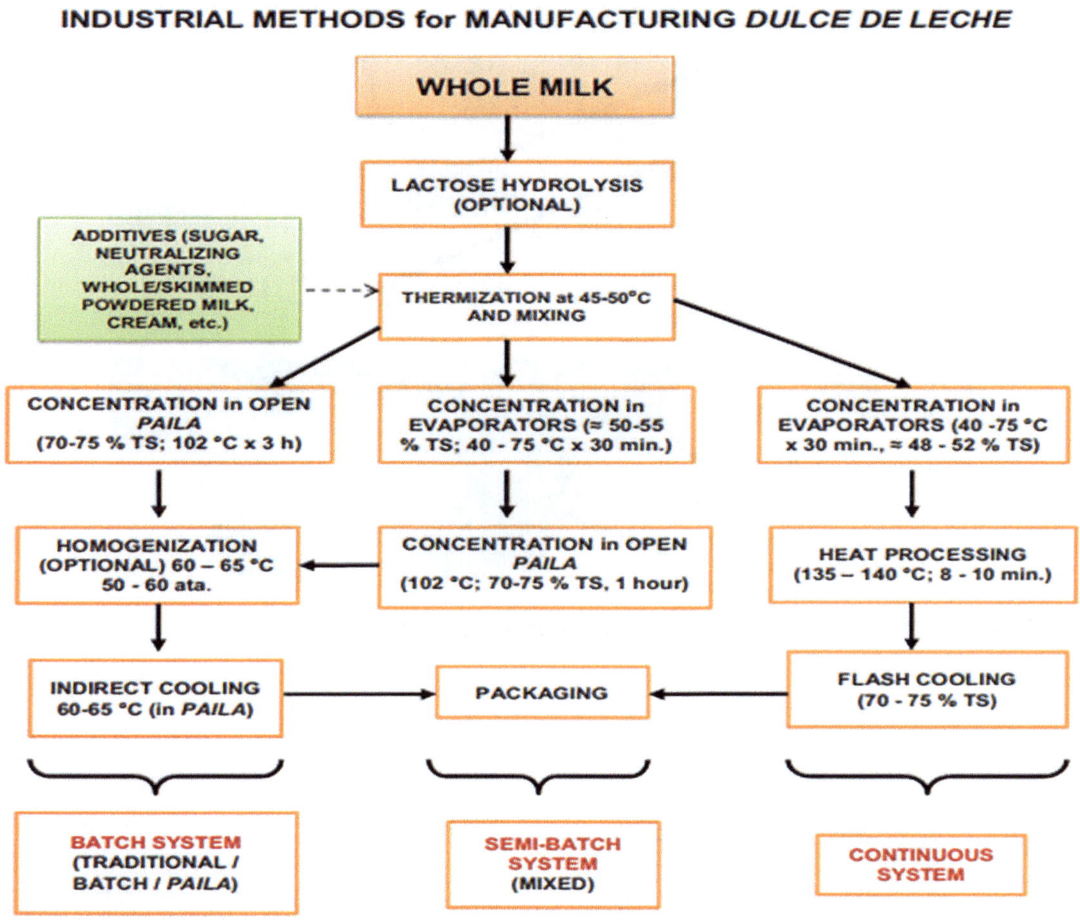

Fig. 3 Diagram of methods for manufacturing *dulce de leche*

Finally, the *dulce de leche* is "flashed" by introducing it at a certain temperature (generally above 100 °C) into vertical tubular equipment in vacuo, where the boiling temperature is lower than 100 °C since the pressure is lower than atmospheric pressure.

Thus, the flashing or sudden partial evaporation of the water in the *dulce de leche* occurs since in such equipment the product is at a temperature higher than boiling temperature, which produces a controlled "extra" concentration and cooling, and the product exits the equipment at the concentration and temperature required for its final packaging.

3 Methods

The general classification for diverse styles of *dulce de leche* and their main attributes are shown in Table 1. Additionally, the *dulce de leche* has an average composition such as the composition shown in Table 2.

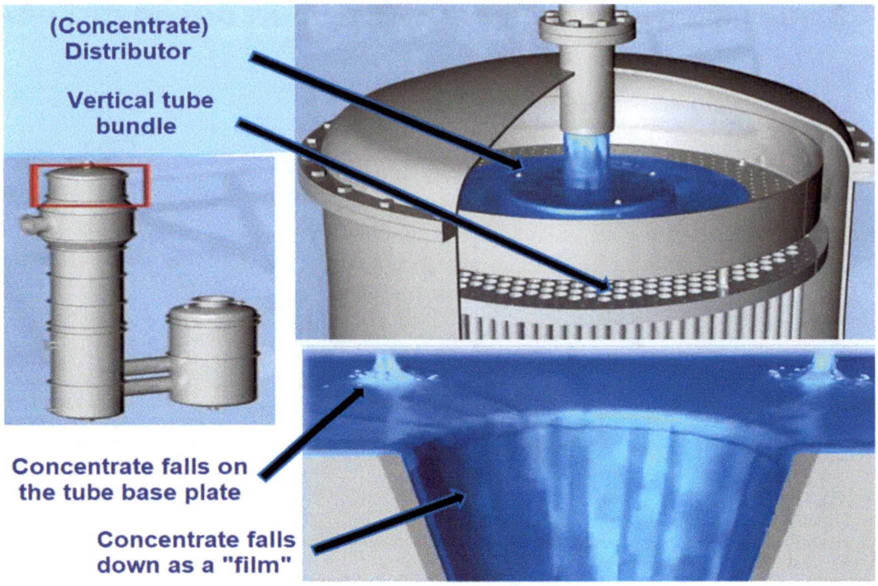

Fig. 4 Diagram for the operation of a falling film evaporator. (Source: GEA Group)

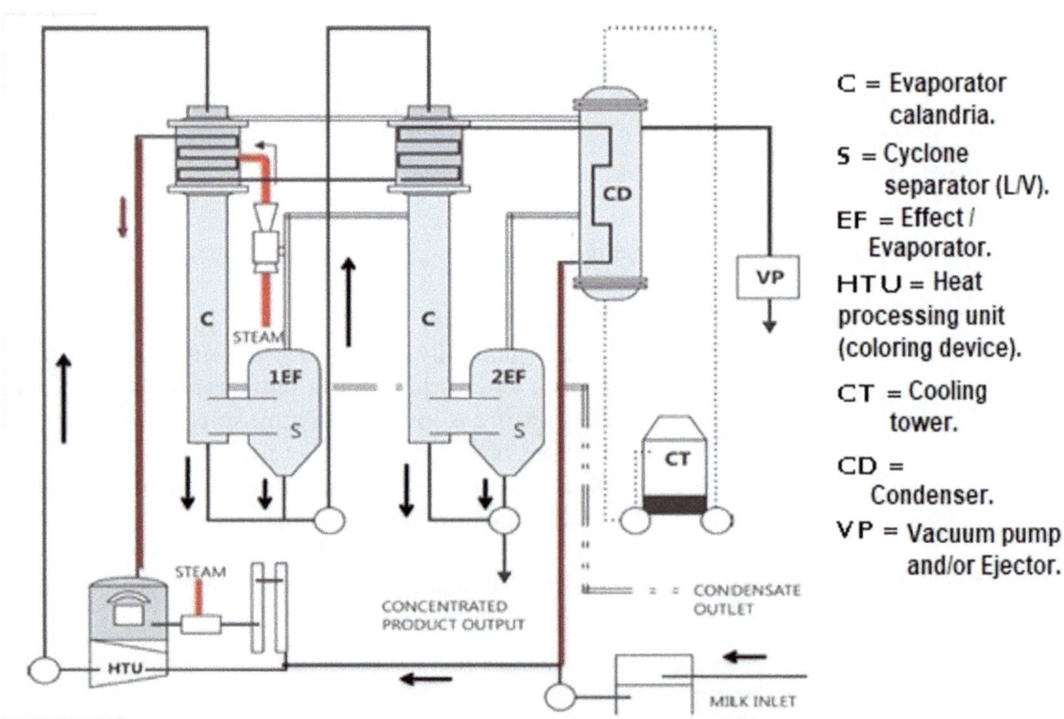

Fig. 5 Diagram for continuous system including evaporator, coloring device, and finisher

Table 1
General styles of dulce de leche, color, texture, gloss, and degree Brix

Style of dulce de leche	Color	Texture	Gloss	Brix
Classic or homestyle	Varies from light to deep brown	Smooth, homogeneous, spreadable, stringy	Intense	67–70
For baking	Varied shades of brown	Thick, cuttable, solid	Moderate	70–75
For patisserie/ alfajor making	Varied	Thicker and more cuttability	Moderate to soft	72–75
For ice-cream making	Deep brown	Minor relevance	Moderate to prominent	72–75

Table 2
Average centesimal composition of dulce de leche

Humidity (maximum)[a]	30%
Dry extract	70–75%
Lactose[a,b]	10–11%
Proteins (minimum)[a]	5%
Milk solids (minimum)[b]	26%
Added sugars	42–44%
Fat content	6–9%
Ash content (maximum)	2%

[a]According to MERCOSUR [12]
[b]According to *Código Alimentario Argentino* (*Ley 18.284*)

As a first step, an initial mixture of fluid milk, sugar, and additives is formed with thermization at 45–50 °C and continuous stirring (*see* **Note 8**).

Nowadays, in the industrial field, it is normal to partially replace sucrose mainly with glucose or fructose (*see* **Note 9**). In addition, the fluid milk can be partially replaced by sweet milk whey. Furthermore, some powdered dairy derivatives are often added, such as concentrated whey proteins, permeates, whole powdered milk, and skimmed powdered milk, that promote starting with an initial mixture with higher TS % which allows reaching the desired final concentration in less time (70 or 75% TS) depending on the style of *dulce de leche*, thereby increasing the company's productivity (*see* **Notes 10** and **11**). In this step, fat content is adjusted to achieve the amount required in the final *dulce de leche*, according to the Mercosur Regulation (*see* **Note 12**).

Table 3
Lactose crystal concentration and size in dulce de leche and their sensorial perception

Texture	Crystal amount/ml	Crystal size (μm)	Mouthfeel
Excellent	400.000.000	9.3	No
Good	300.000.000	10.4	No
Slightly pasty	200.000.000	11.7	No
Finely floury	100.000.000	14.7	Threshold
Floury	50.000.000	18.6	Yes
Sandy	12.500.000	29.4	Yes

As lactose in *dulce de leche* appears as a supersaturated solution and its crystals will appear as a defect during the storage (Table 3), the partial lactose hydrolysis could be optional (or could be necessary to improve shelf life) in this initial step by using the lactase or β-galactosidase enzymes (*see* **Note 13**).

Hydrocolloids are usually added to the initial mixture to provide consistency and viscosity to *dulce de leche*, useful in homestyle *dulce de leche* but indispensable in other styles of *dulce de leche* (mainly for baking and patisserie *dulce de leche*) (*see* **Note 14**).

The neutralization of the initial mixture (using sodium bicarbonate and/or calcium hydroxide) up to an estimated pH of 7.0–7.2 (10–12 °D) is an essential step in the manufacturing of any style of *dulce de leche* (*see* **Note 15**).

Once the initial mixture is ready, thermal and concentration process follow. Figure 3 shows the manufacturing steps of the three manufacturing methods for *dulce de leche* (*see* **Note 16**).

Batch Method The *paila* (open top) is purged by introducing steam into the jacket and displacing the air. Then the initial mixture is introduced, and the steam is introduced at low pressure (0.5–1.0 ata) into the jacket with strong stirring while controlling the foam formation, which is strong at the beginning due to CO_2 generation (due to Maillard reactions and, mainly, if it was neutralized with sodium bicarbonate).

The foam reduces when the first initial mixture has been sufficiently concentrated; then the addition of the initial mixture (dropwise) begins to compensate for the evaporation and maintains the level of the free surface of the initial mixture at the top of the jacket. Optionally, the *paila* can be full, but in order to work under these conditions, it is essential to add an anti-foaming agent (glyceryl monostearate, 0.05–0.1% relative to the volume of the initial mixture) to prevent the mixture from overflowing and falling down the *paila*'s external wall.

Glucose is added (optionally) when approximately 55–60% TS is reached; otherwise, the result would be a very dark *dulce de leche* if the glucose is added at the beginning of the manufacturing process (*see* **Notes 17** and **18**). The steam feeding into the jacket stops when the required TS % is reached (70% for homestyle *dulce de leche* and 75% for other styles of *dulce de leche*), generally measured with a "handheld" refractometer, and the *dulce de leche* is lowered by gravity to a central cooler into which the content of several *pailas* is poured in parallel. Tap or well water (15–20 °C) is introduced into the cooler jacket with intense stirring to cool the *dulce de leche* to the final packaging temperature (65–75 °C) (*see* **Notes 19, 20, 21** and **22**). Previously, it could optionally go through a homogenization step at low pressure (50–60 ata) to not excessively reduce the viscosity, improve surface smoothness, destroy possible lumps, and increase gloss (*see* **Note 23**).

Semi-Batch Method First, the initial mixture is treated in the same way as in the batch method during the mixing of raw materials and additives, and then it is introduced into the evaporator system in vacuo, evaporating and concentrating at temperatures well below 100 °C. This concentrated mixture is introduced into the *paila* when the maximum possible concentration is reached (50–55% TS); the operative conditions are the same as in the batch method, and the product is kept in the *paila* for at least 1 h such that it acquires the desired structure and texture. Then the product is packaged (*see* **Notes 19, 20, 21**) and it can be previously and optionally homogenized, in a way equivalent to the previous method.

Continuous Method No *paila dulcera* is used in this case; instead, a system by evaporation is used, usually a "two-effect" system (two evaporators), the scraped film concentrator, also named "finisher," and a "coloring" device. There are two styles of "arrays": first concentrating and then developing the color, and vice versa; generally, the step involving the scraped film evaporator is at the end in both arrays to obtain the TS % required for each style of *dulce de leche*.

The concentration step is performed under similar conditions to those in the semi-batch method.

The coloring device is the equipment used to provide the final color required for *dulce de leche*, providing that the color reaches a deep brown tone (not the pale color of "*manjar blanco*"). In the first possible "array," where the first step is the development of color, a plate heat exchanger or tubular heat exchanger can be indistinctly used since the initial mixture only has 31–32% TS. Then it concentrates in the evaporator and scraped film evaporator system. In the other type of "array," where the first step is the concentration (up to 50–55% TS), a very viscous product is delivered into the coloring device. This is why a double tube heat exchanger is used (as the coloring device), which does not have as main limitation the increased viscosity of the feeding, as it does in a plate heat exchanger that generates an increased load drop inside the equipment.

Such continuous system usually has limitations for *dulces de leche* other than the homestyle variety since the other styles must have increased viscosity and cuttability (little generation of "strings").

Taking into account that only above 110 °C there is a net coagulation of the system, and that at the operative temperature of the coloring device (135–140 °C) coagulation occurs in less than 3 min compared to the complete working conditions of the coloring device to generate the dark brown color (135–140 °C for about 10–12 min), it is observed that coagulation occurs much earlier than the development of intense brown color (*see* **Note 24**).

In the first "array" mentioned, where the coloring device is placed at the beginning, coagulation of the initial mixture also occurs. Since the initial mixture should then be pumped several times to pass through the evaporator system, where it must circulate inside the tubes, among other mechanical treatments, added to the fact that *dulce de leche* is thixotropic (like most dairy products), the viscosity of the product is reduced. Thus, several industrial operators and equipment manufacturers agree that it is quite difficult to produce *dulces de leche* other than the homestyle variety because they require increased consistency and viscosity. This array is useful for the homestyle *dulce de leche* (classic).

In the second "array," where the coloring step is carried out at the end and after passing through the evaporators, the subsequent mechanical treatment is reduced since, under these conditions, in the coloring device the coagulation also occurs while the *dulce de leche* is at the end of the equipment and ready to exit it, thereby arguing that other styles of *dulce de leche* (such as *dulce de leche* for baking, patisserie, and ice-cream making), which need increased consistency, viscosity, and cuttability, can be manufactured with this "array."

4 Notes

1. *Dulce de leche* has been declared "Patrimonio Cultural Alimentario y Gastronómico de la Argentina" [Food and Gastronomic Cultural Heritage of Argentina] (1998) and "Patrimonio Cultural del Río de la Plata" [Cultural Heritage of the Río de la Plata] (2003), jointly with Uruguay.

2. In many Ibero-American countries, the color is the attribute that most influences the consumers' acceptance of *dulce de leche.*
3. Caramelization reactions developing color do not affect the *dulce de leche* to a great extent [13, 14].
4. Original, or commonly named "traditional," *dulce de leche* is manufactured with only two raw materials: whole fluid milk and sucrose, which form an initial mixture having approximately 31–32% TS.
5. Maillard reactions are favored at high pH with a maximum of approximately pH = 9. Such condition does not apply in the manufacturing of *dulce de leche* since the initial mixture has a pH of approximately 7.0–7.2, which reduces during the manufacturing up to a pH of approximately 6 in the final product.
6. Viscosity is a transport property, which is important to adopt and design product handling systems (pipelines, valves, etc.) and equipment (circulation pumps, heat exchangers, etc.). Therefore, in order to quantify the viscosity by a viscosity analysis (destructive rheometry), design (dimensional) and operative (rotation speed) factors must be known, which are essential to determine the appropriate conditions for the analysis and to perform the correct calculation of such viscosity.
7. It has to be considered that *dulce de leche* is a thixotropic product, which means that its viscosity decreases if the temperature and the amount of mechanical treatment received (pumping, stirring, homogenization, circulation through a pipeline, etc.) are reduced, but also the viscosity decreases if the duration of the mechanical treatment applied is reduced.
8. Temperature (45–50 °C) and continuous stirring favors the dissolution of the powdered ingredients, thereby reducing the possibility of lump formation, which is also reduced by previously mixing the sucrose with the rest of the powders since it is an incredibly soluble sugar that increases the solubility of the mixture.
9. The main alternative material is glucose, which the Mercosur Regulation allows for the replacement of sucrose up to 40%. Such replacement increases the rate of color and gloss development and acts as an anti-crystallizing agent, which is highly relevant for ice-cream making *dulce de leche* [12]. Fructose is also another sugar to be used, and the recommendation is not to replace the sucrose in more than 10% (to avoid *dulce de leche* from becoming "sandy"). Likewise, the inverted sugar (mixture of glucose and fructose) is used in 1–5% relative to milk. It has to be considered that sucrose does not affect the color development since it is a nonreducing sugar and does not

participate in the Maillard reaction, so lactose is the main carbohydrate affecting color development but when monosaccharides are added, they affect the color development to a greater extent than the di- and polysaccharides.

10. Sweet milk whey can replace milk to a maximum of 50% to minimize the development of defects related to lactose crystals, which leave a noticeable mouthfeel when they reach a size greater than 15 microns (Table 3), thus making the product sensorially unacceptable. The addition of milk whey reduces the costs of *dulce de leche*, but it also increases the risk of early lactose crystallization and its perception by consumers. It is important to note that the addition of whey proteins changes color and gloss. Also, the incorporation of milk solids in the initial mixture, such as powdered whey protein concentrate, has the disadvantage that the lactose is in crystalline form, obtained through adequate cooling profiles of the concentrate when it exits the evaporators and before entering the spray dryer, which is essential to sufficiently dry in such spray dryer. This form is present as hard crystals and with a size generally detectable by the palate (it causes a sandy texture). In order to avoid this texture, the initial mixture for manufacturing *dulce de leche* should simmer (boil) for at least 3 min, such that the lactose crystals added to the dairy powders dissolve and do not generate the aforementioned defect.
11. This initial addition of solids can lead to the issue of not achieving the desired color (in the case of deep brown) due to the ostensible reduction of heating time in the *paila*; the final color of the *dulce de leche* also strongly depends on this. In this case, the addition of artificial caramel flavoring is only allowed in ice-cream making *dulce de leche* [12]. In order to technologically counteract the issue of lack of color, the initial mixture can optionally be further neutralized since increasing the initial pH increases the rate of color development [5–8, 13, 15], and/or partially lactose-free milk can be used since the monosaccharides generated in the lactose hydrolysis (i.e., glucose and galactose) significantly increase the rate of color development compared to lactose.
12. Generally, the fat content in the milk or mixture is adjusted to 3% to achieve the minimum required content for *dulce de leche*, that is, 6% according to the Mercosur Regulation [12]. It can be reached by adding cream or butter.
13. The partial lactose hydrolysis of 30–50% can be used as a way to avoid the lactose crystallization and defect which provide a floury or sandy texture (Table 3). The shelf-life of *dulce de leche*, without crystallization, visibly increases from 3 months to 1 year (ideal for export) if 50% of the lactose is hydrolyzed. There are two methods using lactase or β-galactosidase

enzymes: at a room temperature of approximately 30 °C ("fast" method) or at a low temperature of 4 °C ("slow" method), and each step has its advantages and disadvantages. The main disadvantage of the fast method is that there are more probabilities of microbial contamination, which is minimized in the slow method. Additionally, when partially lactose-free milk is used, a smaller number of sweeteners (because the glucose+galactose mixture is much sweeter than lactose) and neutralizers (because it is not necessary to reach a higher pH for the Maillard reaction to occur since the glucose+galactose mixture browns much faster than lactose) is used.

14. The most commonly used ones are modified starch, agar, and carrageenan, alone or in mixtures, which should not exceed 2% relative to milk [12].

15. If neutralization is not performed or is wrongly performed in this step, the *dulce de leche* "breaks" and an extremely sandy mass is formed, which is sensorially unacceptable. The above occurs because, during the manufacturing of *dulce de leche*, the initial mixture becomes more acidic during the heating and concentration due to protons being concentrated as the calcium is exchanged with the micelle and protons exit it, and mainly due to the generation of acids, almost exclusively formic acid [5–8, 13].

    Sodium bicarbonate provides *dulce de leche* with a good "string" formation (Fig. 6), which is desirable for homestyle *dulce de leche* (not for other styles of *dulce de leche*), and it also increases its color and gloss. Calcium hydroxide makes the *dulce de leche* more "cuttable" (no string is formed), which is desirable for other styles of *dulce de leche* (mainly for baking and patisserie *dulce de leche*), thereby reducing the rate of color development and gloss [6].

Fig. 6 "String" formation in homestyle *dulce de leche*

16. It is important to consider that quality control analyses, such as rheological analyses, allow the detection of minor changes in the formulation and/or process, which significantly affect the main attributes of the final product.
17. The more concentrated the system is, the faster the color develops. This results in the *dulce de leche* mainly reaching its definite color in the final manufacturing step.
18. The gelation of the system is produced at temperatures above 100 °C and much earlier than the development of the deep brown color.
19. In order to delay the early formation of lactose crystals (if it is not hydrolyzed), the manufactured product can be cooled with two suitable cooling profiles (the first is slowly cooling up to 93 °C and the second is quickly cooling from 93 °C to packaging temperature), such that the lactose appears as β isomer (not α) since this form is ten times more soluble and much smaller in size, thus much less perceptible to the palate.
    - Crystal size ≤10 μm — Adequately smooth texture.
    - Crystal size ≥15 μm — The palate starts to detect crystals.
    - Crystal size ≥30 μm — Unpleasant sandy texture.

    The maximum rate of lactose crystallization (defect) occurs when the product is at 30–35 °C.
20. Since the osmotic pressure inside *dulce de leche* is elevated, mainly because of the elevated concentration of sugars, microorganisms do not normally grow.
21. *Dulce de leche* is prone to fungal growth on the surface because of the possible humidity that may appear, the water condensation in the lid if the product is packed when it is still relatively hot, and airborne contamination of the working environment.
22. The container is inverted to displace the air between the product and the lid after packaging to avoid fungal growth on the surface. Potassium sorbate is also added to the product mass according to regulations. Additionally, or alternatively, and once packaged, the surface is sprayed with a hydroalcoholic solution of potassium sorbate of different concentrations (even saturated).
23. The use of the batch method (*pailas*) is more versatile than the others, allowing the manufacturing of any style of *dulce de leche*, but it has the disadvantage that there is no energy efficiency, its automation is inexistent or extremely limited, thereby the production standardization is less feasible. The complete opposite occurs with the continuous method, having as a disadvantage its initial cost. It is accepted that the latter method is economically convenient when more than 15.000–20.000 kg of *dulce de leche* are produced per day.

24. In the continuous method, there is heat processing at 135–140 °C but without reaching evaporation, thereby not concentrating the product. This is due to the product being at a pressure well above atmospheric pressure, high enough so that boiling is not achieved. This pressure increase in the product is achieved by introducing pressurized air.

References

1. Zalazar CA, Perotti MC (2011) Concentrated dairy products, Dulce de Leche. In: Fuquay JW, Fox P, McSweeney P (eds) Encyclopedia of dairy science, 2nd edn. Academic Press, Cornwall, UK, pp 874–880
2. Panizzolo L, Rodríguez A, Rozycki S (2016) Reacción de Maillard y caramelización. Efectos sobre el color. Estudios aplicados al proceso de elaboración del dulce de leche. In: Ciencia de Sistemas Lácteos Concentrados, Coordinador S. Rozycki, Centro de Publicaciones, Universidad Nacional del Litoral, Santa Fe, Argentina, pp 167–218
3. Ranalli N, Andrés SC, Califano AN (2012) Physicochemical and rheological characterization of "Dulce de Leche". J Texture Stud 43: 115–123
4. Gaze LV, Costa MP, Monteiro MLG, Lavorato JAA, Conte Júnior CA, Raices RSL, Cruz AG, Freitas MQ (2015) Dulce de Leche, a typical product of Latin America: characterisation by physicochemical, optical and instrumental methods. Food Chem 169:471–477
5. Rozycki SD (2003) Cinética de pardeamiento en sistemas lácteos concentrados: análisis comparativo de modelos. Cienc Tecnol Aliment 4: 95–102-108
6. Rozycki SD, Pauletti MS, Costa SC, Piagentini AM, Buera MP (2007) The kinetic of colour and fluorescence development in concentrated milk systems. Int Dairy J 17(8):907–915
7. Rozycki SD (2008) Tesis de Doctorado en Ingeniería Química: "Estudio de los procesos de gelificación térmica y de pardeamiento no enzimático, de proteínas de leche en presencia de azúcares." Inst. de tecnología de Alimentos (ITA)—Fac. de Ingeniería Química (FIQ)—Univ. Nacional del Litoral (UNL), Santa Fe, Argentina
8. Rozycki SD, Buera MP, Pauletti MS (2010) Heat-induced changes in dairy products containing sucrose. Food Chem 118(1):67–73
9. Pauletti MS, Castelao EL, Bernardi MC (1996) Influence of soluble solids, acidity and sugar on the color of dulce de leche. Food Sci Technol Int 2(1):45–49
10. van Boekel MAJS (1998) Effect of heating on Maillard reactions in milk. Food Chem 62(4): 403–414
11. Hofmann T (1998) Studies on the relationship between molecular weight and the color potency of fractions obtained by thermal treatment of glucose/amino acid and glucose/protein solutions by using ultracentrifugation and color dilution techniques. J Agric Food Chem 46(10):3891–3895
12. MERCOSUR (1996) Resolución No. 137/96 Reglamento Técnico MERCOSUR de identidad y calidad de dulce de leche
13. Rozycki SD (1999) Tesis de Magister en Ciencias y Tecnologías de Alimentos: "Influencia del la Temperatura y el pH Sobre el desarrollo de Color en el Dulce de Leche", Inst. de tecnología de Alimentos (ITA)—Fac. de Ingeniería Química (FIQ)—Univ. Nacional del Litoral (UNL), Santa Fe, Argentina
14. Rodríguez A, Lema P, Bessio MI, Moyna G, Olivaro C, Ferreira F, Panizzolo LA (2021) Characterization of Melanoidins and color development in Dulce de Leche, a confectionary dairy product with high sucrose content : evaluation of pH effect, an essential manufacturing process parameter. Front Nutr 8(November):1–11
15. Rodriguez A, Piagentini A, Rozycki SD, Lema P, Pauletti MS, Panizzolo LA (2012) Evolución del desarrollo del color en sistema modelo de composición similar al dulce de leche. Influencia del tiempo de calentamiento y del pH. Innotec 7:38–42

Index

Adriano Gomes da Cruz et al. (eds.), *Dairy Foods Processing*, Methods and Protocols in Food Science,
https://doi.org/10.1007/978-1-0716-4144-6,

C

D

E

I

J

K

L

Index

N

O

Q

R

S

Index

MIX
Papier aus verantwortungsvollen Quellen
Paper from responsible sources
FSC® C105338
FSC
www.fsc.org

If you have any concerns about our products,
you can contact us on
ProductSafety@springernature.com

In case Publisher is established outside the EU,
the EU authorized representative is:
Springer Nature Customer Service Center GmbH
Europaplatz 3, 69115 Heidelberg, Germany

Printed by Libri Plureos GmbH
in Hamburg, Germany